D0927790

Men on Iron Ponies

Men on
IRON
PONIES

THE DEATH AND REBIRTH OF
THE MODERN U.S. CAVALRY

Matthew Darlington Morton

NORTHERN ILLINOIS UNIVERSITY PRESS   DeKalb

© 2009 by Northern Illinois University Press

Published by the Northern Illinois University Press, DeKalb, Illinois   60115

Manufactured in the United States using postconsumer-recycled, acid-free paper.

All Rights Reserved

Design by Julia Fauci

Library of Congress Cataloging-in-Publication Data

Morton, Matthew Darlington.

Men on iron ponies : the death and rebirth of the modern U.S. Cavalry / Matthew Darlington Morton.

    p.   cm.

Includes bibliographical references and index.

ISBN 978-0-87580-397-5 (acid-free paper)

1. United States. Army. Cavalry—History—20th century. 2. United States. Army—Armored troops—History—20th century. 3. Mechanization, Military—United States—History. I. Title.

UA30.M676 2009

357.0973—dc22

2008052990

For Julie, Macedonia, Mojave, and the Highlander,

no man is as lucky as I.

Contents

Acknowledgments

More than eight years have elapsed since I started this project while attending graduate school at Florida State University. I have been the beneficiary of an enormous amount of assistance along the way and will do my utmost to acknowledge that help.

Historians would not be able to practice their craft without the assistance of research librarians and archival experts. My research odyssey started at Fort Knox, Kentucky, where Lorraine Allen, William Hansen, and Candice Fuller helped me get off to a good start. I received no less help on my visits to the United States Army Military History Institute at Carlisle Barracks, Pennsylvania. At the National Archives, Timothy Nenninger did more to make this project possible, from a research perspective, than anyone else. He quickly oriented me to the best material and was unstinting in sharing interviews he had conducted years before. The staff at Norwich University helped me explore the papers of I.D. White, no small feat for Kay Schlueter was still in the process of cataloging this rich collection of personal papers, documents, and photographs. Back at Florida State, Judy Depew and her staff helped me on a daily basis as I completed my resident graduate work and built the foundation for this final product.

Having finished graduate school, it was my privilege to teach history at the United States Military Academy. I am embarrassed to admit this, but I spent far more time in the library at West Point as an instructor than I ever did as an undergraduate. Susan Lintelmann was the reason. Assisted by Alan Aimone, Debby McKeon-Pogue, Sheila Biles, Alicia Mauldin, and Suzanne Christoff, Susan—with her knowledge of the Special Collections, especially the papers of John K. Herr—put me onto the kinds of material I never could have found on my own. More importantly, Susan has supported the ongoing effort to transform my dissertation into this book during the years since I left West Point in 2004. Words cannot express my gratitude for her support while I was deployed to Iraq; her interest in this project has played an instrumental role in helping me see it through to completion.

Related to the collection of research material, it was my honor to conduct interviews with the real heroes of this story, the men who rode "iron ponies" into combat during World War II. Fred Salter, Keith Royer, Adrian St. John, and Michael Greene provided first-hand accounts of cavalry combat during the Second World War. Some of these troopers have traveled on to Fiddlers Green to refill their canteens, but I will always cherish the time I spent with them. I would especially like to acknowledge the contributions of Paul Willing, whose military career beginning in LeClerc's French 2nd Armored Division is a story unto itself. Robin Sellers and the Reichelt Oral History Program provided the vehicle for conducting these interviews.

Special thanks are in order for Northern Illinois University Press. Melody Herr first contacted me in 2005 to see if I was interested in pursuing this project. I had to ask for an extension at that time, as I was preparing to go to Iraq for a year. Within a month of my return from Iraq, however, and much to my surprise, Melody was knocking at my door again. Without her interest and persistence this project would not have been possible. I would also like to thank acquisitions editor Sara Hoerdeman, managing editor Susan Bean, copyeditor Sandra Batalden, design and production manager Julia Fauci, and the rest of the staff for their assistance. Sara's willingness to work across the Atlantic via email and to accommodate my sometimes strange travel schedule has been a blessing.

I have also been blessed with wonderful mentors. It has been my good fortune that many of these folks also happen to be first-rate historians. Stephen Arata, Keith Bonn, and Dale Wilson saw something in me as an undergraduate that I could not recognize. Their encouragement led me to major in history as an undergraduate. Stephen Arata was at West Point to welcome me back as an instructor more than a decade later. Between West Point the first time and West Point the second time, the history department at Florida State did its best to transform a soldier into a scholar. Donald Horward's willingness to take a chance on me has changed my life. Jonathan Grant and Edward Keuchel lent their hands to the project and I am forever indebted to them both. While serving on the West Point faculty, Robert Doughty, Lance Betros, and Mat Moten encouraged me and provided the necessary support to attain my academic goals. I cannot thank them enough for the opportunity they have given me. The visiting professors at West Point, Dennis Showalter, George Gawrych, and Brian Linn provided wonderful mentorship, but I must single out Joe Glatthaar for special thanks; he has had a bigger impact than he knows. Robert Cameron set me straight very early on as I started this process. I know my work will never approach the level of detail he has achieved with his own research, but I hope he recognizes his influence on my approach. George Hofmann published a book in 2006 that answered the questions I started with in 2000. His encouragement to take my dissertation to the next level upon its completion in 2004 gave me the confidence to follow through.

Like my mentors, the friends who have supported me the most on the project are historians, too. Chris Prigge will ultimately publish the definitive work on this subject, but he has shared everything he has on the topic. Thomas Goss was my company commander the first time I deployed on something more than a training exercise. Since I have known him, he has made a habit of making everything look effortless, to my great frustration as I have tried to emulate his accomplishments. Like Goss, Ben Greene, Sam Watson, Steve Waddell, and Ethan Rafuse have pointed the way and provided the necessary ribbing to keep me moving in the right direction. Mike Runey and Jim Powell are two of the finest warrior-scholars I know; they are also exceptional friends with an unending well of patience for listening to me talk at length about some inane aspect of my research.

Since leaving West Point, many friends have directly and indirectly helped with the project. They are too numerous to list here, but I must publicly acknowledge a few individuals, who stand on a peer level with my family when I consider the people who have mattered most when it comes to this project. Captain Francis "Joe" Myers was a cadet while I served on the faculty at West Point. His remarkable sense of history, inspired by the service of his grandfather during World War II, inspired me. He even provided the introduction to one of the men I interviewed, his grandfather's classmate. Since his graduation he has spent more time in Iraq than almost anyplace else and is now serving with one of my old regiments, the Dragoons. He has been in my thoughts as I have finished this project, because his units have been equipped with Stryker Fighting Vehicles, a purpose-built vehicle that would have made the early pioneers like Charles Scott immensely proud. More important to me, the Stryker shows how the army can get it right, and having done so, save lives on the modern battlefield. Seth Pilgrim always seems to appear in person when I have to deploy to mine-covered mountains or hot, dusty deserts. Thanks for being there. Major David Taylor was my friend while I served in Iraq. I shared with him the hope that I would be able to get this project off the ground when I returned from Iraq, and in return he provided enthusiastic encouragement; that's the kind of man he was. Dave will never get to read this, but I hope he knows his encouragement is still making a difference in my life and in so many of the other lives he touched. Rest in peace.

Gordon Rudd and James Pickett Jones have been the two most influential men in my professional life. Each has touched me as an educator, prepared me to go to war, guided me as a scholar, and provided role models for me to emulate. Gordon Rudd has always been there for my family and me. He has shown me the world and provided sage counsel for the last eighteen years. Nothing I have done as a soldier or scholar would have been possible without his influence. Dr. Jones's deep well of energy, love of life, knowledge, and enthusiasm for teaching is renowned in Tallahassee, Florida, and among the legions of former students who have passed through his classrooms and seminars over more than forty years. I hope have done justice to the children and grandchildren of his own *Yankee Blitzkrieg*. Thank you for offering me a path to success and changing my life in the process.

My family is literally stretched across two continents, so let me use some geographic shorthand in order to mention them all. To the Deerwesters of

Kentucky, thanks for your support and for allowing me to join your family. My brothers in North Carolina and Michigan have seen it all, and yet remain my best friends. My father and the Vermont branch of the Morton tribe have inspired me to continue my education for the sake of always trying to learn something new. My mother in Ohio has provided unwavering support and when necessary a firm push in the right direction when I seem to have lost my way. Words will never adequately express my gratitude. Mom Desmond, thank you for Julie and for sharing your love of history with me.

My wife, Julie, is the guiding force in my life. She has endured many moves, deployments, and even a war and has never uttered a single complaint. She is a remarkable woman and has given me the freedom to pursue my scholarly ambitions. I will never be able to adequately thank her for saying "yes" so many years ago. My children have probably suffered more than anyone else to see this project completed. They have endured countless vacations dedicated to research, writing, and visiting battlefields. Just as being army brats was not their choice, they have had little say in how I have spent time on research that otherwise might have been shared with them in the pursuit of their own interests.

In spite of all the help I have received, I alone remain accountable for the errors of fact and misinterpretation that may lurk here within. I can only extend my humble apologies to those who have done so much to help me avoid these errors of omission and commission.

Abbreviations

Armored Division	AD
Armored Reconnaissance Battalion	ARB
Brigadier General	BG
Captain	CPT
Cavalry Brigade	CavBde
7th Cavalry Brigade (Mechanized)	7thCavBde(M)
Cavalry Division	CavDiv
Cavalry Group	CavGrp
Cavalry Regiment	CavRegt
Cavalry Reconnaissance Squadron	CavReconSqdn
Colonel	COL
Combat Command	CC
General	GEN
Infantry Division	ID
Lieutenant	LT
Lieutenant Colonel	LTC
Lieutenant General	LTG
Major	MAJ
Major General	MG

Men on Iron Ponies

Introduction

★   As darkness descended on the Mojave Desert in early December 1996, it looked like a swarm of lightning bugs on a warm summer night somewhere in the Midwest. But it was cold, cold like it might have been in the Kasserine Pass in February 1943, and the twinkling yellow lights denoted death, just as burning tanks had marked the high-water mark of Rommel's last desperate gamble to check the Allies in North Africa. But here in the Mojave the lights denoted the strikes of laser beams, not high-velocity rounds spit from the barrel of a German 88mm gun, and the blinking vehicles and their associated crews would be resurrected, re-keyed in the military vernacular, to fight another engagement. The battle that had taken place was fought between two proud United States cavalry regiments, each with a rich history and regimental identity. In five short years of service, I had served with each regiment, but on this night I was with the one that had far more lights blinking, wondering how it could have come to pass, the lopsided defeat.

The victorious regiment, the 11th Armored Cavalry Regiment (ACR), Blackhorse!, had assumed the duties of the United States Army's premier standing Opposing Force (OPFOR). Though it no longer resembled the regiment that had fought in the Philippines at the turn of century, across Europe during World War II, and in the jungles of Vietnam, and also defended the Fulda Gap for much of the Cold War, it was still a heavy outfit, with tanks and armored personnel carriers. The regiment's members were the home team and they usually won, but it was their job to exact heavy casualties in the form of blinking lights so that the units they sparred with, fellow brothers and sisters in arms, would be successful on distant battlefields.

I was now assigned to the other regiment, the 2nd ACR, the Dragoons, the longest continuously serving cavalry regiment in the United States Army. Like its opponent in the Mojave, it had departed Europe in the early 1990s, finally settling down in the mosquito-infested swamps at Fort Polk. The regiment was no longer armored, except for the Nuclear/Biological/ Chemical (NBC) reconnaissance vehicles. There were no tracked vehicles, less the bulldozers in the engineer company. Instead the regiment relied on High Mobility Multi-purpose Wheeled Vehicles (HMMWV) to accomplish its missions, none of which had any armor. It was a strange time, even stranger when the "all-wheeled" regiment went to the National Training Center (NTC) to fight the 11th ACR. There, the 2nd ACR proved that any unit can conduct a movement to contact at least once. The Mojave Desert helped germinate the questions that fueled the research that went into this book. In un-armored trucks, were we not unlike Polish cavalry charging German panzers in 1939? Had not the United States abandoned horses after Pearl Harbor to avoid a similar fate? Why had it taken so long and what had transpired between the world wars? What was the story behind the M-10 Greyhound armored car that stood as a silent sentinel at the regimental headquarters, one of the first vehicles young troopers would have used in combat after they traded their real horses for "iron ponies." Why did that Greyhound have more firepower and protection than the modern mounts, HMMWVs, filling the motor stables?

After commanding a cavalry troop, I was fortunate enough to attend graduate school in preparation for teaching military history at West Point; as a cadet I was following the footsteps of men who inspired me to study history, and a company commander who has shaped my career since 1994. Over the years my mentor, Gordon Rudd, had opened my eyes to a variety of issues. Of particular importance, he explained to me how the organizational design of a unit when combined with particular personalities can have a dramatic impact on how a unit ultimately performs. His catch-all phrase for this concept is "chemistry." As a cavalry officer and a historian, I have always been curious about the "chemistry" that led to the decision to abandon horses and form mechanized reconnaissance units. This is the result of my attempt to answer those questions, increase my personal connection to units in which I have served, and improve my understanding of how "chemistry" still influences the profession of arms today.

Men on Iron Ponies: The Death and Rebirth of the Modern U.S. Cavalry, examines the United States Army's development of mechanized ground reconnaissance units between World War I and World War II. The process directly influenced the manner in which these units were used and how well they performed during the Second World War in both the Mediterranean and western European theaters. The evolution of these unique units took place in a complex environment and impacted Cavalry Branch's ability to render effective reconnaissance and security service during World War II. Leaders then, as now, confronted tough questions. What would the nature of the next war be? What kind of doctrine would lend itself to future battlefields? What kind of organization would best fulfill doctrinal objectives, once established, and what kind of equipment should that organization have? How should emerging technology be incorporated into new organizations? For many of the interwar years, the likelihood of fighting in Europe seemed remote; the Panama Canal

Zone and South America seemed like more plausible theaters to the leaders in Cavalry Branch. An ever present constraint, budgetary limitations, confronted army leaders as they sought to carry out their daily missions with an eye to what could be expected of them in the future. Peace did not provide the catalyst to change in the same way that battlefield intercourse with a reactive foe does. The ability of senior military leaders to embrace new concepts and verdant technologies directly influenced the development of new organizations.

One of the most dramatic operations during the waning days of the American Civil War was MG James Harrison Wilson's raid that "cut a path of destruction through the heart of Alabama and Georgia."[1] Leading a corps-sized force of 13,480 men, Wilson set a high bar for future cavalry leaders as they looked to their branch's historic past for inspiration and guidance. The theory behind the practice, at the tactical level of war, was that cavalry existed neither to circulate message and provide security and reconnaissance capabilities for larger infantry formations with small detachments of mounted troopers, nor to charge the enemy with drawn sabers. Rather, Wilson believed, like MG Philip H. Sheridan, that the real potential of cavalry could only be realized in massed formations in which horses lent mobility and speed to the mass. This created a new dilemma for the Confederate civil and military leaders. Ironically, at the heart of the concept was inspiration from a son of the South, Nathan Bedford Forrest, whose maxim was to get there first with the most. Once there, in theory and practice, the fighting was done by dismounted troopers with rapid-firing carbines, not sabers and lances or heavily armored European-style cuirassiers bearing down at the gallop.[2] At the operational and strategic levels of war, Wilson's actions dealt a crippling blow to the rebel cause. Marching 525 miles in two months, his men left a swath of destruction that focused on the very means of making and sustaining war—iron mills, machine shops, rail lines, and rolling stock.[3] Wilson's cavalry was a force to be reckoned with, but as Union victory gave way to continued westward expansion, the concept of cavalry transformed, even if the indelible mark of what had been accomplished remained imprinted on the service's identity.

In 1870 on the battlefield at Mars-la-Tour, Prussian cavalry squadrons thundered across the field in what many historians later cited as the last successful cavalry charge in western Europe. MG Friedrich von Bredow's charge carried the French gun line and eliminated the threat to Prussian positions in the vicinity that day even though the French countered with their own cavalry charge into the flank of the spent Prussian formation. But the cost of success had been high, even before the full-scale employment of machine guns and rapid-fire artillery, with only 420 of the original 800 charging Prussians returning to their lines for future service.[4] Things could not have been more different in the United States.

By the 1870s, Reconstruction in the former Confederacy was no longer a major responsibility for the United States Army and the Cavalry Branch specifically. Rather, the 275,000 Native Americans spread across the American

West garnered the attention of soldiers dispersed over some 116 posts. Stretched thin as they were, often consumed with day-to-day survival on their remote outposts, the idea of assembling forces anything akin to what Wilson had used to sweep through the Deep South or what Bredow charged with at Mars-la-Tour was impossible.[5] Service during this era was characterized by small formations, company-sized in most cases, conducting extended patrols in search of an elusive foe. Like the Civil War, the cavalry horse provided the trooper mobility and extended range, but the operational environment demanded a lightly equipped cavalry force that often abandoned even the trademark weapon of mounted warriors, the saber. It was during this era that the debate about the proper cavalry mount started to emerge, first pitting the Indian pony against the properly bred cavalry horse, each with inherent advantages and disadvantages, which varied depending on what one expected them to do and where it was to be done.[6]

As the Native American threat subsided, orientation toward the Mexican border increased. While Europe was embroiled fighting among the trenches in western Europe, the United States was looking south of the border. In response to Pancho Villa's raid on Columbus, New Mexico, in March 1916, President Woodrow Wilson ordered GEN John Pershing into Mexico at the head of the Punitive Expedition. More than fifty years after Wilson's raid, Pershing's force never accounted for more than 12,000 soldiers from all branches, not just the cavalry, as it plunged into the austere province of Chihuahua. Although airplanes, trucks, and cars made appearances during the expedition, most of the hard work was still carried out by marching columns of men on foot and horseback, employing tactics and techniques learned fighting Native Americans, decidedly not in the roles envisioned by Europeans for their cavalrymen at that time.[7]

During World War I the airplane supplanted the horse for certain long-held strategic reconnaissance missions and forced Cavalry Branch to adapt in order to maintain its ability to contribute to the army's ubiquitous need for information. Machine guns and barbed wire called into question the relevance of horse cavalry on the modern battlefield. From the crucible of battle, tanks and armored cars emerged during World War I and seemed to offer capabilities that might build upon and enhance the existing American cavalry. Because tanks and armored cars could never go all the places a man on horseback could, the issue of tactical mobility sustained the debate that pitted animal against machine. Mechanized reconnaissance units evolved within Cavalry Branch throughout most of the interwar years as the traditional combat arm struggled to incorporate new technology and maintain its ability to perform its wartime missions.

Cavalry Branch developed specialized mechanized reconnaissance forces to serve other mechanized forces developed during the interwar years. These other mechanized forces sought to replicate the combat missions of the horse cavalry, especially offensive combat, exploitation, and pursuit.

Separated from Cavalry Branch in 1940 to become the Armored Force, these other forces were the progenitors of the modern armored divisions. When they departed Cavalry Branch, they took with them their own mechanized ground reconnaissance agencies. Cavalry Branch also developed mechanized reconnaissance forces to support horse cavalry, the branch's premier combat element. Even as World War II erupted in Europe, Cavalry Branch remained focused on the combat role to be played by its horse cavalry divisions, trying to write a new chapter in the history of the U.S. Cavalry in the vein of Wilson rather than the Indian fighters.

Every mechanized reconnaissance unit that served in World War II descended from these interwar units and carried the imprint of the debate that shaped their development, organization, equipment, and doctrine. Ironically, these all-mechanized cavalry units largely defined the contribution of Cavalry Branch to the American war effort in Europe in a manner the branch never would have willingly chosen for itself, in theory acting solely as an agency of reconnaissance and security. The mechanized cavalry formations represented all that was left after the horses disappeared forever. Ironically, after World War II, the men who commanded these all-mechanized cavalry units developed specifically for reconnaissance and security missions—serving armies, corps, and divisions—considered them generally ineffective on such missions, but far more capable of carrying out the traditional combat missions of cavalry than had been thought possible before World War II.[8]

How could this have happened? Two full decades separated the world wars, yet America fought World War II with ground reconnaissance agencies deemed less than satisfactory. This resulted from the competing perspectives of men destined for service in the Armored Force and traditional horse cavalrymen. Each of the sides in the struggle that shaped mechanized ground reconnaissance doctrine, organization, and equipment had a more important priority than specialized reconnaissance units. For each side, the real issue was combat; reconnaissance was merely a means to that end. Common to both were the realities of the Depression and its effect on the interwar environment, an effect that constrained practical development, but did not constrain the minds of all men involved in the process of preparing for the next war. Unfortunately, the prejudices held by those who sought to constrain mechanization, so as not to see it eclipse their beloved horse cavalry in the next war, directly contributed to the flawed doctrine and organization that characterized the World War II experience of mechanized cavalry units. Fortunately, the men who commanded these units drew on a doctrinal background that emphasized combat. They were able to overcome the shortcomings of their organizations with creativity and the application of traditional horse cavalry doctrine, which emphasized combat, rather than the doctrine developed specifically for mechanized ground reconnaissance units. The efforts of the men who served in mechanized cavalry units during World War II led directly to the reconciliation of Cavalry Branch with its Armored Force offspring and have contributed directly to the doctrine and design of contemporary cavalry units.

The same challenges face Army leaders today as they contemplate the nature of the next war, a hypothetical undertaking that begs myriad questions about how to equip, train, and organize for that war, all the while having to retain some flexibility to deal with the war that actually happens. These issues are at the heart of this book. Two decades of peace dividends did not result in the proper solution for how to conduct ground reconnaissance. The nature of World War II did not completely match interwar expectations. High hopes for technology were dashed, as they had been in the past, by the reality that war remains a human endeavor, a contest between competing wills, in which one must inevitably close with the enemy to determine his intentions.

The very term "cavalry" can easily lead to confusion, and thus it is important to establish a doctrinal understanding of the subject. Well into World War II, the June 1944 edition of *FM 100-5, Field Service Regulations, Operations* still recognized two forms of cavalry: horse cavalry and mechanized cavalry.[9] Cavalry, in total, was "characterized by a high degree of battlefield mobility" and derived its "special value" from its ability to rapidly move firepower from "one position or locality to another."[10] Doctrinally, *horse cavalry* retained all of its traditional combat functions, including offensive and defensive combat, as well as the missions more commonly associated with cavalry, reconnaissance and counter-reconnaissance. The same 1944 doctrine characterized *mechanized cavalry's* mission as conducting "reconnaissance missions employing infiltration tactics, fire and maneuver."[11] The army defined reconnaissance as "the directed effort in the field to gather information of the enemy, terrain, or resources" and its purpose was to "gain the information upon which to base tactical or strategical operations"[12] Reconnaissance had always been an important function of cavalry, but throughout the interwar years and well into World War II, Cavalry Branch continued to doctrinally define itself as a branch capable of performing offensive and defensive combat missions with its horsed elements, although it expected to conduct most combat operations dismounted. Mechanized elements remained doctrinally limited to the mission of reconnaissance. Why then had the branch broken with its past traditions of homogeneity of force composition to develop a subordinate arm for an important, but narrow function?

At the end of World War I, GEN John J. Pershing conducted an extensive staff study of the Great War that culminated in a report by the American Expeditionary Force (AEF) Superior Board. Although few American cavalry units saw service in Europe, a board of cavalry officers contributed to the Superior Board report.[13] They concluded that the role of cavalry had "changed but little when considering a war of movement," but "that the aeroplane will do the major part of what was formerly believed to be the role of cavalry in its strategic nature, so far as very distant reconnaissance is concerned."[14] The board was impressed with another piece of technology, the armored car used by the French for reconnaissance on roads, for pursuit and delaying actions, and for helping to overcome "machine gun resistance."[15] The board was so impressed that it recommended the inclusion of twelve of these cars in the horse cavalry division.[16]

By the early 1920s, doctrinal literature began to reflect the Cavalry Branch's new role in reconnaissance. There was no question that the airplane now carried the bulk of what was defined as *strategic reconnaissance,* but because of weather conditions and inability to cover terrain in detail, the army still considered there to be a reconnaissance role for the cavalry in the gap between opposing armies.[17] Into this gap the army anticipated the employment of an entire horse cavalry division.[18] To help the cavalry division extend its range, Cavalry Branch introduced motorization and mechanization.[19] Motorization focused almost exclusively on the logistics support of horse cavalry units. Mechanization took on ever greater roles in the arena of "distant" reconnaissance, with the emphasis on saving "horseflesh." Still, the horse's all-terrain, all-weather mobility had been and remained the central pillar of the argument for retaining horse units. As the mechanical reliability of the vehicles increased, a different question began to emerge: How would the horse and the mechanized vehicle work together to accomplish Cavalry Branch's mission? The army also sought the answer to the question of whether the cavalry mission could be completely performed without the horse.

In 1931, Chief of Staff of the Army GEN Douglas MacArthur made an important decision about how the army would experiment and develop mechanization. He authorized all branches to develop and implement mechanization and motorization as they saw fit. No longer was Infantry Branch solely entitled to experimentation with tanks. This paved the way for Cavalry Branch to experiment and led to the parallel development of mechanized reconnaissance units during the interwar years, since the all-mechanized forces envisioned under MacArthur's directive also needed an agency of reconnaissance. This resulted in the establishment of the first permanent all-mechanized force. The 1st Cavalry Regiment (Mechanized)—CavRegt(M)—was equipped with a variety of vehicles including the "combat car," a way of saying "tank" that allowed Cavalry Branch to circumvent existing federal statutes, which reserved tank experimentation to Infantry Branch.[20] Shortly thereafter, the cavalry's mechanized force expanded into the 7th Cavalry Brigade (Mechanized)—CavBde(M).

The 7th CavBde(M) presented a serious possibility for confusion in the study of the development of mechanized ground reconnaissance between the world wars and during World War II because of the nomenclature associated with its subordinate components, the 1st and 13th CavRegts.

These were fully mechanized cavalry regiments that sought to replicate all cavalry missions, including combat. To this end, they were organized with all arms, especially tanks. In 1940, the 7th CavBde(M) became the 1st Armored Division—1st AD—and left Cavalry Branch to become part of the Armored Force. The development and evolution of this organization, which led directly to the American armored division, has been well documented and researched, but must be traced in outline to show the evolution of the reconnaissance units that served it. The mechanized cavalry units that saw service in World War II were not descended from the mechanized cavalry regiments that gave rise to the armored divisions. Rather, they grew out of the specialized units developed solely for reconnaissance, not the full scope of combat missions.

The perception of what kind of war Cavalry Branch might be called upon to fight in the future also influenced the integration of mechanization. The findings of the cavalrymen in their report to the Superior Board predicated their conclusions on how they expected their branch to be used in a "War of Movement," since a return to trench warfare would also see a return of cavalrymen to the role of infantrymen, thus requiring no special modifications to training, doctrine, or organization of the branch.[21] The tales of Allied uses of horse cavalry and the endurance of the horses sustained an entire generation of American cavalrymen and reaffirmed their confidence in the relevance of horse cavalry as a wholly modern combat arm. Lord Edmund Allenby's exploits in the deserts of the Middle East were easily transferred in the minds of American horse soldiers to their own remote, desolate theater of operations, the Mexican border.

With peace in Europe, America quickly withdrew behind the Atlantic. The citizens of the United States sought "normalcy," which meant thinking about the army little and spending even less.[22] The realm of probable action in the United States, Central and South America, and colonial holdings in the Philippines suggested a viable role for horse-mounted warriors. This was another important and complex factor that must be examined in the overall development of the mechanized reconnaissance forces of the United States Army. After World War II, it was all too easy to criticize those key leaders who resisted the trend toward mechanization in favor of retaining what they viewed as proven technology, the horse. The American political landscape, upon which the debate over and the development of mechanized reconnaissance forces took place, did not suggest that the United States would return to Europe to fight in another war.

The Great Depression had minimal impact on the army as a whole, since as an institution it was already in terrible condition, starving financially from the 1920s and on.[23] The army's leadership was still thinking in terms of World War I manpower and rifle strength, and as Thomas Collier points out, what the army really lacked was vision, not money.[24] Fortunately for the army and the nation as a whole, the 15 percent reduction in military pay that President Franklin D. Roosevelt instituted with his New Deal had little impact on officer and enlisted retention and desertion.[25] The army as an institution may have lacked vision, but there were enough men with that quality who persevered and developed mechanized reconnaissance tactics, doctrine, and equipment in conjunction with the same process that resulted in the formation of armored divisions in 1940. Even with their vision, innovation, and creativity, it was hard to ignore the economic constraints of the Depression and its retarding impact on the growth of mechanized reconnaissance. Throughout this period the Ordnance Corps had a minimal share of the army's overall budget to invest in new equipment. From 1922 until 1935, its allotment consisted of a mere 3.5 percent of the overall army budget, in stark contrast to the 25 percent share the corps enjoyed in 1939.[26] This lends context to the firm belief of those involved in the pioneering process of the early 1930s that they had the right ideas but were not able to see them executed on the proper scale. General Robert W. Grow, one of the mechanization pioneers at Fort Knox, said it best, remarking, "Under Hitler they were in a position to act while we were denied funds to more than improvise and experiment."[27]

As the likelihood of a renewed war in Europe grew, so did the backdrop that shaped the issues of mechanized reconnaissance. As Hitler's military arm spilled over the restraints of Versailles, Americans took notice. Even though the process of mechanization had started out as a "homegrown" idea, it could no longer resist the influence of events abroad. At the end of the interwar period, and on the eve of American involvement in the new European war, the means of conducting ground reconnaissance were transformed. The debate over cavalry's role changed again and gathered steam as Europe moved toward and became embroiled in war. Hitler's *panzers* lent new urgency to questions about the organization of the United States Army. Cavalry Branch suffered a major blow when GEN George C. Marshall allowed the 7th CavBde(M) to become the nucleus of the Ist Armored Corps in June 1940, just before the massive expansion of the peacetime army.

Cavalry Branch fought hard to retain a combat role for its horse cavalry divisions, but placed increased emphasis on its ability to serve corps-sized units with "corps reconnaissance regiments" built with horse and mechanized elements. As the branch strove to maintain relevance for the coming war, it focused on retaining a role for the horse and the mounted man. This last effort played up the horse's unique capability to perform reconnaissance better than any mechanized trooper. The last effort to secure the horse's future took place during the 1941 General Headquarters Maneuvers. Only weeks after the completion of these maneuvers, the nation was at war and the games were over. In 1942, Army Ground Forces eliminated all horses from the horse and mechanized corps cavalry regiments.[28] Even so, mechanized ground reconnaissance units continued to evolve and proponents of the horse continued to campaign for its restoration to fighting and reconnaissance units.

The United States entered World War II with a completely mechanized ground reconnaissance force. The 1st Cavalry Division—1st CavDiv— fought in the Pacific without its horses, serving as an infantry division.[29] This meant that as Cavalry Branch grew from 13,000 troopers in 1940 to its peak strength of 91,948 in 1945, the vast majority of its men served in seventy-three mechanized ground reconnaissance units.[30] This should have meant that the Cavalry Branch's contribution to the war effort largely consisted of reconnaissance and security missions, yet it did not. A board of officers who had commanded mechanized cavalry groups, squadrons, and troops concluded that in total, their units had rarely conducted purely reconnaissance missions. Rather, they concluded, their units performed all of the traditional cavalry missions despite their doctrine and organization.[31]

This work attempts to explain the development of the specialized mechanized ground reconnaissance units that provided the basis of today's cavalry units, but follows these units through World War II as the first to serve in a transformed United States Cavalry. The horses were no more after 1942, but the men on iron ponies kept alive traditions reaching back to the American Revolution. When World War II ended, members of these units emerged as the leaders calling for reconciliation with the former members of Cavalry Branch who had fought with the Armored Force. They did so in the finest traditions of the men who had led the interwar effort to master new tasks with emerging technology. They did so with an eye to the future while acknowledging their roots in the past.

1

The Lessons of World War I

The next war will begin as the last war ended . . .

WITH MOVEMENT![1]—*Ferdinand Foch*

★   During the opening days of World War I, the Russian army honored its commitment to France and thrust two massive armies into East Prussia. GEN Maximilian von Prittwitz commanded the German Eighth Army and nearly panicked when faced with the onslaught of forces from the east. His task was to hold in the east while the bulk of the German army sought decisive victory in the west under the Schlieffen Plan. Prittwitz's panic resulted in his being relieved of command, and generals Paul von Hindenburg and Erich Ludendorff were appointed to steady the situation. By the end of August, their famous partnership directed the massive German victory that erased the Teutonic failure of 1410 and restored German glory to the name Tannenberg. During the ensuing orgy of violence and death, one scene haunted a German officer for the rest of his life. More than the thrashing sounds of thousands of Russian soldiers drowning in the many lakes that dotted the landscape, it was the image of white horses that gnawed at his soul. Five hundred men mounted on five hundred white horses stood silent and dead in the image forever burned in his mind. The horses and riders were packed in a formation so tight they remained transfixed, unable to fall where the modern means of war and death found them.[2] If such a tragedy could befall mounted men in so gruesome a fashion as this, on the very front heralded as a model of effective maneuver—in contrast to the static western front of putrid trenches, what then was the future of cavalry?

Even tragedies such as this did not deter the belief that there was still a future for the horse on the modern battlefield and that the role of the cavalryman was secure. Almost ten years passed, however, before experimentation with emerging technology, doctrine, and organizational design set the conditions that ultimately displaced the horse. The first American clash of horse and mechanized cavalry did not occur until 1934 at Fort Riley, Kansas, touching off a struggle that lasted eight additional years. American views on World War I shaped the contours of debate as did attempts to predict what might happen in the future. Advocates for the continued use of horse cavalry embraced technologies during the 1920s insofar as these technologies enhanced the ability of horse-mounted units to remain viable and relevant. During the same time, advocates for the full mechanization of all cavalry functions began to emerge. Common to both positions was the introduction of rudimentary mechanized reconnaissance organizations that served horse and machine alike.

The United States entered World War I with seventeen regiments of cavalry, but left most of these units on the Mexican border to safeguard against the threat of German agents combining their efforts with Mexicans to "unite against the hated gringos."[3] The austere nature of the border region, lacking in infrastructure for communication and transportation, demanded the employment of large horse-mounted formations and changed little during the interwar years.[4] Even though the 2nd Cavalry Regiment (CavRegt) was the only cavalry regiment to see service in Europe during World War I, its single mounted attack, conducted by CPT Ernest Harmon, as well as actions in other theaters, provided rich material to sustain the belief that horses remained relevant on the modern battlefield. Themes of proper tactical deployment and dispersion in conjunction with the use of terrain recurred throughout the interwar years as justification for retaining horse cavalry.[5]

Following the armistice of November 1918, the U.S. Army remained in Europe until 1923. During this five-year period, the army did more than just demobilize; it also gave serious consideration to the laws regulating its organization and composition and to how such factors might affect its ability to respond to future threats.[6] GEN John J. Pershing established a series of boards to review the American experience during the war. He also convened a Superior Board to review the findings of subordinate boards. Although the primary objective of the Superior Board was to draw conclusions about the performance of the World War I square infantry division, it also commented on the future of cavalry.[7]

The American cavalrymen deployed in Europe took advantage of their unique opportunity to investigate and learn from the experiences of the other Allied cavalry forces that fought in the war. Pershing, who chased Pancho Villa in 1916, directed that a board of officers investigate and report on the "Armament, Organization, the role of Cavalry and Cavalry tactics."[8] The board prepared a report that focused on the traditional role of cavalry as a combat force, but key observations within its pages held the seeds of future changes in the composition of the U.S. cavalry, including new ways to conduct reconnaissance. To this end, the board considered the addition of light armored cars carrying 37mm or 47mm as desirable, but did not see them as an integral part of the cavalry division. The decisive campaign conducted by GEN E.H.H. Allenby in Palestine and Syria attracted the board's attention because mounted troops there inflicted large losses on a crumbling Turkish army. The use of armored cars with the mobile Arab column operating east of Amman merited a single sentence. During a visit to the French 5th Cavalry Division (CavDiv) at Vincennes, members of the board met a commander who saw in the armored car the future of cavalry. In contrast, the British did not favor the further development of the armored car, seeing more potential in the development of a light tank to accompany the cavalry.[9]

Impressed with the French armored car, the board saw merit in its ability to project firepower across a mile-wide front, conduct reconnaissance operations on roads, assist in delays and pursuits, and facilitate the reduction of machine gun resistance. With an eye toward employment in America, the board was concerned about the car's mobility and thus pinned higher hopes on the development of the light tank as tractor technology increased. Even so, the board envisioned as many as twelve armored cars in each cavalry division.[10] It also recommended the inclusion of airplanes, motorcycles, and pack radios. Motorcycles would save "horse flesh" by moving messages behind a screen of horses, and radios mounted on horses could also keep messages flowing.[11] Board members did not view airplanes as "an integral part of the organization," because "aeroplanes are met by aeroplanes, requiring them to keep very high and preventing any close reconnaissance of troops on the ground." Even so, the board saw the need to practice more night marches and to scatter formations in an effort to conceal them from observation.[12] Night operations continued to gain importance in the debate over what the best means of accomplishing reconnaissance would be in the future.

With respect to the cavalry's role in the support of infantry divisions, the board recommended one squadron or one regiment per division. It seemed to favor temporary cavalry attachments, thereby retaining the ability to form these squadrons or regiments into divisions and corps for large-scale employment. The board also acknowledged that cavalry troops on the western front had often fought dismounted in the trenches as infantry. Calling for the discontinuation of the mounted attack in close order, the board recommended additional dismounted training for those times the cavalry was expected to fight like infantrymen. With all the focus on the ground, the board still saw a viable role for mounted troops, especially in the pursuit.[13]

The board concluded that the strategic role of cavalry had changed as the airplane assumed the majority of cavalry tasks that pertained to strategic reconnaissance. The endless trenches of the western front did not dissuade members of the board from their belief in the need or ability to maneuver; they saw instead an entire front that "could be included within the confines of Texas alone." Further, they viewed the ability of a cavalry division to move cross-country over extended distances faster and arrive fresher than other types of units as an important quality that ensured their future. The board also commented on the proper means of employing the cavalry's force so that it not be "frittered away or wasted unnecessarily." This message continued to be a common theme when commanders were forced to decide how best to use their specialized reconnaissance units as they began to emerge. In 1919, centralized command and control of cavalry forces seemed the best means to prevent their misuse. Only when needed would cavalry be "sent or loaned to lesser commands as occasion may demand."[14]

While the United States attempted to learn the lessons of the Great War from the experience of others, the "others" drew their own lessons. Great Britain entered the interwar period with an undefined continental commitment that in some ways hampered its ability to reform or experiment.[15] J.F.C. Fuller remarked that cavalry officers in Great Britain led the fight against any expansion of the armored force and described these officers as "this equine Tammany Hall, which would rather have lost the war than have seen cavalry replaced by tanks."[16] The Germans looked on the issue with clear and rational analysis. As losers they were willing to search for answers in new techniques, and they saw great promise for their cavalry with the continued incorporation of aviation, radio technology, and mechanization. Moreover, GEN Hans von Seeckt supported efforts to advance these measures.[17] Heinz Guderian, another German general associated with interwar mechanization theory, may have drawn his ideas from the actions of GEN Hermann Francois, Ist Corps commander at Tannenberg and at the first battle of the Masurian Lakes, where deep encirclements enabled dramatic results.[18] Regardless of where he got his ideas, Guderian and the Germans were experimenting with motorization and mechanization in the early 1920s.[19]

In the aftermath of the war, the American people wanted an army that was small, inexpensive, and out of the public eye.[20] Having a small budget, the army deferred modernization, but the National Defense Act of 1920 did restructure the cavalry arm of the army. The National Defense Act also created new jobs inside the War Department, and these jobs had a lasting impact on the development of mechanized reconnaissance units during the interwar years, specifically the creation of the chief of cavalry. Along with the chiefs of infantry, artillery, and air service for their respective branches, the chief of cavalry wielded great influence at the War Department in regard to all cavalry matters, including the integration and implementation of new technologies.[21] Going into effect in 1921, the cavalry service retained one active cavalry division, the 1st CavDiv commanded by MG Robert Lee Howze, and one inactive division.[22] Thus, in 1921, Congress reduced the seventeen active regiments to fourteen, and the fourteen active duty regiments were skeletonized by placing selected troops and squadrons on 'inactive' status, resulting in a cavalry force of half-strength.[23]

The first chief of cavalry, MG Willard A. Holbrook, was not pleased with the strength of his branch and begged for a larger force to patrol and secure the Mexican frontier. He believed that "the three and a half years of struggle in the trenches" inspired faulty conclusions and lamented the dearth of observers on the eastern front of the Great War "where cavalry found its greatest usefulness." Holbrook did not doubt the role of the horse in the next war and cautioned against too much faith in emerging technologies. Trucks were "slaves of the road and subject to its condition" and were unable to reach the battlefront, leaving those dependent on them unprepared. In his view, tanks needed "the support of friendly troops" to occupy terrain, and he realized weather conditions constrained the use of airplanes. Holbrook conceded that all of the emerging technology could serve the horse under specific conditions, but the lack of developed roads in "North and South America, Asia, Africa, and Australia" continued to favor the retention of the large horse cavalry units.[24]

Even with such radical reductions affecting the force as a whole, the 1922 edition of *Rasp,* the Cavalry School's annual yearbook, provided a snapshot of how the force viewed itself in the present and how it saw itself relative to the future. MAJ George S. Patton Jr. associated with the first U.S. tank forces in World War I, prepared a short piece emphasizing veterinary science and equine care.[25] COL Hamilton S. Hawkins, "Mr. Cavalry," provided a succinct explanation for why the cavalry could expect to see future service in its present form. Entitled "Why Is the Cavalry Indispensable?" Hawkins's essay explained the role of cavalry, from the perspective of a senior officer who had seen service both in the recent world war and on the border. For him cavalry remained an integral member of the combined arms team that also included infantry and artillery. As a member of this team he listed ten functions the cavalry was expected to perform: obtain information, guard against surprise, hold terrain until infantry arrives, hide the movement of the infantry, strike suddenly and swiftly against points holding up the advance of the infantry, delay to allow the infantry time to escape, exploit success of the infantry, pursue the beaten enemy, attack and delay the enemy's attempt to commit his reserve, and keep away enemy cavalry.[26] To perform these missions, Hawkins saw a place for the motor truck to help improve the cavalry's "radius" of operation by increasing the speed and distance that logistics could be transported. Airplanes would more effectively direct the movements of the cavalry. Tanks would serve as supporting cast members and modern methods and technology must be integrated, but the cavalry was still the cavalry.[27] The conclusion of Hawkins allowed for interpretation. Was cavalry a mission to be accomplished regardless of the equipment used?

Hawkins's cavalry was horse mounted. There was room for technology and modernization as long as no one misunderstood the fact that these modern conveniences were only present on the battlefield to support the horse-mounted soldier. Horses and the men riding them were expected to defeat troops equipped with machine guns and modern rapid-firing artillery by quickly attacking across open areas with the support of their own machine guns and artillery. Open-order movements aimed at the enemy's flanks were viewed as other means to reduce the impact of the enemy's firepower.[28] The basic history instruction offered at the Cavalry School reinforced Hawkins's views by emphasizing the Middle Eastern theater and the eastern front.[29]

While the army of occupation in Coblenz filled its time fox hunting, racing, and playing polo, the stateside cavalry shrunk and the cavalry community as a whole entered into a long lull. Modernization and technological innovation were useful as long as they served the horse. The newly formed air corps assumed the responsibility for providing strategic reconnaissance once filled by far-ranging mounted formations. It would be unfair to say the force was obsessed with horses, but it was clear that those with equine interests were not uncommon and carried with them a certain degree of institutional momentum. Even so, the force's willingness to embrace modernization, recognition that it would continue to need increasing amounts of firepower to remain relevant, provided the opportunity for the future employment of specialized formations to serve and expand the traditional roles of the horse-mounted cavalry.

United States Secretary of War Dwight Davis visited Great Britain in 1927, where he saw a smaller mechanized force beat a larger non-mechanized force.[30] The exercise Secretary Davis witnessed may have caused Rudyard Kipling to remark, "It smells like a garage and looks like a circus," but what Davis saw did inspire change.[31] The general staff issued a memorandum in March 1928 that explained how a mechanized force was intended to fit into the existing organization of the army.[32] COL Adna R. Chaffee Jr. was working in the War Department's Training Section of the Operations Division, G3, when the chief of staff of the army, GEN Charles P. Summerall, received the directive from the secretary of war to begin mechanized testing at Fort Leonard Wood, outside Washington, in July 1928.[33] Approximately 3,000 troops were to take part in a test of light and medium tanks, motorized artillery and infantry, support troops, and cavalry mounted in armored cars.[34] The first test involved a road march from Fort Leonard Wood to Aberdeen Proving Grounds to Carlisle Barracks and back to Fort Leonard Wood. The vintage World War I equipment was barely capable of making four miles per hour, and the visits of foreign military attaches were cancelled since the experiment was viewed as an embarrassment.[35]

CPT Harold G. Holt commanded Provisional Platoon, 1st Armored Car Troop, which was activated on 15 February 1928 at Fort Myer specifically for the experiment. This unit represented the birth of specialized reconnaissance formations in the modern army. Holt had been a horsemanship instructor at Fort Riley, Kansas, but filled his platoon with troopers from the 3rd CavRegt at Fort Myer. Holt's new unit received its motorization instruction at Camp Holabird, Maryland, at the Motor Transportation School and then traveled to Camp Meade for instruction in the use of machine guns and "one-pounder guns." Trucks came from Fort Bragg, North Carolina, and Fort Benning, Georgia, but the platoon had to build its own "imitation armored car bodies" before moving on to take part in the Experimental Mechanized Force. Classified as Armored Car, Light T-1, and Armored Car, Medium T-2, the new steeds cruised 150 miles on a single "drink" of fuel and carried .30-caliber machine guns.[36] The *Cavalry Journal* article that introduced this unit remarked:

> Armored cars, like aircraft, are of special value to cavalry in facilitating reconnaissance and thereby making it possible for the main cavalry forces with supporting troops, including artillery and tanks, to concentrate their chief efforts in the most advantageous direction.[37]

Cavalry once again had a stake in strategic reconnaissance. A specialized force was now needed to maximize the impact of larger formations that might include tanks, artillery, and supporting troops. The idea of reconnaissance was not new, but the use of armored cars to perform reconnaissance tasks was an important step.

> When the cowboy down here is herding cattle in a Ford we must realize that the world has undergone a change.[38] —*BG Van Horn Moseley to MG H.B. Crosby*

★   Experimentation was not limited to the east coast during the summer of 1928. In 1927, the 1st CavDiv completed its first division-level maneuvers since 1923.[39] Large-scale maneuvers, with their interesting constraints, provide a glimpse of how peacetime biases influenced doctrinal debate about how best to perform reconnaissance. Although the 1st CavDiv conducted, on its own initiative, a number of innovative experiments during the late 1920s, these experiments were almost always done with an eye toward preserving a role for the division's beloved horses.

Trucks provided extensive support to the maneuvers. The 8th Engineer Battalion used them to move about and prepare the maneuver area by marking gates, repairing roads and bridges, and improving the ever important water facilities, a major limiting factor in planning the divisional maneuvers.[40] Ironically, little was said in 1927 about the limiting factor that water placed on horses and men alike, yet mechanization's ubiquitous need for fuel was always at the top of the list for those seeking justification for retaining horses. In later years, as the debate developed, the *Cavalry Journal* went to great lengths to describe cavalry operations from World War I in which mounted units traversed great distances for days on end with only minimal water resources.[41] Infantry troops used trucks to travel 223 miles in three days to join the division for the exercise. A Battery, of the 1st Field Artillery, was organized as a porté battery utilizing trucks and trailers to move the cannons and tractors to Fort D.A. Russell, Texas, from Fort Sill, Oklahoma, and the division conducted an important exercise with porté cavalry.[42] More important, the horse division motorized its entire rear echelon for the exercise. The decision to motorize the logistics trains proved crucial. Distribution points for the units in the field were some thirty-six miles from the railhead and "the use of animal-drawn trains exclusively under this situation would have been impossible," noted the closing remarks on supply and logistics. Cavalrymen embraced trucks for their ability to extend the radius of operations for the division's logistics apparatus. Prior to including the use of trucks, animal-drawn trains were only capable of supplying the division eight miles from the railhead—a sixteen-mile daily round trip for fodder, ammunition, and other supplies.

During the exercise, WHITE and BROWN forces conducted reconnaissance. The WHITE reconnaissance squadron, which was camped at a well, did not dispatch patrols until 0800 hours, but discovered a BROWN column. Unable to notify higher headquarters by radio, the squadron dispatched a motorcycle messenger to report the incident. This small excerpt from a much larger exercise illustrated critical inconsistencies in later arguments pertaining to reconnaissance. Other than predictably encamping at or near a well, the reconnaissance squadron also confined its reconnaissance to daylight. When the pack radio failed, it took hours to notify the WHITE higher headquarters.[43] Armored and unarmored cars equipped with radios later helped solve some of these problems. Ironically, those most opposed to mechanized reconnaissance would argue in favor of the horse because it was not limited by fuel or darkness, when these early examples provide insight to horse cavalry performance under field conditions.

CPT Charles Cramer's F Troop, 5th Cavalry, conducted the maneuver's porté cavalry exercise at the conclusion of the larger maneuvers, and thus, no tactical lessons could be drawn from his troop's experience. Using twelve Liberty trucks and make-shift horse carriers, the troop moved forty-seven men and forty-eight horses 288 miles to Fort Clarke.[44] The horses and men were no worse for wear than had they moved as far by train, and in the process the cavalry community gained its first large-scale (certainly a relative term) experience with porté cavalry.[45] This concept surfaced again as the chief of cavalry went to great lengths in the late 1930s and early 1940s to preserve the role of the horse.

When the 1st CavDiv conducted maneuvers again in 1929, its different appearance reflected organizational changes that had occurred in 1928. The chief of cavalry in 1928, MG Herbert B. Crosby, in an early bid to increase the firepower of the cavalry division while reducing personnel, reorganized the division's four cavalry regiments. Aside from redistributing the brigades' machine guns by distributing them to each regiment, he authorized the addition of an armored car squadron.[46] The 1929 maneuvers included the first incorporation of armored cars and anti-tank guns, and the division revisited the use of porté cavalry. The armored car troop, Troop A, 1st Armored Car Squadron, participated in the maneuvers with its "motley" assemblage of CPT Holt's homemade armored cars from the previous year's mechanized experiment. The mechanized scouts earned high marks for their ability to conduct delaying operations, but their good mobility was attributed to dry weather and the lack of fences and ditches along the Texas roads that otherwise would have prevented them from gaining any degree of off-road mobility.[47] There was some surprise at the relative "invisibility" of the cars until they moved. Even with these generally positive comments, a bias was forming that armored cars were best on roads and that they would be countered by units operating off the roads.[48]

During the regimental phase of the maneuvers the platoon conducted reconnaissance ten miles forward of the main body across a five-mile front. Radio sets mounted in the vehicles allowed the men to send reports every two hours. The platoon was generally successful in delaying the opposing force with the use of ambuscades and effective long-range machine gun fire. Opposing forces learned to get off the roads and, using their own towed anti-tank weapons as a supporting base of fire, maneuvered to the flanks of the armored cars.[49] The horse-mounted troopers opted to attack the cars with mounted charges that umpires deemed successful and whose "opportune" arrival was all that prevented the untimely death of the armored car crews by "saber thrusts."[50] Armed as they were with .30-caliber and .50-caliber machine guns and an assortment of small arms, it seemed odd that the armored cars would have succumbed so readily to such flamboyant attacks, yet this notion of vulnerability persisted.[51]

MAJ Patton and MAJ C.C. Benson wrote an article in 1930 that clearly outlined the concerns and possible advantages for the increased mechanization. The article acknowledged that since foreign nations were proceeding in the development of "fast tanks, armored cars, self-propelled gun mounts, and their auxiliaries," it would be "ostrich-like," to ignore these developments.[52] The authors expressed concern that machines, unlike men and horses, required full rations of fuel, oil, grease, and spare parts to remain effective. Additional concern about the loss of immobilized vehicles and the ever present issue of battlefield mobility reinforced these detractors. Horses were still attributed the accolades of all-weather, stream-swimming, forested hillside-traversing, nighttime operating, any-condition-one-can-think-of mobility.[53] Even with these limitations, the authors saw real reconnaissance potential in the armored vehicle's ability to cover long distances at high rates of speed. Security operations, such as flank guards, could be enhanced by the ability of vehicles to move faster and farther to the flanks to warn of potential threats. All these positive attributes were couched in the terms of "wherever the terrain permitted the use of machines."[54]

Patton addressed the same theme in another article that year. Again the issue of mobility was paramount in his argument noting that "in any theater of war save Western Europe," the U.S. Army and its cavalry could expect no better than the road conditions of the United States, which at the time were only 6.5 percent improved.[55] Justifiably, he expressed concern that the assumption, "Weather is cool, roads are dry and hard, all bridges two-way and up to fifteen tons," all too often reflected the conditions offered in map problems.[56] Even with these concerns, his theme thereafter was one of inclusion and combination of the best attributes of the horse and the machine to fulfill the roles assigned to the cavalry arm.

Patton correctly stated that distant and strategic reconnaissance was still the function of cavalry and that the mounted arm had not been supplanted by the airplane with its lack of ability in "storms, fogs, darkness, [and] forests."[57] He viewed the role of armored cars for this role with skepticism. He believed the loss of a single car would end the mission, whereas the loss of a single lame horse only amounted to the loss of a cavalry scout. In his opinion the cars were far too susceptible to barriers on the roads and would be unable to penetrate enemy screens to gather information. He saw some use for them as messengers to keep patrols linked, but saw no utility beyond simplified logistics for combining armored cars into single units. The same man who had written about the relative "invisibility" of armored cars in tactical exercises only months before now wrote that men mounted on horses were less conspicuous. According to Patton, the horse-mounted soldier even had the advantage of his horse's keener hearing to aid him in the accomplishment of his mission.[58] This work provided an excellent example of Patton's vacillating position on utility and on the possibilities presented by mechanized reconnaissance units.

The article continued in the same vein, finding only limited roles for mechanized forces in the other assortment of cavalry missions. Horse cavalry still held the advantage in Patton's mind even when pitted against opposing mechanized forces. He concluded: "In consideration of the foregoing it is our firm belief that the independent employment of mechanized forces is so largely illusory that it will never be seriously employed. Certainly not after a few trials."[59] Perhaps the pictures used to illustrate the article stated the case more plainly. The frontispiece of that issue of *Cavalry Journal* depicted horses negotiating the "water jump" during the Cavalry School Graduation Race Meet, and the very next page was the beginning of Patton's article. The only other illustration for the article filled the unused portion of the last page of Patton's remarks. It depicted a truck mired to its axles while in the background a mule-drawn logistics train moved on.[60]

Limited experimentation in the late 1920s and early 1930s left a great deal of doubt about the real utility of mechanization in the role of reconnaissance. Field Marshal Viscount E.H.H. Allenby wrote a letter to the *Cavalry Journal,* accompanying his article "The Future of Cavalry," in which he focused on the limitations rather than the positive possibilities of mechanization and motorization. Specific to reconnaissance, Allenby still held that the airplane would do the distant work, and that the increased use of mechanization in the performance of tactical reconnaissance would never "supersede" cavalry (from this, one is to assume he implies horse cavalry, for the article draws heavily on historical accounts of horse cavalry operating in the Middle East during World War I). Allenby's view was that armored cars were "purblind" and the more "invulnerable the machine, the blinder the crew."[61] A car dominated the photo that accompanied the article, which referred to the action in Palestine. The photo's citation reads as follows: "This photograph was taken as the Field Marshal was making a reconnaissance immediately before the attack of September 19, 1918, on the stabilized front north of Jerusalem, which resulted in the destruction of the enemy's army."[62] No one, unless it happened to be the photo editor, recognized the irony.

The comments of one of the most recognized and successful cavalry leaders of the last Great War could not deter what was slowly taking place any more than the shifting views of one prominent American writer on the subject, MAJ Patton.[63] Armored cars allowed the cavalry to reclaim a stake in the area of distant and strategic reconnaissance. The establishment of a second armored car squadron at Camp Holabird with its projected permanent home at Fort Riley, home of the Cavalry School, all but insured a continued if not expanding role for some form of mechanized reconnaissance force.[64] Changes occurred at the front and rear of the cavalry division with the introduction of motorization and mechanization, but what had not changed was the notion that the horse would continue to dominate the cavalry force. This unquestioned truth only allowed for continued development of motorization and mechanization in ways that directly or indirectly served the horse. Progress, learning, and experimentation thus far had been voluntary and tinged with curiosity. Along the way, a list of arguments developed that sought to retard the expansion of motorization and mechanization, often focusing on mobility and logistical constraints. These themes took on an even more important life in the next level of development, just about to begin.

2

The 1930s

New weapons are useful in that they add to the
repertoire of killing but be that tank or tomahawk
weapons are only weapons after all.[1]

—*MAJ George S. Patton Jr.*

★ Events of the 1930s directly impacted what type of army the United States would deploy to Europe during World War II. Doctrine was decided, weapons selected, and the role for horses and their mechanized competitors largely decided. Mechanized reconnaissance developed along two tracks, in service of horse regiments including the army's only horse cavalry division, and in service of the all-mechanized forces that emerged. This twin development influenced the views of key decision makers as they approached questions relating to the employment and equipping of mechanized reconnaissance units, views that ultimately affected the wartime performance of these units.

The early part of the decade saw the continuation of what had begun in the late 1920s, but soon the use of mechanization for reconnaissance extended beyond the service of horse-mounted units and into the realm of fully mechanized units established on an experimental basis. As mechanized reconnaissance began serving fully mechanized units, it shared the common radius of action and speed, but still lacked the mobility of the unit it served, because tracked vehicles, like horses, had greater cross-country mobility. It also remained mounted on different platforms to perform its missions, scout cars and armored cars, while the units it served used combat cars (tanks) and half-tracked vehicles. The mechanized troopers at Fort Knox saw their role the same as that of their horse-mounted brethren, namely, one of mounted combat, even though mechanization afforded them large amounts of firepower and protection that horse cavalry could only gain by deploying limbered or packed guns and by occupying suitable terrain.[2] Such differences—between the horse and mechanized units—directly impacted the development of the reconnaissance units that served them. With fire-power, mobility, speed, and radios that enhanced command and control, horse cavalry commanders saw their mechanized reconnaissance assets as a means to an end, sustained viability. The all-mechanized troopers at Fort Knox developed more aggressive techniques and called for the concomitant levels of firepower and protection that such techniques demanded. On the whole, they were less concerned about the loss of their reconnaissance assets, given the availability of additional vehicles in their organization. Furthermore, the success of the mechanized reconnaissance men posed no threat to those they served at Fort Knox, whereas the horse cavalry was just beginning a death struggle between horse and machine.

The Great Depression restrained modernization and growth within the army, yet failed to constrain the minds of men set on improving the army's capabilities.[3] By mid-decade, the role of mechanized reconnaissance continued to gain acceptance within Cavalry Branch.[4] Even so, the horse and mechanized divisions within the branch drifted further apart as they attempted to define the proper place for these specialized assets. With the debate over technique also emerged more spirited discussion about what kind of equipment the scouts should use to accomplish their missions. Debate with respect to equipment focused largely on the need for both mobility and protection. Compounding the difficulty of arriving at the correct solutions for the organization, doctrine, and employment of mechanized reconnaissance units was a shift in the attitudes of those occupying the key position, chief of cavalry. Over the course of the decade, leadership in this post devolved from those who might be characterized as optimistically progressive to a man who at times seemed almost paranoid and delusional. The rift within the branch grew larger as events in Europe thrust the world closer to war. The use of mechanized reconnaissance units was accepted as long as an important role for horse cavalry remained. To what extent the army should expand the mechanized force as a whole, at the possible expense of the horsed regiments, became the basis of a larger debate embroiling mechanized reconnaissance units and influencing their future organization and how they would ultimately fulfill their missions during World War II. Fighting in Poland in 1939 provided the final catalyst for change, but final decisions about how the United States would conduct mechanized reconnaissance was not achieved for two additional years.

The army chief of staff, GEN Charles P. Summerall, directed in October 1930 that a permanent mechanized force should be stationed at Fort Eustis, Virginia.[5] Using $284,999 allotted in the fiscal budget for 1931, Summerall created the mechanized force. The same month he submitted an article to the *Cavalry Journal* that made no mention of the planned force, but instead summarized the view of cavalry's future, pointing out in the very first line that the lack of transportation prevented its overseas service during World War I and that this had led to the "misunderstanding . . . as to the value of cavalry."[6] The army chief of staff believed the presence of cavalry would have helped to "expedite decisions and change the course of battle."[7] Believing that the role of the army in 1930 was largely defensive, he saw great utility for the continued use of cavalry along the borders of the continental United States. Summerall viewed armored cars as a means to extend the range of patrols that would "effect a great saving in the exhausting efforts of horse and men on such duty."[8] He championed the use of large mounted formations on the order of corps and discouraged the detachment of small cavalry units to support others. He also doubted the practicality of moving large bodies of men and horses by trucks in the future. In every sense Summerall's vision was that of the status quo. His views persisted in the Cavalry Branch long after his tenure as chief of staff, but were not held by the man who replaced him, Douglas MacArthur.

MacArthur decentralized the process of mechanization throughout the United States Army and in so doing changed the process by which the army mechanized and motorized.[9] He brought new vision to the office of the chief of staff and was not afraid to put forward radical ideas. MacArthur believed that past military reorganization had focused too much on the equipment, at the expense of looking at the missions expected of certain branches. Simply put, equipment only provided the means of accomplishing the mission and for MacArthur the mission was the bigger picture. He believed, contrary to much of what was being written in the *Cavalry Journal,* that "Modern firearms have eliminated the horse as a weapon, and as a means of transportation he has generally become, next to the dismounted man, the slowest means of transportation."[10] The cavalry spirit was to be maintained, but the new chief wanted the branch to be able to carry out its missions in the context of a modern battlefield. Summerall's newly created Mechanized Force was to be reorganized as a "reinforced cavalry regiment" with appropriate supporting arms.[11]

Building on his theme, MacArthur tasked the cavalry with: long-distance strategic reconnaissance, fighting for control of the theater of reconnaissance, tactical reconnaissance, seizing points of strategic and tactical importance, pursuit, delay, exploitation, and being part of a general reserve to be used tactically or strategically. MacArthur, even with his lack of confidence in the future utility of the horse, did allow that in order to accomplish all of these tasks, two types of cavalry might be required, should horses be retained for operations in terrain unsuitable for mechanized vehicles. He further allowed for the development of reconnaissance formations and doctrine on twin tracks, but he saw cavalry as one branch. He expected that officers and enlisted men would gain experience in each type of unit through normal periodic personnel rotations.[12] As army chief of staff, MacArthur's vision of the future, unlike Summerall's, reached farther, yet was constrained by the realities of the day.

MacArthur found a mechanization ally in the current chief of cavalry, MG Guy V. Henry Jr.[13] Henry assumed leadership of the branch in March 1930 and later expressed a great deal of pride in the modernization that occurred during his tenure, despite resistance on many fronts.[14] Henry was a talented branch chief and operated within the constraints of a tight budget, cooperating with the navy to procure the .50-caliber machine gun for both services.[15] Henry recalled his first meeting with MacArthur, who pointed out the window at a passenger car and remarked, "Henry, there is your cavalry of the future."[16] Henry was no stranger to cars, traveling to every one of his cavalry outposts by car, crisscrossing the United States eight times during his tenure.[17]

Like MacArthur, Henry embraced technology as a means of improving the ability of Cavalry Branch to accomplish its missions, including reconnaissance, but also allowed for the continued role of the horse in some capacity. He believed that not one, but two of the fourteen existing cavalry regiments should be mechanized. The boldness of his thinking can only be appreciated in the context of Henry's overall concern that the proper balance between horse units and mechanized units be maintained in an army that he felt already had too few cavalry units.[18] Such a balance would provide the ability to conduct brigade-level operations and serve as a mobility lab for continued testing of emerging mechanized equipment. Horse units were to be improved with the addition of anti-armored vehicle weapons and scout cars in each regiment, and trucks were to replace all animal-drawn trains and be given increased firepower with the addition of semi-automatic rifles and light machine guns. Henry also saw the need for better radios throughout the branch. The National Guard was to benefit from the dissemination of better radios and enjoy the advantages provided by scout cars.[19] Henry's views on how technology could be used to assist the cavalry in performing its missions were influenced by his expectation of the nature of the next war.

Henry anticipated maneuver and movement. This kind of warfare required forces to maintain their security while they moved to keep contact with one another as they sought the enemy, denied his reconnaissance forces the opportunity to gain intelligence on the American main body, and found the best routes for the friendly maneuvering forces. Looking beyond the Mexican border and the threat found there, one can only believe he had greater expectations for the cavalry when he wrote: "Due to the development of the automobile, armored fighting vehicles, (mechanization), the radio, improved roads and the air-plane, the next war will be characterized by increased rapidity of movement for the main forces."[20]

Such conditions demanded more vigilance for what he saw as expanded flanks, fronts, and rears, and the need for larger mobile reserves. Like MacArthur, he too allowed that these expanded fronts required two kinds of cavalry, mechanized and horse. He held the common prejudice of the day that mechanization promised good road mobility and strong offensive power, but that it lacked the cross-country mobility and defensive holding power attributed to horse cavalrymen equipped with rifles and machine guns.[21]

Thus, generals MacArthur and Henry, the two men in the most important positions to determine the nature and role of the cavalry force were generally synchronized in their views and beliefs.[22] They both believed that the future would demand more than the army and the cavalry were doing, while at the same time each was willing to accept the continued role of the horse in some capacity. Therefore, the stage was set to continue the trend in the specialization of units performing reconnaissance.

Late in 1931, in an effort to save money, MacArthur disbanded the Mechanized Force at Fort Eustis, formed only a year before.[23] The new unit was commanded by BG Daniel Van Voorhis. It drew on all branches of the army but had only attained a degree of concentration in November 1930. The ground reconnaissance functions had been performed by Troop A, 2nd Armored Car Squadron, with ten cars, built by various manufacturers with varying degrees of armored protection and armament.[24] Testing of equipment and doctrine had been limited to a two-day exercise in June 1931 when the platoon reconnoitered routes for the Mechanized Force and provided the commander with continuous updates by radio, motorcycle messenger, and even by messages snatched by airplanes off a wire and dropped to higher headquarters.[25]

The Virginia exercise had been largely scripted, providing a fairly accurate view of what the expectation was for ground reconnaissance while at the same time exposing one of the fundamental flaws of peacetime maneuvers. The armored cars were to travel the routes to determine the best ones for the advanced guard and main body of the Mechanized Force, and to gain and maintain contact with enemy forces. The constraints of peacetime training dictated that the reconnaissance force commander receive a map with the enemy force positions and general disposition.[26] The armored car troop was thus able to perform its designated function by keeping the Mechanized Force's commander updated and allowing him to direct the employment of the remainder of the force, but it had not been forced to "fight" for the information. Had the men in the troop been forced to fight for the information, they would have had to depend on some degree of protection from their armored vehicles, a protection no mounted man could ever claim. They also could have demonstrated the ready firepower of their vehicular mounted machine guns. Even so, the general expectation was that the ground reconnaissance element in the form of the armored car troop would be able to find the enemy and develop the contour of his flanks. This general theme continued even through some of the first action of World War II.

With the Mechanized Force disbanded, each branch was to develop its own mechanization policies and conduct its own projects. MacArthur expected Infantry Branch to focus on its ability to attack strongly held positions. In keeping with his earlier expressed views, Cavalry Branch was to focus on "reconnaissance, counter reconnaissance, flank action, pursuit, and similar operations."[27] To accomplish this mission MacArthur directed that: "The mechanization of one regiment is the first step in determining the application of modern machines to Cavalry missions in war and in developing the technique and basic tactical principles applicable to Cavalry in which the horse is replaced by machines."[28]

The decision to disband the Mechanized Force at Fort Eustis, Virginia, spawned a new debate inside the cavalry community about where the new all-mechanized cavalry regiment should be stationed. Some were in favor of the Mexican border, where the regiment could work with the horse cavalry division and others focused on proximity to the industrial heartland. In the end, Fort Knox won, and Van Voorhis, accompanied by the remnants of Cavalry Branch's Mechanized Force arrived there in late 1931 and waited for the arrival of additional troopers.[29] MG Henry notified the commander of the 1st CavRegt, then stationed at Marfa, Texas, that "with a feeling of sadness" the "oldest mounted organization" was about to undergo a dramatic change.[30] The troopers moved north without their horses in January 1933.[31] Before the regiment even arrived at its new home, the War Department, in 1932, had created the parent headquarters for what was to become its first all-mechanized cavalry brigade by creating the 7th CavBde(M).[32]

Having secured the War Department's approval for a mechanized cavalry regiment, Henry's task was to develop it in a manner consistent with independent operations.[33] The regiment would perform "distant missions covering a wide area," but not hold terrain without the support of infantry, field artillery, or horse cavalry.[34] On paper it had 42 officers and 610 enlisted men filling out a covering squadron and the combat car squadron. A "combat car" was the equivalent of a "tank."[35] Emphasizing speed, combat cars were to be light and fast, not the plodding type one might expect to accompany an infantry attack. The force's designers expected the covering squadron to operate well in front of the regiment while conducting air-ground liaison and searching for enemy positions.[36] The regimental commander expected the squadron to fulfill its role of keeping him informed by employing the ten armored cars in the armored car troop and the seven scout cars in the scout troop.[37]

The 1st CavDiv in Texas continued to experiment with its armored car squadron, too. The men developed new techniques and equipment to facilitate their ability to accomplish missions, including the development of antennas that allowed them to communicate on the move during their 600-mile road march from Fort Bliss, Texas, to San Antonio.[38] They also received new Marmon-Herrington "4-wheel, 4-wheel drive" trucks, which they tested under a variety of conditions.[39] In small ways the mechanized reconnaissance men serving the horse cavalry division could now communicate better via radio than the horse-mounted units they served and were narrowing the gap in the race of tactical mobility still claimed by the horse.[40]

The 7th CavBde(M) received visitors from Germany in 1933. Two officers from the German General Staff toured Fort Knox, while back home men with names that would become famous during World War II—Gerd von Rundstedt, Ewald von Kleist, and Fedor von Bock—conducted war games. Beginning in late 1932, these games exercised concepts involving mechanized reconnaissance. American military observers watched from the sidelines.[41] Back at Fort Knox, Van Voorhis remarked that the visitors, "were not particularly interested in our equipment, which was certainly not formidable at the time, but were interested in our views on the proper tactical and strategical employment of mechanized forces."[42] Van Voorhis's operations officer, Robert W. Grow, who accompanied his commander from the 12th CavRegt through the trials with the Mechanized Force at Fort Eustis and on to Fort Knox, echoed the sentiment. With the benefit of hindsight, Grow saw the linkage with the rise of Adolph Hitler in Germany the same year when he wrote, "Under Hitler they were in a position to act while we were denied funds to [do] more than improvise and experiment."[43] One of the visitors, COL Adolph von Schnell, who paid a return visit to Fort Knox in 1936, later became the chief of motorization in Germany.[44] Even if the Germans were interested in doctrine under development at Fort Knox, those serving at Fort Knox were still fighting for a chance to use their new force in the field for the first time. The post commander, BG Julian R. Lindsey, helped with decisions that revolved around money and with Civil Conservation Corps (CCC) obligations that allowed the mechanized cavalrymen to exercise their new steeds for the first time, including a practical test of their newly acquired mechanized reconnaissance vehicles.[45]

The first major test of the reorganized 1st CavRegt(M) took place at Fort Riley in April and May 1934. It was hoped that such an exercise at the home of the Cavalry School would build confidence in the idea that the two strains of cavalry could exist in one common branch.[46] It was also the same month that the saber was discontinued as a cavalry weapon.[47] COL Adna Chaffee, who now commanded the 1st CavRegt(M) and oversaw its road march from Fort Knox to Fort Riley, proved his continued interest in horses by serving as a judge at the annual Fort Riley horse show upon arriving.[48] But the regiment had not traveled to Kansas to watch horse shows, rather it was there to provide demonstrations on mobility and firepower before taking part in a relatively large force-on-force exercise. Broken into two phases, the 1st CavRegt(M) was pitted against the Cavalry School's provisional brigade, built around the 2nd and 13th CavRegts during the first phase. They were to carry out coordinated operations against a notional enemy during the second phase.[49]

The demonstration phase gave the umpires for the upcoming exercise, and all others in attendance, an opportunity to see the effectiveness of the weapons and equipment used by the 1st CavRegt(M). Firepower and driving demonstrations culminated in a mock reconnaissance mission that orchestrated all of the techniques developed to date.[50] The distinction between the armored car platoon's distant reconnaissance mission and the close reconnaissance tasks delegated to the scout troop became apparent during the full regimental demonstration as the armored car troop deployed forward of the regiment to gain and maintain contact with the enemy while the scout troop secured the immediate advance of the main attack delivered by the combat car squadron. Later the reconnaissance men would withdraw to the flanks as the fighting elements—combat cars and machine guns—closed with the enemy.[51] Notional enemies gave way to force-on-force maneuvers, but heavy rainfall hampered the 1st CavRegt(M)'s mobility during the initial phase. The armored car troop completed distant reconnaissance during hours of darkness that enabled the main body to avoid destroyed bridges.[52] Not only had it accomplished its mission at night, but it had done so largely without the assistance of its radios, which were unreliable during the rain experienced that night. Most observers focused on the loss of small bridges, which altered the mechanized force's plan, while seeing merit in the horse-mounted force's ability to manage some nineteen miles over muddy roads in two hours. What they seemed to miss was the mechanized regiment's ability to cover approximately sixty miles that evening as it shifted its entire thrust to what it thought was an intact bridgehead.[53] The mechanized ground reconnaissance effort during the exercise proved the ability of scout cars and armored cars to operate with speed in both day and night and in poor weather.

The Cavalry School's horse brigade also used scout cars in an emerging race for the bridge, which they won, forcing the mechanized regiment to detour seventy-five miles before bivouacking, refueling, and conducting further reconnaissance. The conclusion drawn from this occasion was that it exposed the weakness of the mechanized force because of its reliance on fuel, as well as the vulnerability the force experienced from the undue fatigue of such long marches.[54] No comparison was even suggested to reflect how a similar body of mounted men might feel after covering such a great distance in what would have certainly been a longer amount of time. It was clear in the minds of many that as long as the mechanized reconnaissance served the horse units by racing ahead to blow up bridges it was a good thing, but in the larger scheme of things mechanization and motorization seemed to be too dependent on terrain and fuel.

The emerging ground reconnaissance organizations serving both the horse and the mechanized units had performed well on both sides and were gaining acceptance in both communities. The exercise also further convinced the horse-mounted community that there was no turning back from the requirement to deploy machine guns capable of deterring the combat cars and tanks whose role remained debatable. Unquestionably, these weapons were required to defeat the now fully accepted presence of the armored and scout cars in the arena of reconnaissance and counter-reconnaissance.[55] The importance of bridges was not lost on the participants of the 1934 exercise.

Addressing the Army War College in September 1934, the new chief of cavalry, MG Leon B. Kromer, presented his own observations and lessons learned from the recent exercise. He believed the 1st CavRegt(M) was good for employment during daylight hours, mainly in delaying engagement and holding key terrain until horse cavalry could arrive to relieve it. Kromer pointed to the 1st CavRegt(M)'s considerable amount of firepower in the form of some 155 .30-caliber machine guns and 53 .50-caliber machine guns. He also recognized that communication was limited to simple codes on three voice channels, one of which was entirely dedicated to reconnaissance reporting and the dissemination of reconnaissance information. Kromer was confident that mechanized forces would be capable of countering other mechanized forces but "should not be able to stop horse patrols."[56] Therefore, one might argue, great faith was still placed in the horse's ability to penetrate the enemy's defenses to gain information, especially under cover of darkness. But Kromer realized the mechanized threat could not be discounted, hence his emphasis on the importance of defending streams and water obstacles and the increasing role of demolition to deny the use of bridges. The new chief also called for the complete motorization of all cavalry logistics, the addition of six light machine guns in every line troop, and the assurance that each of the army's remaining thirteen horse cavalry regiments would be equipped with a platoon of six scout cars.[57]

One of the students present, MAJ R.I. Sasse, introduced another important and interesting theme that took a different view on the issue of reconnaissance by mechanized units protected by armor. He acknowledged that personnel in such units would have to be highly trained in such skills as map reading and observation, but he failed to draw the direct connection that reconnaissance tasks were moving beyond the scope of the generic cavalry trooper.[58] All troopers learned to ride a horse and fire a rifle or machine gun, but now the men who were expected to range far to the front could no longer be drawn from any line troop at a moment's notice. Sasse, like Patton, recognized the value of armor protection and agreed with Patton's assertion that this armor gave them a degree of immunity, which should allow for more aggressive reconnaissance.[59] Patton, typically straddling the issue, saw a need for combat cars to accompany distant reconnaissance, but at the same time espoused a continued role for horses when operations were taking place closer to the main body.[60] While the senior leadership seemed to focus more on the shortcomings of the recent exercise, some of the army's junior field-grade officers were beginning to debate the best way to employ the advantages provided by mechanized reconnaissance.

Kromer focused on the reality of the times, remarking that "[t]he situation is not unlike our drilling Infantry with wooden guns in 1917," as he acknowledged the tenacity and resourcefulness of the troopers in creating the equipment they needed.[61] But his true feelings were best expressed when he said, "We must not expect too much of our mechanized regiment pioneering with inadequate equipment in the maneuvers against our highly developed horse cavalry."[62] His comment on the "highly developed" force of cavalry was typical of army mentality in general during this period. Testimony to Congress often found experts asserting that American equipment was superior to foreign competitors, when in fact the United States lacked the required technical intelligence-gathering abilities to make a good assessment. Coupled with a "can do" mentality, such false pride all but ensured that the United States would be doomed to a game of equipment "catch-up," regardless of how good the theory was.[63]

The 1934 maneuvers exposed an important trend in the role of reconnaissance. Mechanized units working to the front of dismounted or horse-mounted units were able to use their speed and range to provide the main body with the maximum amount of time to make a decision. A motorized reconnaissance unit serving a motorized or mechanized force did not have the same luxury of time since its parent unit was equally fast. Hence, both the mechanized and horse-mounted regiments were growing increasingly dependent on their specialized reconnaissance formations. No unit could afford the time required for horses to march to the front. The detail provided by horse-mounted units was greater, but getting there first seemed to be more important.

By the end of 1934 there was little question that a lasting role for mechanized reconnaissance was secure. Troop A, 1st Armored Car Squadron, operating with the army's only organized cavalry division, had definitely assured a role for itself. The armored car troop operating to the front of 1st CavRegt(M) did not have to compete with any other unit, especially horse-mounted units, for its role leading the regiment's combat cars. Even with the future of some form of mechanized reconnaissance secured, there was still much to be learned and many unresolved issues. The notion that armored cars would operate in excess of 100 miles to the front of the main body seemed to be diminishing, and certainly this type of mission was absent from any of the exercises carried out in the first half of the 1930s. Commanders were growing more reliant on the command and control capability provided by scout and armored cars. Some in the cavalry community were beginning to raise questions about the combat and survival capabilities of these specialized troops. With such questions came debate over what could and should be expected from the troops. This debate carried over into the realm of how the troops should be equipped to perform their missions. It was clear at the end of 1934, there was no going back. There would hereafter be a role for mechanized reconnaissance units, but there was still plenty of room for debate about how great a role the men on machines could expect to play in the next war. Even though the cavalry had discontinued the use of the saber in 1934, troopers continued to practice shooting at targets from the backs of galloping horses.[64] As 1935 dawned, Adolph Hitler restored conscription and acted on his own grandiose thoughts that had little to do with target practice from the backs of horses; he was bent on creating three panzer divisions before the end of the year.[65]

> As to the desirability of having them operate together there can be no question. Both arms are powerful. Both of them have their weaknesses. In a coordinated effort the powers of one remove the weakness of the other. Therefore, the combination of force is not only twice as strong as either one alone but immeasurably stronger.[66]—*COL Bruce Palmer to the Army War College, 1936*

★   On 5 April 1935, the army reaffirmed the mechanization and motorization policy first promulgated on 1 May 1931.[67] In 1935, the Cavalry

School at Fort Riley published an entire section of its *Cavalry Weapons and Material* manual dedicated to motor vehicles.[68] Not only did the manual provide detailed explanations about automotive principles such as wiring, cooling, axles, ignition systems, and maintenance, it also suggested that the United States must utilize motor vehicles as much as possible. Specifically, it called for increased use of motors to replace horse-drawn transportation, carry weapons and personnel to battle, and close with the enemy in the assault.[69] Mechanization and motorization were not merely passing trends.

COL Charles L. Scott commanded the 13th CavRegt during the Cavalry School maneuvers in the fall of 1934. He espoused the belief that "the service of information through reconnaissance is the paramount mission of these cars," and that they "must be considered the direct information agency of the regimental commander."[70] He was convinced that mechanized reconnaissance assets were required at the regimental level and that mechanized forces could accomplish these missions better than men mounted on horses or the infantry. Summarizing his conviction, he noted that "we shall need armored cars as well as scout cars (or reconnaissance) cars, in the regiment."[71] This proposal represented one of the next major steps in the evolution of mechanized reconnaissance. Scott proposed a combined arms approach to reconnaissance, and although it would not be realized for years, it was at least being considered at this early stage in the process of developing reconnaissance doctrine and organizations.

In the past, the detachment of a troop, squadron, or regiment might be expected to help it conduct distant or strategic reconnaissance. The airplane largely eliminated that role, but the cavalry responded to the challenge with the concept of the armored car, capable of ranging out in front with a sufficient amount of protection and firepower to operate independently. The threat of an enemy similarly equipped gave rise to a need for scout cars to operate somewhat closer to the horse regiments and, among other things, to provide enough time for what was becoming an increasingly heavily armed horse cavalry to deploy its heavy machine guns. Now, Scott proposed to give the regimental scout cars a more robust traveling companion to help them accomplish their missions.

There were three major field exercises in 1936 that tested the army's now varied cavalry formations. The 1st CavDiv, led by the soon to retire but still vocal horse advocate, BG Hamilton S. Hawkins, took the field in April and May to test and train with its modern equipment. As the army's largest cavalry formation, the division contained armored cars at the division level and scout-car platoons in each of its four horse cavalry regiments. The only piece of cavalry equipment not available to the division for this exercise was the combat car.[72] The other two major exercises, collectively known as the Second Army maneuvers, included the 7th CavBde(M), first during the V Corps Area maneuvers held at Fort Knox, followed by the VI Corps Area maneuvers at Camp Custer, Michigan. These maneuvers gave BG Van Voorhis the opportunity to exercise his unit with the changes made after the 1934 Fort Riley maneuvers; they also provided future army chief of staff, George C. Marshall, who commanded an infantry brigade during the Michigan phase of the Second Army maneuvers, an opportunity to see the impact of mechanization firsthand.[73]

Going into the Second Army maneuvers in 1936, the 7th CavBde(M) still consisted of the 1st CavRegt(M) and the 68th Field Artillery(M). The cavalry regiment abandoned the two-tiered reconnaissance structure seen in 1934, which incorporated the use of armored cars for more distant reconnaissance and held scout cars closer to the main body. During the Kentucky phase of the exercises, the armored car platoons sought routes around the flanks of the enemy force and into his rear areas and remained undetected throughout most of the exercise.[74]

Having completed operations as part of the V Corps portion of the Second Army maneuvers on their home turf, the 7th CavBde(M) moved north to Camp Custer. This was the most extensive and first modern exercise of the United States Army since 1918 and involved 26,000 troops from the National Guard and the regular army.[75] The accounts of the first day of activity rendered by George C. Marshall, who commanded the 12th Infantry Brigade, and COL Palmer, who commanded the 7th CavBde(M), were very different. Marshall remembered successfully screening the corps he supported and capturing eight of eighteen armored cars from the 1st CavRegt(M) that opposed him on that day.[76] Palmer wrote in the *Cavalry Journal* that his use of the horse-mounted 106th CavRegt in the north and Troop A of the 1st CavRegt(M) gained the complete "contour" of the opposition. Using its ability to conduct long movements, one armored car platoon located the enemy's corps headquarters and other critical logistics nodes well in the enemy rear.[77] Palmer made no mention of what would have nearly amounted to a 50 percent casualty rate as the price of this information. Perhaps knowing that he could only use Troop A and the 106th CavRegt on 13 August because of exercise limitations, and with the full knowledge that these forces would be returned for the next day's mission, Palmer acted boldly by sending his reconnaissance force on this extended mission without the benefit of support. Marshall and Palmer's differing interpretations about the performance of the armored cars became largely irrelevant as their forces joined for an action that pitted them against two National Guard divisions.[78]

On 15 August, the 12th Infantry Brigade continued its attack, whereas the mechanized cavalry sought to envelop the enemy from the north rather than the south as it had on the previous day. Palmer teamed one platoon of armored cars from Troop A, 1st CavRegt(M), with 3rd Battalion, 2nd Infantry Regiment, the only motorized infantry formation at the exercise, for an independent mission. The armored car platoon preceded the truck-mounted infantrymen until it made contact on the enemy's extreme left flank. The infantry dismounted and attacked while the armored cars continued the envelopment in an effort to attack the enemy artillery supporting the forces now under attack by the motorized infantry battalion.[79] In this example, the armored car platoon far exceeded its expected role of reconnaissance by seizing the initiative on an open flank to strike at a slow-moving and poorly defended unit operating in the rear area.

Palmer concluded at the end of the Second Army maneuvers that it was better to organize the resources found in the covering squadron into a reconnaissance troop and a machine gun troop. He believed these supporting arms and the mortars in the regimental headquarters could be easily attached to the combat car squadrons as needed. He also called for the inclusion of more infantry to perform dismounted patrolling and outpost duty.[80] Although there were no mortars in Troop A at this time, they were held in high esteem as an important addition to the mechanized regiment, and the mechanized cavalrymen enthusiastically embraced the ability to obscure the enemy anti-tank gunner's aim with mortar-delivered smoke rounds.[81]

One month later, COL Palmer addressed the Army War College at Fort Humphreys, Washington, D.C.[82] His remarks on this occasion captured the essence of what was expected of the ground reconnaissance agencies in the mechanized regiment and also spoke to the proper roles they should be performing in the more numerous horse cavalry regiments. Palmer opened on a balanced note for what would have been a conservative audience at the war college by remarking that the recent maneuvers had been driven by the need for a force capable of reconnaissance and combat with a high degree of mobility and striking power. He also noted in the introduction that the role of the horse was not gone yet, but that the horse cavalry was "in need of repair and building up."[83] This building up was to include the addition of more anti-tank guns to each of the horse regiments and the introduction of chemical mortars into the horse cavalry division.[84] Palmer's comments on scout and armored cars were also conservative.

Addressing the issue of scout cars, an entity common to both the horse and mechanized force, Palmer reiterated the standard argument that the cars were there to help preserve horse flesh and to allow the commanders of horsed units to disperse these mechanized assets across a broad front to locate the enemy main body.[85] Deeper in his remarks he went to great lengths to spell out the proper role of scout cars in the horse cavalry:

> . . . the sole purpose of the reconnaissance vehicle with horse cavalry is to gather information by observation. Its use for combat will be accidental, emergency or self protective. It is employed in small groups which are enjoyed to avoid fighting. The information it returns will be indicative rather than detailed, and positive rather than negative.[86]

Palmer couched his comments on the use of the combined nineteen armored and scout cars in the mechanized brigade in somewhat different terms. Here he praised the wheeled vehicles for their range, speed, and quiet. Quiet had never been a positive attribute assigned to the mechanized reconnaissance units in the past and had in fact been a constant detractor. One can only assume that in just a few years of tactical operations, if not the first combined use of wheeled and tracked combat cars, the stealth of the wheeled platform in contrast to its tracked companion was immediately apparent. Speed, in Palmer's view, was not an asset for the pursuit; rather, it afforded the reconnaissance men the ability to "thrust out rapidly" on their missions while the combat element of the mechanized regiment marched to their support. He expected armored and scout cars, like their brethren in the horsed units, to use their .50-caliber machine guns during the time that intervened between their first contact and the arrival of reinforcements and to counter enemy mechanized reconnaissance forces.[87]

This view of the use of mechanized reconnaissance reflected the mechanized cavalry regiment's interpretation of how it was to accomplish its missions. Rather than waiting for the detailed reconnaissance that most agreed could only be accomplished by a man on horse or on foot, the regiment expected to act like its own reconnaissance agency on a larger scale. Not willing to wait for a detailed reconnaissance report, it too would use its mobility to "thrust" out and seize key terrain only cursorily examined by the scouts forward of the main body. This not only allowed for the rapid projection of direct fire weapons, but it also envisioned the ability to move artillery forward even faster while at the same time affording the forward observers good locations to call for and adjust indirect fire. The mobility and armored protection of the regiment helped to safeguard it against the possibility of surprise associated with less than detailed reconnaissance. Palmer emphasized that each position occupied by the regiment had to afford more than one way in and out, and that the unit could at a minimum displace rapidly.[88]

The ubiquitous issue of mobility, a constant companion to every discussion of cavalry and reconnaissance, was not absent from Palmer's remarks at the war college. During the body of his lecture he did not fail to acknowledge the standard criticisms of mechanized reconnaissance and the movement of larger mechanized combat formations. In the question and answer session following the lecture, however, Palmer somewhat contradicted his earlier remarks and rather supported the general feeling that mechanized units were dependent on good roads, bridges, and a steady supply of fuel, oil, and lubricants.[89]

QUESTION: How does the lack of good roads and mud and water affect operations of the mechanized force?

ANSWER: As I said, we have been a little bit astonished ourselves at the excellent results we have gotten from our vehicles off the roads. I would answer your question by saying that beyond slowing us down slightly, it has no effect at all. We go right ahead and keep going at good speed.

The next question was even more specific, regarding the defense needs found along the U.S.-Mexican border, then home to the army's largest concentration of cavalry forces.

> QUESTION: Would that gumbo mud in West Texas affect it much, that sandy black mud?
>
> ANSWER: It will affect us, but it won't stop us. Our present four-wheel drive scout car will negotiate practically any condition. Its cross-country mobility is pretty close to that of the combat car and the combat car is pretty difficult to stop.[90]

Both answers spoke to the possibility of more mobility by mechanized reconnaissance forces than had previously been allowed for in the debate. Few questioned the greater mobility of tracked vehicles, but these vehicles remained unacceptable for reconnaissance in most minds because of the guiding fundamental of stealth. Palmer's answer to the second question dodged the matter on one major point. His response referred to the scout car and excluded the armored car, then in use by the mechanized regiment to move at the farthest point forward of the regiment. With its additional armored protection, and the weight that came with that protection, the armored car certainly would have lacked the mobility of its lighter cousin, the scout car. Palmer chose to focus on the positive aspect of the scout car's mobility while neglecting its lesser degree of protection.[91] Although Palmer presented the war college audience with a view of mechanized reconnaissance from the perspective of the all-mechanized force, the horse soldiers of the army's only cavalry division spoke volumes about their thoughts on mechanized reconnaissance in the methods they chose to use as iron horsemen during their divisional maneuvers that same year.

Maneuvers in 1936 provided an opportunity to exercise the horse division's new equipment and capabilities, including an increased number of radios, mechanized elements, and anti-tank weapons. The exercise also served as BG Hawkins's swan song as he neared retirement, although he would continue to influence the future of the horse in cavalry operations by using his editorship of the *Cavalry Journal* to espouse such ideas. For this exercise he not only wanted to test the new equipment and exercise the division as a whole, but also to provide his brigade commanders the opportunity to test their skills against one another.[92] A motorized infantry company from Fort Sam Houston, Texas, joined the exercise to provide Hawkins with additional firepower and the advantages of a more mobile infantry force.[93] Like previous exercises, the availability of land and water restricted much of the scenario planning. While the horse-mounted soldiers planned their operations with an eye toward this constraint, their mechanized elements in the division's armored car squadron, still only a troop in size, and the scout platoons in each of the regiments had little to worry about in terms of the terrain they would have to traverse in the conduct of their missions, since it was considered nearly ideal.[94]

During the first phase of the exercise, mechanized elements displayed more cavalier action than their division commander might have liked as suggested by his observations about overexposing themselves to hostile fire, but the horse soldiers' actions were no less flamboyant.[95] The men on horses conducted mounted attacks in waves against deployed machine guns, while noon armistices allowed each side time to water its horses and men at wells before resuming the fights in the afternoon.[96] The divisional phase of the exercise held little for the mechanized reconnaissance men as, "Armored and scout cars are assumed to have finished their reconnaissance duties prior to the action and are employed during the action for flank protection and for liaison duties."[97] In retrospect, for an exercise whose stated purpose was to test and integrate new equipment and rehearse the orders process, it seemed odd that no attempt was made to integrate the use of the division's and each regiment's distant reconnaissance agencies. Unlike at Fort Knox, the mechanized men of the 1st CavDiv were there only to serve the horse and only if the horse wanted to be served. Not surprisingly, the entire division staff abandoned its cars for this portion of the maneuvers as Hawkins elected to command from horseback.[98]

During one portion of the exercise, Hawkins selected COL John K. Herr to command the delaying regiment and gave him the use of the motorized infantry company, a platoon from the division's armored car troop, and a battery of field artillery from the division's artillery regiment. Hawkins issued a written field order to the remainder of the division that sought to push through the resistance offered by Herr and his reinforced 7th CavRegt. Hawkins's use of his mechanized reconnaissance elements left some room for interpretation about how he really viewed their capabilities. He directed the division's remaining armored car troop to reconnoiter a specific route and area. Once there, they were to report on the presence of the enemy and his disposition. Hawkins directed that the regimental scout cars operate to the flanks while the field order directed that each of the marching columns establish an advanced guard that was to be equipped with two .50-caliber machine guns.[99] Unlike the same kind of machine guns carried on scout cars, these two machine guns packed on the backs of horses had to be deployed upon contact. Again, the division commander elected to command and control the two marching columns from horseback allowing his radio truck to follow as closely as possible. As the report on the maneuvers suggests, "Unfortunately, this was not as close as might be desired."[100] It was clear that by deploying the armored cars well in advance of the division, Hawkins gave himself some flexibility to deploy once contact was established with the opposing force. Yet, he minimized his ability to act on the reports offered by his radio-equipped mechanized scouts by placing himself at an "unfortunate" distance from the receiving end of those reports, his own radio truck. In regard to the use of regimental scout cars, Hawkins showed himself as a micromanager reaching over the heads of even his brigade commanders in telling the regimental commanders how to organize their advanced guards. Perhaps his desire to see the use of packed machine guns blinded him to the fact that the same machine gun was mounted on each of the regimental scout cars. Not only were these weapons on the regimental scout cars, they were also immediately deployable, far more mobile than a tripod-mounted machine gun, and the operators of the scout car were at least provided some modicum of protection from hostile fire by the armor plating on each of their vehicles. All of this was apparently lost on Hawkins, based on an interpretation of his field order and actions during the exercise.

COL Herr also used armored cars on the flanks and to counter the thrusts of Hawkins's armored cars. Herr used a section of cars to get himself around the battlefield to check the readiness of his units as well as to command and control each of his squadrons and the attached infantry and field artillery. He held the other section in reserve for future reconnaissance roles or to be used to reestablish lateral communications. There was some indication that the mechanized units were used to assist units in withdrawing from their initial positions to subsequent prearranged locations on the battlefield. Herr believed that his portion of the operation provided an excellent demonstration of the use of the radio and mechanized elements to command and control horse elements.[101]

Like Hawkins, Herr failed to maximize the potential of his mechanized assets. Instead of deploying the armored cars well forward across his front to report and delay the marching columns of horsemen, so deployed that all they could have countered with was packed machine guns, Herr relegated them to one flank. The diagrams accompanying the article seemed to indicate that this might have been prearranged with Hawkins, thus allowing the armored cars to battle in their own arena out on the flank so as not to disrupt the main event. Herr used the scout cars to increase his ability to command and control his unit, but in so doing prevented them from fulfilling their intended mission of reconnaissance forward. Later, it fell to Herr to oversee the planned expansion of the mechanized cavalry forces, often at the expense of his beloved horsed units. He lost control of the men at Fort Knox in 1940, seeing them permanently taken away from Cavalry Branch, and he was the last chief of cavalry when all the branch chiefs were eliminated in 1942.[102]

Thus in 1936, even the use of scout cars was not consistent in the horse cavalry community. The men in Texas might be interpreted as being overly concerned for the possible loss of this important emerging technology; hence their decision to relegate scout cars to the flanks on most occasions. This would be more plausible had the commanders in Texas ever committed their mechanized resources to a mission deemed important enough to merit the risk of losing them, following the old precept that great risk can offer great gains in return. But as the 1936 1st CavDiv maneuvers indicated, this was never the case. The men at Fort Knox, with their mechanized regiment, had little choice but to use their armored car troop to gather the information they needed. Not surprisingly, it was the same men at Fort Knox who started to become even more vocal about what type of scout cars should be used to accomplish these missions.

COL Scott's article in the *Cavalry Journal* of July–August 1936 reflected the thinking of those who had spent the most time working with mechanization. It also reflected the lingering economic effects of the Great Depression and the role it played in how ground reconnaissance units should be equipped. Scott argued against the urge to equip the army with a fleet of scout cars that could be purchased straight off the assembly line, thus eliminating the cost of special modifications and the research and development involved in such a process. He cited the testing that had been conducted, which favored the use of specially designed cars over commercially produced models. With some flair, Scott suggested that those who opposed the acquisition of specially designed reconnaissance vehicles were the same kind of people who thought the infantry could be equipped with any rifle and that the air corps could accomplish their missions with any type of plane.[103] Having put forward this line of reasoning, Scott was equally quick to point out that should war come tomorrow the army would probably have to resort to such measures, but that they would never "meet all the cavalry needs by any means."[104]

One of the most important reasons a car right off the assembly line would not meet the needs of the mechanized scout car was the simple requirement for features such as a four-wheel drive, radio and weapons' mounts, and the minimum amount of armor required to protect the crew from medium- to long-range fire from .30-caliber weapons.[105] It was the issue of armor protection, a generally accepted requirement for armored cars conducting more distant reconnaissance, that now entered the debate over how scout cars should be designed. Scott did not shy away from answering the question that probably dated all the way back to antiquity when the first man of more means donned some form of protection while the men around him had none: Why armor when the man in the ranks or on horseback has none? Simply, a man on foot, or even mounted on horse back, can make more ready use of terrain than the man mounted on a vehicle, unless as Scott suggested, one wanted the mechanized scouts to immediately stop and abandon their cars on first contact. In Scott's view, the protection provided by a minimal amount of armor gave the scout-car men a sense of boldness that translated into the effective performance and accomplishment of their assigned missions.[106]

Scott recognized the potential of America's industrial might and did not suggest that the army continue in any way the tradition that had started with CPT Holt in the late 1920s with his hand-crafted cars. Rather, he saw the goal of the cavalry and the army as this:

> . . . the cavalry problem of today is not to be satisfied with commercial designs unsuited for wartime use, but to work up, the use of major commercial parts, a simple effective *fighting vehicle* with armor and armament so that it can be produced from blue prints by the automotive industry in about the same time as a standard commercial car.[107]

Perhaps he reached even more deeply within the cavalry community when he couched the issue in terms they were all familiar with.

> To adopt a policy for taking any cheap ill-bred scrub iron horse obtainable in quantity on the streets of the nation is certain to greatly reduce the effectiveness of mechanized cavalry units in the performance of their missions. It is a makeshift poorly planned procedure.[108]

The next addition of the *Cavalry Journal* contained an article written by MAJ Robert W. Grow, who picked up the same theme that was becoming a more common issue in the pages of the mounted arm's professional forum.[109]

Grow acknowledged that the continued improvements and advances in automotive technology allowed for a vehicle with almost any capability, just not in a single machine. To those who called for a "light, cheap, commercial vehicle for cavalry," he pointed out that once they added the requisite radio mounts, front-wheel drive, weapons mounts, and a little bit of armor they had in effect the M2A1 Scout Car that was then in use.[110] A cavalryman like Scott, Grow had a professionally imbued belief in the superiority of mobility, but saw it as an equally important cavalry characteristic, a "fighting power" that had to be carried out mounted or dismounted. For him, "protection" rounded out the needs of the mechanized man to accomplish his mission.[111] Whereas Scott used the horse-breeding analogy to make his point, Grow reverted to football terms: "Cavalry operations are comparable to open field running in football. Tacklers are likely to spring up from anywhere, flank or rear."[112] Armor provided the "temporary protection" for such chance meetings in the "open field."[113]

Grow went on in the same article to delineate the cavalry's vehicular needs into two basic categories: those designed primarily for fighting, thus requiring maximum mobility, fighting power, and protection; and those designed primarily for reconnaissance, needing high road mobility at the cost of cross-country capability, maximum fire power, and minimum protection. This seemed very similar to his introductory thesis that the cavalry could have a vehicle that did almost anything, but not all in the same vehicle. His conclusion exemplified the optimistic attitude of those who had been associated with the mechanization of ground reconnaissance from the very beginning. Grow admitted that "vehicles must conform to engineering capabilities as they exist today, always with the hope that the future will offer a 'lighter, cheaper, commercial vehicle' whose characteristics we can accept and apply to the accomplishment of cavalry missions."[114] Rather than wait on Detroit, men like Scott and Grow knew they had to continue to forge ahead, while at the same time not sacrificing what they saw as essential needs because of those clamoring to save a few dollars or perpetuate the utility of the horse for similar missions.

MAJ Grow was doing more in 1936 than simply writing articles for the *Cavalry Journal*, he was also busy with his duties as office chief of cavalry.[115] The 1936 maneuvers of the army's only mechanized brigade inspired a new round of discussion about what types of changes should be made based on what new lessons had been learned. The same type of activity had occurred after the 1934 maneuvers at Fort Riley. The issues of the addition of organic aviation and increases in the attached rifle strength of the brigade were revisited. In this round Grow made it clear that any additional infantrymen should be mounted in "cross-country" carriers, since the ability to maneuver off-road would be key to the unit's success. He went on to say, "I would like to see the mechanized cavalry made completely road free," hoping to avoid the inclusion of any more trucks.[116]

The idea of stationing a permanent air component at Fort Knox also emerged from the 1936 maneuvers. It was hoped that such an arrangement would facilitate air-ground coordination in the forms of "air-ground combat, observation, and communications."[117]

> Combined training by Air Corps and Cavalry units is recognized as being of the greatest importance. The Air Corps-Cavalry team in highly mobile operations, particularly in reconnaissance, is analogous to the Infantry-Artillery team in the division.[118]

There was a strong belief that the relationship between the cavalry and those in the air above was applicable to horse units, but even truer of those in the mechanized cavalry brigade.

Grow also believed that it was time for the office chief of cavalry to recommend to the War Department that a mechanized cavalry division be formed using the men and units at Fort Knox as its basis. The division staff would be an outgrowth of the current brigade staff. The division itself would require the addition of engineers, quartermasters, medical units, and a divisional reconnaissance unit. In Grow's view, they should all be mounted on cross-country carriers.[119] In his proposal were the seeds of future controversy. Grow's solution was to "[m]odify present mechanized regiments by reducing somewhat reconnaissance and as much as possible supply, probably adding rifle powers and adjusting it with assault power, the regiment to be designed to *fight*."[120] Though just a theory at the end of 1936, Grow's proposal sought to place the burden of reconnaissance more firmly on the shoulders of the division commander, freeing the regimental commanders to concentrate on employing their increased combat power. It was not surprising, since all along the advocates of mechanization saw their role as the same as that of the horse cavalry, only mounted on different platforms.

LTC Alexander D. Surles, a cavalry officer attached to the army's General Staff, wrote an article for the *Cavalry Journal* about the possibilities of mechanization of the cavalry. GEN Malin C. Craig, the army chief of staff, even included a short endorsement that accompanied the article, in which he admitted that though "much of the experimentation is theoretical, no modern, progressive army can afford to neglect its potential power."[121] Even so, the general's very next words were, "However, it is likewise true that motorization and mechanization have not driven the horse and mule from the battlefield." He then closed his brief remarks with a statement that supported Surles's theme, offering that, "Both horsed and mechanized cavalry will be present in future warfare, but in what proportion, only circumstance can tell."[122] Surles put a positive spin on the British decision to motorize eight of its twenty-eight cavalry regiments. He suggested that rather than focusing on the decrease in the number of horse units, instead, attention should be directed to the fact that these newly equipped cavalry regiments remained under the control of Britain's cavalry branch. For Surles the real issue facing the United States was in determining the proper ratio of horsed to mechanized units. His personal views on the matter were somewhat illuminated when he pointed out that, except for Russia, the United States had a reservoir of some eighteen million horses to draw on compared to Europe's nine million.[123]

This notion of finding the correct ration of horse units to mechanized units confronted the current chief of cavalry, MG Kromer. Late in 1936, in a letter to BG Daniel Van Voorhis, then in command of the army's only mechanized brigade, Kromer addressed this issue. He pointed out what was going on in England and France. He commented that in Germany they now had an entire mechanized and motorized corps divided into three divisions, which totaled some 4,000 tanks and 17,000 vehicles, and that there were no horses in the organization. He believed it was true that the German cavalry branch had passed up its chance back in 1933 and was now relegated to the role of local reconnaissance and security.[124] At home, Kromer saw infantry regiments being equipped with the same light tanks in service with the units at Fort Knox and felt they were "too light for combat in the infantry attack" and that they were being held in reserve for exploitation. Kromer felt, "This indicates their intention to take over the Cavalry mission of exploitation."[125]

Acceptance of the role of mechanized ground reconnaissance was complete at the lowest echelons. With platoons and troops serving both regiments and the cavalry division, the branch now placed increased attention on how to equip these special-purpose troops, and how and if these organizations should grow beyond the size of platoons and troops. If the role of mechanized reconnaissance was accepted, the demise of the horse was far from settled. Its utility continued to enter the debate that exceeded reconnaissance problems, as the army and Cavalry Branch continued to seek the proper ratio of horse and mechanized units required for the nation's defense.

I frankly admit that if the sole purpose of our existence is to fight Mexicans we can dispense with all pack weapons and do our work very handsomely with the rifle, the pistol and the SABER, but if we are organizing for the purpose of conducting war in Europe or in Asia, or of fighting Europeans or Asiatics in this country, we must have both the light and heavy machine gun, even in Mexico the light 37.[126]—*George Patton Jr.*

★   The subject was machine guns, but LTC George Patton's statement summed up the tone of the late 1930s and begged important questions: what kind of war was the United States preparing for and did the nation have the right equipment on hand to respond to threats from abroad? Patton's own experience in another, somewhat unrelated way, provided the background scenery for the American stage on which the question was posed. As the commander of the 3rd CavRegt at Fort Myer, Virginia, the Pattons' family dog went missing in the middle of every month. Patton, a man of means, posted a reward of a few dollars and the dog was always returned safely.[127] "Dognapping" proved a consistent way to extend a trooper's minimal paycheck to the end of the month in an army already short on funds, in a nation and world of economic depression. The problem of funds did not completely go away in September of 1939, but the question, "What are we getting ready for?" took on even greater importance and in the process continued the debate on how best to integrate mechanized reconnaissance forces in what was soon an expanding army.

The change in the army's priorities started in the mid-1930s when it was generally recognized that the nation's financial situation was not adequate for the current mobilization plans. GEN Craig, who replaced MacArthur in October 1935, recognized that the nation would never be able to buy the massive amounts of equipment needed to support a "million-man" army under the current mobilization plans.[128] Craig's leadership resulted in the Protective Mobilization Plan issued in 1937, which was to provide an initial force built around four regular army divisions and fifteen National Guard divisions, with a complement of corps and army troops bringing the total strength to around 400,000.[129] The nation needed at least a small force ready and equipped to fight immediately. This led to the end of meaningful research and development as the army became more interested in securing the needed equipment at once, and the need to spend funds on what was "functional today" came to exceed the search for "even better" in a few months or years. There followed a downturn in innovation that was exacerbated later as America found itself supplying European allies when war did break out.[130]

As chief of staff, Craig was not hostile to research and development, but his sense of urgency to equip the army was more important and led to purchases of equipment such as the 37mm gun built on the German design.[131] At the time of its purchase, the 37mm was an adequate anti-tank weapon, but also a weapon the Germans soon abandoned for larger and more powerful weapons, like the famous 88mm. Thus, the threat of prematurely committing to any given new technology or system was quickly exposed. On Craig's watch, the Italians invaded Ethiopia in 1935; the Nazis reoccupied the Rhineland in 1936; and the Spanish civil war gave the world a glimpse of new weapons and the techniques of employing them that soon exploded across the world. In 1938, the Nazis marched into Austria and Czechoslovakia. All of these events supplied the motivation to "revitalize" the nation's mobilization planning.[132]

The sense of urgency that infected the chief of staff spread to Cavalry Branch. Expected to fight on M-Day (Mobilization Day), not days, weeks, or months later, the chief of cavalry, in 1938, petitioned to have the branch revitalized with the immediate creation of a corps headquarters, four division headquarters, and their associated reconnaissance elements, and to bring all regiments up to at least two-thirds of their wartime strength.[133] Inside the 7th CavBde(M) the feelings of at least one senior officer mirrored the mood that permeated from above. COL Scott wrote, "Now that the brigade is fully equipped I think the time has come to call a halt on all theocratcial [sic] studies that tend to throw a monkey wrench in the machinery."[134] More practical work, less theory, seemed to be the mantra of those involved in the mechanization process for some time. Not only did the men at Fort Knox feel they had come to acceptable answers on technique and equipment, they watched the events in Europe for confirmation that they were on the right path.

Even while the men at Fort Knox, and other advocates of mechanization, felt they had and were continuing to develop the proper organization, equipment, and techniques, including the use of mechanized reconnaissance forces, efforts to maintain the perception that Cavalry Branch was one big happy family persisted. As good family members they continued to participate as team members, but were also willing participants in the heated debates involving their branch's future. MAJ Grow, a contributor to the *Cavalry Journal* on mechanized reconnaissance issues, prepared an article in 1937 that detailed just how far the branch had come, by creating a very telling chart. The 1937 Horse Regiment was devoid of mules, substituting sixty-eight pack horses to move around its extensive number of machine guns. Both regiments' machine gun totals included sub-machine guns.[135]

Colonels Scott and Palmer, commanders of the mechanized regiments at Fort Knox, sent letters to COL Innis P. Swift, commander of the 8th CavRegt, congratulating him on a recent 150-mile endurance ride conducted by a lieutenant in Swift's regiment, and these letters were published in the *Cavalry Journal* for all to see. Yet no amount of note writing could paper over the growing rift within the branch and the siege mentality that was beginning to manifest itself in those desperately trying to retain a role for the horse.

MAJ Grow captured the cause of much of the concern and made a suggestion on how to deal with the threats to the branch's future in a particularly revealing intra-office memorandum titled "Thought for the Day." He traced the feelings of concern for the future of American cavalry to the demise of large horse units in Europe. Grow's solution was to emphasize that only the *"horse"* had been replaced and that the mission of cavalry remained. To this end he insisted that any unit performing a cavalry type of mission be designated as such. All of this was important because he expected that during the next war an independent tank corps would reemerge and "gobble up" any "loose" mechanized units. His last paragraph summed it up: "MORAL: We must not let the Army lose sight of the fact that the mechanized units designed to carry out cavalry missions *are* Cavalry and, obviously, should be designated 'Cavalry.'"[136] In 1940, events revealed Grow's prescience, for in 1937 he had offered a solution for the branch to retain control over what continued to grow in importance, the mechanized elements already inside the branch.

Whereas Grow's vision was realized, those who fought the hardest during this period to retain all roles for the horse in the cavalry community sought to reduce the number of combat cars in the mechanized brigade at Fort Knox. They hoped that by adding scout cars to replace the diminished number of vehicles dedicated to combat they might push the organization into more of an overall reconnaissance role.[137] Not surprisingly given the tone of his command at the 1st CavDiv, BG Hawkins provided a loud voice backed by years of horse cavalry experience to bolster this general argument, even as he saw a limited role for the mechanized men in reconnaissance. One of his initial points of attack involved the role of cavalry in the opening days of a war, when opposing armies groped across space and time to find one another. Covering forces, that is a body of troops operating independent of the main body, had been used in the past not only to locate the enemy, but also to deny his covering force the ability to locate friendly troops. Hawkins argued that it was impossible to expect a mechanized unit to perform such a role and maintain its mobility over any type of terrain.[138] To this he added the well-worn litany of other reasons mechanization could not be depended on to perform such roles. He included the lack of stealth, the inability to get around roadblocks, and an exaggerated depiction of a mechanized section driving blindly into a town to determine the presence of the enemy. There was also the constant issue of supply: "We cannot block the roads with passenger cars, nor could we supply the gas and oil to move, not only supplies of all kinds, but also all the troops."[139]

For Hawkins the proper place for mechanization was just as he had used it, at the flanks of large bodies of mounted or marching men, since "opposing armies do not rush toward each other in motor cars."[140] If not being used to allow large horsed bodies to march unmolested by enemy mechanization, then the mechanical monsters should find a role closer to the field army, leaving the distant missions to the men on horses. In this vein of reasoning, he predicted that an "independent mechanized force is almost certain to be a failure except in the easiest kind of situation with a very inferior enemy."[141] No, the cavalry must retain its horse-drawn artillery and pack-mounted radios and avoid the possibility of giving up mobility on any type of terrain.

This was the overall tone going into the last years of the Great Depression in the United States while Europe edged closer to general war. The Protective

Mobilization Plan dominated the thinking of the senior leaders in the army and filtered its way down into the smaller arguments most closely related to mechanized reconnaissance. Starting with the belief that mechanization was now a permanent fixture within the army and the cavalry, Van Voorhis believed the best way for the nation to maximize its industrial potential in the next war was to continue enlarging the role of mechanization. "In Germany where there are newly built eight-lane highways leading to the frontiers it may be possible to maneuver the German Panzer Corps with its 4,000 tanks and 13,000 other vehicles."[142] Reflecting awareness of the world in which he lived from a financial perspective, he acknowledged the army could expect to spend millions. Van Voorhis was adamant that the money be spent in a manner that best served the defensive needs of the country. To this end he recommended the expansion of the 7th CavBde(M) into a full-fledged division of not more than five hundred vehicles.[143] Somewhat in deference to critics' arguments that centered on where the army would be expected to fight, Van Voorhis admitted that the nature of the road network must be considered when organizing such a division.

The key component Van Voorhis sought was an additional armored car troop that would provide the division commander with his own asset to reconnoiter the flanks and rear of the division. Van Voorhis also saw the additional troop as a ready pool of replacements for the armored car troops in each of the regiments. In this respect, operating to the flanks and rear of the division was similar to the employment of armored cars in the horse division, but the expectation that mounted on an identical vehicle they could replace losses incurred in the armored car troops of the regiments illustrated a major difference. He anticipated and expected his mechanized regiments to continue pushing their armored cars forward of their regiments, even at the risk of loss. A replacement pool of these important reconnaissance assets allowed the division to conduct extended missions. This was in contrast to Hawkins's view of how mechanized units should operate in close proximity to the main body of the army. Van Voorhis also proposed the addition of a rifle troop that, when paired with the divisional armored car troop, would form the basis of a division's reconnaissance and support squadron.[144] The rifle troop provided the ability to conduct detailed reconnaissance and patrolling, protect divisional installations, and hold ground. Rifle troops mounted in half-tracks with mounted machine guns were able to keep pace with the other mechanized assets in the division with comparable protection. Although the horse never entered into the equation, Van Voorhis's inclusion of a rifle troop sought to fill the void of manpower required to conduct detailed dismounted reconnaissance and hold ground in a fashion similar to what might have been expected from a horse-equipped troop.

Van Voorhis also sought to increase institutional change through education, politics, and, indirectly, through voters. He made sure West Point cadets had the opportunity to experience the mechanized cavalry experience, because expanded mechanized formations would require more officers to lead them, and who better than newly recruited enthusiasts. Beyond the subtleties of exposing West Point cadets to the increasing possibilities of mechanization for both reconnaissance with the armored cars and combat with the combat cars, the leaders at Fort Knox sought help for their vision of expansion on Capitol Hill. A field-grade officer at Fort Knox called on Senator Henry Cabot Lodge to submit an article to the *Cavalry Journal*. Lodge held a reserve commission in the cavalry and had seen service with horsed, porté, and mechanized cavalry units and also served a tour in the chief of cavalry's office. He was a member of the Senate Military Affairs Committee, and there was a feeling at Fort Knox that, "Because someone is liable to touch off this mechanized sky rocket at any moment it would appear to be the mission of us here on duty with the Mechanized Cavalry Brigade to be conservative about the matter of possible extensive expansion."[145] LTC Willis D. Crittenberger offered to supply the information and pictures that Lodge would build his article around in the hope that the senator's words would be helpful "at this particular time when mechanization is climbing into the saddle."[146] One way which the 7th CavBde(M) "climbed into the saddle" that year was in using its expertise in mechanized reconnaissance to introduce, guide, and inculcate the National Guard in the proper use of such assets, and in doing so the guard made an indirect appeal to the voters who had a vested interest in the future of the army.

The National Guard brigadier used his scout cars effectively and was able to locate the 7th CavBde(M) by not "fritter[ing] away these highly specialized vehicles on command, staff, and messenger missions no matter how urgent."[147] COL Bruce Palmer, who commanded the 7th CavBde(M) for the exercise also remarked on the proper use of the scout cars by his National Guard opponent and further addressed the role of mechanization in scouting by remarking, "Reconnaissance mechanization is purposed for reconnaissance, not for combat, when combat can be avoided, nor for command."[148] Van Voorhis, who observed the exercise, reinforced this theme with his own extended remarks on the subject.

These maneuvers again demonstrated the usefulness of armored reconnaissance vehicles to gain and maintain contact with a highly mobile enemy. Van Voorhis wrote:

> I am convinced that this is the primary mission of our reconnaissance elements in mechanized cavalry as well as the scout car platoon in horse cavalry, regardless of any desire to use these vehicles for command or messenger service, no matter how urgent this desire may be.[149]

By offering these comments publicly in the *Cavalry Journal*, the leaders at Fort Knox made it perfectly clear what they thought about the use of mechanized reconnaissance assets outside of Kentucky. They may have attempted a conservative approach to the expansion of the brigade into a division with its entailed increases in mechanized reconnaissance, but there was no subtlety in their view of how everyone in the cavalry family, Regular or National Guard, horse or mechanized, should be using these specialized units.

Outside the public and professional gaze, COL Chaffee continued to stay abreast of the happenings at Fort Knox. LTC Crittenberger wrote Chaffee about the same exercise. The 7th CavBde(M)'s mortars, artillery positions, and machine gun troop had been effectively engaged by the rifle troops of the National Guard. This, in combination with proper equipment and command, allowed the National Guard to "take issue with mechanized cavalry on suitable terrain."[150] The guardsmen had effectively moved and employed their notional .50-calibers, which continued to "demonstrate the usefulness of the horse to quickly transport fire power from one point to another."[151] Crittenberger's letter also revealed that in addition to COL Scott, ten other regular army officers bolstered the leadership of the National Guard unit during the exercise.[152] For the time, this was nothing less than an incredible injection of mechanized reconnaissance expertise. The mechanized men at Fort Knox went out of their way to insure the National Guard turned in a quality performance using modern equipment, specifically the use of scout cars.

Crittenberger's letter to Chaffee revealed another factor that further illustrated a different interpretation of how the scout cars had not been "frittered" away on command and control, a common abuse in the regular army horse regiments. In addition to the six scout cars, the 7th CavBde(M) augmented the 54th's meager ability to command and control their units with six horse-packed-radios by providing them with three command cars for the brigade headquarters and nine vehicular radios.[153] The National Guard did not misuse their scout cars because GEN Van Voorhis and his regimental commanders removed the temptation. What might be viewed as a short-term solution for promoting what they viewed as the proper employment of a valuable asset was built on as shaky a foundation as the National Guard's fictitious .50-caliber machine guns. Just as the machine guns did not exist, neither did the number of radios required to turn in a similar performance in war. How then can the actions of the Fort Knox leaders be justified?

Machine guns and radios might eventually be procured in the event of war, even for National Guard units. The men of the 7th CavBde(M) realized that more important than the equipment was the doctrine; the ideas behind how best to maximize the on-hand technology. The men in the horse units limited their horizons to preserving the role of their mounts, and mechanization only played a small role in that effort, as long as it served the horse in some role, if not the proper role. Viewed in this way, the actions of Van Voorhis, Palmer, Scott, and the many others who assisted them must be applauded. They must have recognized that were war to happen the next day they could not supply the men and equipment the National Guard needed to fight effectively, but they gave them the more lasting gift of knowledge.

TABLE 1—Advances in Cavalry Armament, 1917–1937

	1917 *Horse Regiment*	*1937* *Horse Regiment*	*1937* *Mechanized Regiment*
Horses and mules	1,028/166	174	0
Sabers	890	0	0
Radios	0	7 on hand 10 planned for in future	36 on hand 44 planned for in future
Machine guns	6	69	522

The growing schism within the branch also came out during the visit from Cavalry School faculty during a presentation by CPT I.D. White. White, a former mechanized reconnaissance leader at Fort Knox, now served at Fort Riley at the behest of the chief of cavalry's office. Probably the first instructor of mechanized tactics, White later recalled that the subject of mechanization "wasn't popular," and that the horse enthusiasts were hostile. Still, he took part in all the horse sports and played on the department polo team, in the finest tradition of the officers at Fort Knox who had long sought by every means to maintain their "horsey" ties to Cavalry Branch.[154] Back among friends, COL Bruce Palmer praised him for his service during the Michigan maneuvers only a year before, remarking that the mechanized regiment's success had rested on White's ability to get the information needed for successful operations. White's remarks were rich in explanation for why the agencies of mechanized reconnaissance had developed as they had and why they were organized and equipped as they were.[155] His remarks also revealed the building tension within the branch, at one point acknowledging his lack of full agreement with the precepts of the Cavalry School he now served.

As many good speeches often start with a joke, White reported on the two words recently "coined" by a bright student at Fort Riley. Intended to be humorous, the words distilled the essence of the ongoing reconnaissance debate. Speaking of the student, "He said when we refer to horse cavalry we should say 'equinoiter,' and when we refer to mechanized cavalry we should say 'mechanoitor.'"[156] Throughout the speech, White repeated the Cavalry Branch party line of there being only one branch that was learning to accomplish its missions by different means. Now as the Cavalry School's instructor on the subject of mechanized reconnaissance, White reminded his audience that the "general principle" of reconnaissance remained the same. Before shifting to the subject of his real expertise, use of the armored car troop used to gather information for the mechanized regiment, White said something that must have raised the hairs on the backs of the necks of the early mechanized pioneers at Fort Knox. White, or perhaps White under the influence of the Cavalry School, was of the opinion that "in the mechanized cavalry regiment we have a unit whose principal duty is reconnaissance and its very organization and equipment and the very fact that it specializes in reconnaissance results in a very definite set of rules governing its employment."[157] Chaffee, Van Voorhis, Scott, and others at Fort Knox were not interested in reconnaissance beyond its ability to serve their greater desires to fulfill the traditional combat missions of the cavalry.

White began his remarks on reconnaissance with the uncontested assumption that the purpose of the armored car troop, then the mechanized cavalry regiment's agency of reconnaissance, but similarly replicated on a smaller scale in all the horse cavalry regiments with a platoon of scout cars, was to gather information. He briefly reiterated that because of the unit's equipment, wheeled vehicles by design, commanders expected them to find the location and composition of the main bodies of enemy forces. This was fully within the capabilities of the unit since "large forces [were] confined to main routes."[158]

White offered the Cavalry School solution: "We say that the best reconnaissance can be performed by stealth. That is a very definite tactical principle. Move as quietly as possible and avoid contact with the enemy."[159] This made sense, especially given the attributes of the wheeled vehicles then in use. White further emphasized the need for much of the reconnaissance to be carried out on foot in such places as defiles and before crossing bridges. This then led to the issue of combat. Armored cars carrying machine guns spoke to the basic understanding that mechanized reconnaissance units expected engagements with the enemy and fought in those situations for self-preservation. White did not challenge this, nor did the Cavalry School. What the Cavalry School did insist on was "a very definite policy . . . that reconnaissance units on reconnaissance missions would not engage in combat except to accomplish their missions."[160] This message was issued by the same school that anticipated mechanized units transitioning from reconnaissance to flank and rear security missions faster than horse cavalry units, "because of the comparatively greater combat strength" of horsed units.[161] White offered the following statement regarding the avoidance of combat: "I don't believe that principle should be ironclad for mechanized reconnaissance elements. I might say that those are my own personal ideas and that they do not meet with the approval of the Department [Tactics] of the Cavalry School."[162] White did not deny the primacy of the reconnaissance mission, but he envisioned that "the initial stages of contact with large hostile forces" would prevent the furtherance of the reconnaissance mission. When this occurred, White believed it important that mechanized reconnaissance units be prepared to "further the general plan of the commander by engaging in combat on their own initiative without specific orders from higher authority."[163] Stealth and reconnaissance remained paramount, but in conflict with the Cavalry School, CPT White, like others at Fort Knox, anticipated a greater need to be able to fight for information.

The faculty members of the Cavalry School were not the only visitors to Fort Knox in 1937. Nazi Germany's MG Heinz Guderian also visited, bringing with him a small party of officers. The group was treated to a series of rehearsed demonstrations and participated in roundtable discussions on mechanization. Some felt this visit helped the Germans "solidify" their thoughts on blitzkrieg warfare.[164] This may have been true, but the growing German mechanized juggernaut already had plenty of its own momentum. During September 1936 it conducted an exercise in the state of Hesse that involved some fifty thousand soldiers. Hitler's army was growing with the reintroduction of compulsory service, and the expansion caused a shortage in the officer corps.[165] The exercise featured a revamped infantry division with a "ubiquitous" distribution of anti-tank guns, even in the divisional reconnaissance battalion.[166] This battalion was unlike anything in the American army at the time, whose infantry division still lacked an organic reconnaissance unit, but it did have some common features with what was still the independent mechanized cavalry brigade at Fort Knox.

The Soviets were also providing some insight to their views on mechanized reconnaissance in their professional journals. They too concluded that mechanized forces must be supported by constant reconnaissance on the ground and from the air to gather information on the enemy and on the terrain the larger mechanized force would have to travel over. The Soviets also expected ground reconnaissance units to carry out security functions by denying the prying eyes of the enemy access to the friendly main body.[167] The article, nominally on the training of mechanized forces, reveals far more about the political climate of the Soviet Union as it developed its own doctrine.

> Each member of the tank corps, however, must thoroughly understand that all problems to be encountered by them in the course of their forthcoming exercise may be properly solved only with the aid of efficient political activity, by unswerving loyalty to the cause of the party of Lenin and Stalin and by strict observance of all instructions of the People's Commissar of Defense governing military training.[168]

There was no commensurate faith in the United States that President Roosevelt alone held the key to all the questions troubling the cavalry community.

As the end of 1937 approached, the chief of cavalry addressed the Army War College. He opened by telling the audience that the German army no longer contained any cavalry divisions; all that remained were eleven regiments with the limited function of performing reconnaissance only. He quoted a member of the German War Office at length: "Our decision in this matter is a clear-cut one. We are accepting the principle of mechanization in toto. For the mounted soldier there would be only one field of activity in the future—reconnaissance."[169] This was very different from the American army where to date mechanization had performed some of the reconnaissance in the horsed units, but was considered either too valuable to be risked in combat or, more often, ignored. Only in the mechanized brigade at Fort Knox were machines fulfilling all cavalry roles including combat. Kromer agreed that, as a branch, cavalry "welcomes any mechanization that will increase its effectiveness" but concluded that, "If, in the future, we develop any breed or breeds of the iron horse that can prove superior to the thoroughbred for *all* cavalry missions, we will abandon the horse."[170] While the Germans retained horses for limited reconnaissance duties close to their foot-infantry regiments, the American army clung to its horsed units with the expectation that they still filled a vital role in combat. Thus, mechanization might lend assistance, but there was a lack of consensus that the "iron horse" was bred well enough to supplant the "thoroughbred."

If 1937 appeared as a year in which those who sought the most from mechanization in all aspects of cavalry missions, including reconnaissance, were doing their best to advance their ideas, then one must view the arrival of the new chief of cavalry, MG John K. Herr, as a sign of resistance to the thoughts of those at Fort Knox. The arrival of Herr marked the end of the tenure of balanced views on mechanization as a whole that had marked the service of the two recent chiefs, Henry and Kromer. Now there was no question that the horse would be supreme and that the voice of BG Hawkins, Herr's former division commander, would be more influential than it had in the past.[171] In all fairness, Herr visited Fort Knox as he traveled to Washington in March 1938 to assume his new command, but his actions and words, until his position as branch chief was eliminated in 1942, identified him as the unquestionable champion of the horse.[172] He did not deny the importance of reconnaissance, but did not want to see his beloved horse cavalry relegated to this task alone.

During an exercise that suggested hostile forces had landed on the Atlantic seaboard, the 7th CavBde(M) marched south to respond to the intrusion. As it entered Chattanooga it was met by the scout car platoon of the 6th CavRegt. The men from Fort Knox presented an appearance that was "somewhat different from the arrival of the Union's forces seventy-five years" earlier.[173] The mechanized men may have lost the polo matches that ensued after their arrival, but they had once again demonstrated their ability to conduct mechanized reconnaissance across a broad front. In this exercise the armored car troops of each regiment covered a two-hundred-mile front. BG Van Voorhis experimented with the use of light aircraft to command and control the vastly dispersed force. COL Patton, the Cavalry School's observer, noted that this was "the obvious answer" for "maintenance of liaison between the several columns of the mechanized cavalry command."[174] Patton also saw the utility of aircraft in directing the ground reconnaissance effort to new routes and bridges as required.[175]

Again, the link between ground and air observation was recognized, but continued to remain unresolved.

After a brief visit at Fort Oglethorpe, Van Voorhis turned his iron steeds to the north. This time the opposing force took the form of a hostile infantry brigade that had penetrated south of the Ohio River. Aviation provided the mechanized brigade commander with some early and detailed reports about the enemy's disposition, but it was up to the ground reconnaissance men to develop the situation in greater detail. The armored car troops moved forward of the brigade after the entire force reached Nashville. The next morning, the entire brigade went on the attack, in what they must have considered their own back yard by now, prompting one observer to remark with surprise that such a feat would be attempted without preliminary reconnaissance.[176] What the observer had not seen or heard were the twenty radio-delivered messages supplemented with motorcycle courier messages that had been delivered to the brigade commander as his unit marched north through the night. Even if it must be granted that the infantry brigade that opposed them was operating on the cavalrymen's "stomping grounds," which certainly facilitated the collection of reconnaissance information from both the air and ground, it was a considerable accomplishment. An entire body of mechanized warriors had marched through the night and into the attack without pause. No horse unit could have carried off any aspect of the operation, even the reconnaissance.

The march through the hills of Kentucky and Tennessee reinforced Van Voorhis's desire to see his reconnaissance agencies augmented. He exchanged letters with a mechanization-friendly staff officer assigned to the chief of cavalry's office, LTC Willis D. Crittenberger. In his letter to Crittenberger about how the future mechanized division should be organized, Van Voorhis expressed hope that Herr would "sacrifice the necessary amount of horse cavalry to meet the War Department's wishes in expanding to a division."[177] One of the major demands on personnel remained Van Voorhis's desire for the addition of a reconnaissance and support squadron so that he could "at once work them into the combat team."[178] Just like MAJ Grow had warned, Van Voorhis saw the possibility that Cavalry Branch would lose its control of an expanded division, especially as other branches contributed large units to fill out the structure of the new unit.[179]

Crittenberger used his exchange with Van Voorhis to draft a document that he presented to Herr. Specific to reconnaissance, it suggested that there should be no reductions in the number of soldiers in the existing armored car troops and that in fact these troops should be increased. The need of an entire squadron for reconnaissance and support was justified by the inability of the existing reconnaissance troops to "function efficiently for more than 24 hours' straight run."[180] Van Voorhis argued that those suggesting that a mechanized cavalry division only required a single troop for reconnaissance were thinking "more in terms of strategical reconnaissance and are not giving any consideration to battle reconnaissance."[181] Strategic reconnaissance had always envisioned long, independent movements, such as those carried out by the armored car troops, but Van Voorhis's attention on "battle reconnaissance," fighting for information, may have been based on a combination of factors. It may have been a means of getting more reconnaissance assets than he had any intention of using to "fight for information," since he had already expressed a desire to use them to supplement the capability of the existing armored car troops. The more probable explanation was his recognition that a division would contain far more units than the existing brigade did; therefore, a division needed more reconnaissance assets to deploy safely.[182]

Deciding on what types of vehicles to procure for mechanized reconnaissance inspired more debate within the cavalry community. Inside the chief of cavalry's office, LTC Crittenberger prepared a memorandum for the chief entitled "Improvised Mechanized Reconnaissance," which outlined many of the same points that others associated with the mechanization process had already made. Crittenberger started softly with the horse-minded chief, but finished strong. If a war occurred on the morrow, Cavalry Branch would have to maximize the availability of commercial vehicles, especially for reconnaissance.[183] This would be problematic, he reasoned, for there was an expectation that the scouts would make contact with the enemy and then need some means of fighting, at minimum, machine gun mounts. He urged Herr not to allow cost to drive the decision since the less expensive alternatives lacked the combat strength for offensive or defensive action. Crittenberger lent emphasis,

> NO OTHER COUNTRY IN THE WORLD WITH ANY KIND OF MECHANIZATION PROGRAM IS GOING SOLEY WITH A LIGHT COMMERCIAL CAR FOR MECHANIZED RECONNAISSANCE MISSIONS.[184]

Having reviewed the "Improvised Mechanized Reconnaissance" memorandum, Van Voorhis sent a letter to Crittenberger and agreed that "improvisation" would be a mistake.[185] In contrast, members of the horse-minded community were less interested in developing better-designed reconnaissance vehicles, because they operated from the mindset that the ultimate vehicle was already in their inventory, the horse.

COL Scott, an increasingly important figure in the development of mechanization, prepared an article for the *Cavalry Journal* in response to a series of articles critical of the existing vehicular technology. He urged the readers to demand that the technology be brought forward to meet their needs rather than adapting their needs to that which was currently available.[186] This attitude of always reaching for and expecting more seemed to dominate the mind-set of those involved with mechanization, whereas those in the horse cavalry community seemed to be tailoring their expectations to meet the existing capabilities presented by the horse. Unfortunately for many a casualty in the next world war, it was the attitude of the latter that ruled the day in the interwar army.[187] For all their efforts to head off the introduction of a light, commercially produced vehicle that lacked armored protection, the "jeep" had already started to undergo testing with the Cavalry Board at Fort Riley. For decades to come reconnaissance men conducted their work from the back of this nimble machine and many would wish Scott's reasoning had won out.[188]

The stress of maintaining a single-branch identity continued to grow in late 1938, as a new round of cavalry maneuvers entered the final stages of planning. Considerations for the exercise were not limited to the problems from within the branch, but also examined the threats from outside. Crittenberger recommended that Herr give consideration to the integration of aviation assets as a way to show the mounted arm's willingness to be progressive.[189] He also advised Herr to maximize the publicity for the exercise and get many of the War Department General Staff officers to attend, writing, "Unless an early and determined effort is made to publicize these maneuvers, they will be shoved into the background by the Anti-aircraft GHQ Air Force maneuvers now scheduled to be held at Fort Bragg during the first two weeks of October."[190] A successful showing at Fort Riley would help show that the horse cavalry could not be reduced any more and that none of the remaining horse units should be converted into air defense units.[191] Leaving nothing to chance, Crittenberger asked his fellow mechanized cavalryman COL Scott to make sure 1LT Henry Cabot Lodge received "a good look at the mechanized cavalry."[192] Senator Lodge requested that he be brought onto active duty status for the Fort Riley exercise and be attached to the staff of the chief of cavalry.[193]

A month prior to the Fort Riley maneuvers, Herr asked Crittenberger to assist him in editing remarks he was to deliver to the Army War College. Herr planned to make the following comment on the use of mechanized forces: "Most of them advancing alone will be waylaid and captured and will deliver to the enemy the fine arsenal of weapons with which they are equipped." Crittenberger advised the chief that as the head of both horse and mechanized cavalry, this comment might make it seem as if he had little confidence in the "development of mechanized reconnaissance to date." In addition, he doubted "the wisdom of the Chief of Cavalry emphasising [sic] any weakness of mechanized cavalry, any more than a weakness of horse cavalry."[194] Herr also wanted to remark that the 7th CavBde(M) should be moved away from Fort Knox, eliciting Crittenberger's response that this would be inconsistent with the ongoing request for more funds to purchase more land at Fort Knox and sustain ongoing construction, and would undermine the role the mechanized cavalrymen played in defending the "bullion depository."[195] Herr's original text also seemed to suggest that it was unlikely the United States would find itself fighting in western Europe, which Crittenberger again reminded Herr was inconsistent with recent remarks by President Roosevelt.[196]

Herr enlisted help from retired officers in his effort to maintain the prestige of the horse cavalry. In a letter to retired MG Harold B. Fiske, Herr asked for the senior leader's input on the utility of cavalry in the Panama

Canal Zone. Fiske replied that his five years of experience in that strategically vital part of Central America had left him confident that cavalry would be of assistance. Fiske provided a sound argument that met Herr's needs and had plenty of details. The retired general emphasized the lack of roads on both the Atlantic and Pacific coasts and stated that what few roads and trails did exist were so narrow as to only be used by men on foot or horseback. He then reasoned that any enemy attacking the Canal Zone would bypass the forts that dominated either end of the canal, thus forcing American defenders to control the network of trails and small roads. Having said that, he provided the cavalry chief the words he really wanted to hear: "To depend entirely upon infantry for reconnaissance and delaying action away from the vicinity of the canal means the expenditure of enormous labor in excessive heat by men and long delayed information of, and slow action against the enemy."[197] Fiske suggested that small bodies of men mounted on native ponies would be the best tactical solution to the defense of the Canal Zone's undeveloped periphery.[198] This last line of reasoning reappeared in a memorandum to the chief of staff a little more than one month later. In the memorandum were outlined the requirements for cavalry forces in the United States.

> Considering the difficulty of transporting the horses of invaders across the Atlantic or Pacific Oceans into this hemisphere, which difficulties unquestionably would restrict the number of hostile cavalrymen brought over here to fight us, the advantages of an American campaign of movement resulting from the horses available to our army would be tremendous.[199]

Herr believed that "the cavalry had already been bled white" and marshaled all his resources to justify why any increases in mechanized forces should come from an overall enlargement of the army, not through the reduction of more horse cavalry units.[200] He centered much of his argument on the questions of where the army would be expected to fight and what were the resources America had to offer to support his vision of how the cavalry should look. Underlying Herr's assumption was his belief that given the international situation the army's size would increase in the near future and that the next war would be mobile.[201]

At the time the memorandum was submitted to the chief of staff of the army, the on-hand cavalry strength to fill its required role in a 280,000-man army was a mere 50 percent, and the army as a whole had only 59 percent of its total requirement. Therefore, Herr, using statistics, showed that the cavalry was already at a manpower deficit in respect to the rest of the army.[202] He noted that in European armies, cavalry composed some 7 percent of the active forces and that in the United States, this figure only amounted to 6 percent. The Poles had forty regiments of cavalry and the Soviets some 112 regiments, but with the exception of Russia, no other country in the world could match the animal resources of the United States. Herr acknowledged that no other country in the world had the same motorized production capability either.[203]

Herr answered the question of where the cavalry might be expected to fight in his proposal to form all U.S. cavalry units into a single corps. Herr saw such a corps as the logical next step, because the nation was faced with two, long land frontiers that lacked adequate communications and were given to extremes in weather and climate. The cavalry's position was that given these realities, the regular army, the National Guard, and Reserve cavalry components should all be formed into a corps structure as part of the Protective Mobilization Plan.[204] The regular army cavalry corps was to be stationed in the American Southwest where it would have access to large training areas, thus eliminating the current situation of regiments being scattered all over the United States. He hoped that the corps would contain at least two horse divisions and one mechanized division.[205] Stationing this large body of mounted men in the Southwest would put U.S. forces in sync with the report's material on the arrangements and strength of other nations' cavalry. The report's summary illustrated Mexico's sizeable mounted arm.[206] The cavalry chief expected that a corps of this size would initially conduct reconnaissance with the two horse divisions abreast, while the mechanized division was sent to some important, but possibly distant, location. Both types of cavalry were expected to penetrate the enemy's screen line, conduct delays, and fill gaps between friendly units.[207] Were the army not to adopt the corps idea, it was feared that "this glorious opportunity" might allow the current trend in cavalry employment to continue "frittering away our cavalry in small units designed only for reconnaissance," and that this "would be the height of folly."[208]

In regard to the state of mechanized cavalry in the United States, the memorandum to the army chief of staff reflected the views of those most closely associated with the process of the development of mechanized cavalry. Mechanization advocates, and the cavalry branch as a whole, felt that the American force was the equal of any similar force in the world.[209] Even so, the official position held that the mechanized men were in no position to replace the traditional horse cavalry.[210] Horse advocates envisioned mechanized support of offensive actions involving mounted units by drawing the enemy's fire, thus "enabling the horse cavalry in close support to seize the objective without great loss."[211] The cavalry chief was willing to agree to the expansion of the 7th CavBde(M) into a division with at least a troop-size reconnaissance organization, but he also wanted a mechanized unit stationed at Fort Riley. In addition, he expressed a need for a new horse cavalry unit in Panama, and for some consideration to be given to creating a cavalry unit at the Infantry School. With all that was wanted, the bottom line was clear: "No further horse units should be converted to mechanized units. Horse units are already at the irreducible minimum."[212]

Abroad, other armies considered some of the same issues and what bearing they would have on their own cavalry forces. In 1938, the Germans, already well on the road to blitzkrieg, recognized that their shortage of domestic fuel and petroleum products placed an inherent limitation on their continued mechanized development. They also recognized that their agricultural economy was still animal based, offering a total of 3,400,000 horses. This large number of horses and the number of men experienced in the care of animals as a result of their importance in the farm economy brought about a compromise. The Germans eliminated all cavalry regimental headquarters and reorganized each of the regiments into two battalions. One battalion was composed of five horse squadrons, and the other battalion contained three motorcycle squadrons, one anti-tank gun squadron, and one light gun squadron. All of these units now worked for infantry divisions in the role of divisional reconnaissance.[213] This was the same role Herr hoped to avoid for his own units.

In Great Britain one of the complaints offered about the country's development of mechanization was the lack of maneuver space in the "entire British Isles."[214] One place in Europe where there was no complaint about room to maneuver was Spain. The poor performance of tanks in the fighting there provided material for horse advocates. The *Cavalry Journal* treated its readers to many accounts of the meaningful and effective use of horse cavalry. One Soviet observer offered that, "intelligently employed and using proper tactics, the cavalry is a modern arm in every respect and should be widely employed both for strategic missions and combat."[215] Mechanized proponents asserted that the poor performance of the tanks in Spain was directly linked to the tactics used, which were deficient in the employment of combined arms and proper reconnaissance, one of the hallmarks of the American mechanized experience.[216]

Another place with no shortage of space was Poland. A cavalry officer from this nation that was wedged between two larger nations, neither of which might be considered a true friend of the Poles, offered his thoughts on the role of cavalry in the next war. After a long examination of the ebb and flow of technology and its impact on the mounted arm through history, he summed up the current situation that faced Poland in regard to potential cavalry threats. In the West, the Germans and the French had opted for mechanization and motorization. In the East, the Soviets retained large horse-mounted units because of the type of terrain. The Polish colonel's conclusion took the form of a question, "Could we abolish cavalry and then stand hopelessly confused when the mobility of armored units proved fiction?"[217] He bolstered his position with the same arguments heard in the United States: the columns are too long, vehicles are too sensitive to the terrain, motorized and mechanized units are too susceptible to attack from the air and by other ground forces at night. In the final analysis, the Polish colonel was "of the opinion that the judgment of the so-called 'East' is correct and cannot be proved false." Less than a year after his article appeared in the United States, events in Europe proved the colonel wrong.

Some in the United States were already beginning to hear the drumbeats of the war that was about to engulf the European continent, and they hoped

America would be ready when it started. Two army officers, one a cavalryman and the other a coastal artilleryman, teamed together for an article that captured the current state of mechanization in England, Germany, France, Italy, and the Soviet Union. The authors framed their story in the fictitious tasking from the chief executive of Atlantis to his secretary of defense to develop a proposal for the creation of a mechanized force. The authors suggested the people of Atlantis had realized that,

> modern means of transportation, communication, size of navies, and, especially, the capacity, potentialities, speed and range of airplanes today and those planned for future use, have taken from Atlantis the security that she has enjoyed from her birth due to her geographic position.[218]

After analyzing what the aforementioned nations had pursued in regard to their mechanized policy, the secretary of defense of Atlantis arrived at what he thought would be the ideal mechanized force for his country. In regard to mechanized reconnaissance, the secretary was in favor of the inclusion of a battalion-size organization for each of the mechanized divisions. The larger conclusions pertaining to mechanization as a national issue were more important. The fictional secretary of defense focused on Atlantis's ability to manufacture more goods than any other country in the world. The report recognized that domestic oil reserves in Atlantis far exceeded anything found in Europe, as illustrated by the fact that the Germans were unable to complete their mechanized movements to Vienna during the *Anschluss* for a lack of fuel. Unlike the Germans, who recognized the potential of men accustomed to farming with animals to transfer that knowledge into military service, the citizens of Atlantis were viewed as being "thoroughly accustomed to the use of mechanical devices." This fact would assist in the process of bringing the army of Atlantis into the age of mechanization.[219]

Maneuvers throughout the 1930s exposed important indicators about the future of reconnaissance. Mechanized units working forward of dismounted or horse-mounted units could use their speed and range to provide the main body the maximum amount of time to make decisions. Mechanized reconnaissance units serving equally mobile forces could not provide as much decision-making time, but were absolutely essential to fill the gap between fast-moving motorized/mechanized forces and air reconnaissance. Thus, mechanized and horse-mounted regiments grew increasingly dependent on their specialized reconnaissance formations. No unit could afford the time required for horses to march to the front. The detail provided by horse-mounted units was greater, but getting there first was increasingly important, especially in the army's only horse cavalry division.

Even with the future of some form of mechanized reconnaissance secured, there was still much to be learned and many unresolved issues. The notion that armored cars would operate in excess of one hundred miles to the front of the main body seemed to be diminishing, and certainly this type of mission was absent from any of the exercises carried out in the first half of the 1930s. While commanders grew more reliant on the command and control capability provided by scout and armored cars, others questioned the combat and survival capabilities of these specialized troops. With the questions came debate over what could and should be expected from the mechanized forces. The debate carried over into the realm of how these forces should be equipped to perform their missions, even as troopers continued to practice shooting at targets from the backs of galloping horses.[220]

As 1935 dawned, Adolph Hitler restored conscription and acted on his own grandiose thoughts that had little to do with target practice from the backs of horses; he was bent on creating three panzer divisions before the end of the year.[221] Even if the members of the horse cavalry community could not achieve a common consensus on the proper employment of mechanized reconnaissance units, as indicated by major exercise throughout the 1930s, the men at Fort Knox took an increasingly important role in shaping the future of this emerging capability. Their role was important in the light of events beginning to transpire in Europe, but also led to more divisiveness that in turn threatened the integrity of Cavalry Branch. Having begun the decade with pragmatic and progressive leadership, Cavalry Branch under John K. Herr drifted in a different direction as the very survival of horse cavalry came to the fore, inextricably linked to the issue of mechanized reconnaissance.

3

The Big Maneuvers and War

The magnificent horsemen, with their lances

lowered to the charge and their saddle leather

creaking, died with their steeds in droves with

the drumbeat of hoofs at the gallop in their ears.

They pitted man and animal and courage

time and again against tanks and

gunfire, and it was no contest . . .[1]

—Poland, September 1939

★   At the beginning of 1939 the world was still at peace while Americans considered their defense needs in splendid isolation. Relying heavily on the navy and emerging fleets of long-range bombers, it seemed that the army was to remain small and professional, barring two "major calamities": becoming involved in a European war or suffering an invasion of the United States.[2] MG Herr no longer doubted "that we are about to arm," but his focus remained in the Western Hemisphere. To this end he continued to advocate massed bodies of mounted men formed into entire corps that would train in the American Southwest.[3] Blitzkrieg on the plains of Poland later that year did not dissuade Herr from his conviction that the horse remained a viable instrument of war. The Polish disaster inspired compromise, the formation of corps reconnaissance regiments composed of both horse and mechanized squadrons. Hybrid as they were, the inclusion of horses reflected the continued belief that mechanization alone could not satisfy all the army's needs while also signaling an increased responsibility for reconnaissance missions in a branch that still prided itself on its combat role.

War in Europe also triggered the revitalization of the army. Large-scale exercises conducted in 1940 and 1941 forced decisions on reconnaissance topics that had been debated throughout the previous decade. Cavalry

Branch reached a crossroads as army-level leaders made their decisions about the force they would send to war. The conduct of ground reconnaissance was one of the key issues, especially after the Cavalry Branch lost control of 7th CavBde(M) when it merged with Infantry Branch's tank force at Fort Benning to form the Armored Force. Efforts by those who sought to retain an important combat role for the horse had done so at the expense of serious reflection on what was needed to increase the capabilities of the existing mechanized reconnaissance formations already fielded, yet these individuals now found themselves with a single mounted division and an expanding reconnaissance mission. Between the attack on Pearl Harbor and the Allied invasion of North Africa, GEN George C. Marshall resolved the reconnaissance debate. Although horse and machine would continue to exist in a single branch, reconnaissance henceforth was a mechanized affair marking the culmination of all that had been done during the interwar years. With a firm decision, doctrine and organizational design that would see wartime service emerged, bearing the tell-tale signs of interwar debate.

In 1939, MG Herr testified before Congress, where he expressed the belief that "American horse cavalry today constitutes the most effective highly mobile fighting force in the world for operations on the American continent or similar areas."[4] When congressmen posed questions about what "Stuart" or "Sheridan" might have done, Herr assured them that "the cavalry is not going to charge machine guns," but would still conduct attacks in dispersed waves.[5] Vehicles and airplanes were useful for seeking distant information, but horse cavalry had "become more the fingers than the eyes of the army, feeling out what the Air Corps and men in mechanized vehicles have seen."[6] Herr remained as unconvinced about mechanization as he had been unconvinced when he was a regimental commander.

While Herr testified to Congress, Marshall, then working in the army chief of staff's War Plans Division, wrote a letter to BG Lesley McNair about the needs of the army in a larger context. Marshall told McNair that it was time to start preparing officers and men for the tasks required of units and troops in an expanding army that needed "highly trained and experienced" men.[7] Herr's testimony ignored this requirement even as Cavalry Branch continued to field new technology.[8] Herr's testimony also contradicted lessons still being learned in the field. During a recent exercise, a regiment of motorized infantry embarrassed the horse soldiers by moving more than one hundred miles over rough terrain in one night only to appear in the division's rear at dawn. The motorized infantry were confronted by the division's armored reconnaissance troop, but they had only been dispatched to intercept this broad envelopment after it was too late.[9]

During the exercise, mounted attacks had been the norm and, as one observer recognized, "suicidal" against an infantry force equipped with modern weapons.[10] Even more damning, officers were criticized for their lack of initiative and leadership. Officers neglected their men and horses by missed waterings and feedings because they feared what might develop tactically if they took the time to perform these vital functions.[11] Scout car patrols were probably substituted for local security patrols because the horses and men in the rifle troops were unable to perform these vital tasks. Lacking the programmed "armistices" that had been a feature in previous divisional exercises, the horse faired poorly against the untiring machine.

The First Army maneuvers held near Plattsburg, New York, in August 1939 involved 50,000 participants in regiment versus regiment, brigade versus brigade, and corps versus corps exercises. During the corps versus corps phase, vintage World War I "square" infantry divisions fought smaller "triangular" divisions that also included the 7th CavBde(M).[12] COL Patton lobbied GEN Hugh Drum for maneuver-area boundaries that would allow the horse and mechanized cavalry to demonstrate their full potential.[13] The maneuver box was quite small and size mattered. Engineers sought to contain mobile forces with obstacles where the army had not secured trespass rights on either side of the road.[14] However, even a cavalry force using sound mechanized reconnaissance doctrine would find it hard to overcome such gamesmanship. As the exercise ended, the mechanized men marched south to West Point and on to New York City to participate in the world's fair being held at Flushing Meadows.[15] They arrived on 31 August 1939; understandably, interest in their presence increased markedly the next day. GEN Marshall's interest in mechanization increased too as he was sworn in to his new job as chief of staff of the army, while German tanks poured into Poland.[16]

As the German forces perpetrated a swift and brutal invasion of Poland, they touched off the bloodiest war in history, a war that claimed the lives of tens of millions before it ran its course, only ending in the mushroom clouds over Japan.[17] In retrospect, the *Cavalry Journal,* in its own strange way, contributed to the brutalization of the Polish military establishment, specifically the Polish army's vaunted arm of cavalry. The July–August edition of the *Journal* featured a translated article titled "Training of Modern Cavalry for War, Polish Cavalry Doctrine."[18] The Poles recognized that "there can be no mistake about it that any spark in the present political situation in Europe may well set off a conflagration."[19] Like Herr, the Poles placed the highest priority on their horses' ability to get them to the battlefield, where they would fight dismounted, and they saw weakness in the rear echelons of mobile forces whose supplies were tied to road networks. Defensive stratagems induced the Polish cavalry to seek refuge from mechanized forces by operating in swamps and forests and teaching their troopers how to shoot at the chinks in the enemy's tank armor at ranges of 40–50 meters.[20] When Germany struck Poland, only the Soviets fielded more mounted men than the Poles did, but they were no match for the lightning blows of six panzer divisions and four light divisions (largely composed of motorized infantry and tanks), each led by mechanized reconnaissance units equipped with light tanks and armored cars.[21] Herr viewed the Poles' defeat with disappointment, which forced him reluctantly to accept the role mechanization would play in the new European war.[22] The Poles, using doctrine not dissimilar to that which Herr advocated, failed.[23] Herr believed German success was a function of "good roads," and saw his countrymen's clamor for more mechanization and motorization as a reflection of a "tendency to rush *en masse* to extremes."[24]

One man who did not miss the opportunity to reach what must have been becoming a more receptive audience was BG Adna R. Chaffee who addressed the Army War College in September 1939, only weeks after leading his brigade at Plattsburg, while the Nazis were finishing their brief campaign in Poland.[25] He provided war college students an overview of how the mechanized brigade had developed through the years and spoke of lessons learned from specific exercises that influenced its current organization. Although most of Chaffee's remarks focused on the larger issue of the entire mechanized brigade, he did explain the current ground reconnaissance agencies employed by the brigade and continued his call for the creation of a separate brigade reconnaissance agency that would serve.[26] He emphasized the importance of the work being performed by the reconnaissance men, commenting that he usually tuned one of his radios to one of the regimental reconnaissance radio networks.[27] The proliferation of radios in the 7th CavBde(M) was incredible when compared to other units. Chaffee's men used 158 radios during the Plattsburg maneuvers to keep the brigade connected.[28]

Communication had become so important that Terry Allen, an instructor at the Cavalry School, prepared a small manual on the subject. Its text revealed the growing disparity between the horsed and mechanized units and the reliance of the horsed units on the radios in their limited number of vehicles, mostly concentrated in the regimental reconnaissance platoon. Since communication was a "vital factor" in the proper functioning of the regimental scout platoon, it was well equipped with radios. Allen's manual allowed that, "although the platoon normally operates by section on reconnaissance and security missions, its sections may also furnish means of communication with distant units." Furthermore, he allowed that scout car platoon sections could be used to augment the communication capability of a horse cavalry squadron on an independent mission or, in an emergency, could transport messages or function as a relay station.[29] The need for the assistance was found in the organizational capabilities of the regiment's organic communication assets. The Regimental Communications Platoon only had two mounted messengers in peace and four in war, as well as one motorcycle messenger with sidecar in peace and two in war. These usually accompanied the radio-equipped regimental command car. Other than this, all the squadron could boast of was three pack radios in peace and four during war, of which one was usually dedicated to the regimental commander when he commanded from the saddle.[30] Since the mounted squadrons of the regiment may or may not have been assigned a pack radio for any given mission, they still relied on mounted messengers and bugle calls to command and control their rifle troops.[31] It was easy to see why there was a great temptation to divert the reconnaissance platoon from the reconnaissance mission.

The recent maneuvers in Plattsburg, Chaffee believed, had proved that mechanized cavalry could operate on difficult terrain, could make strategic moves at night, and should not be divided. He also believed the exercise demonstrated, in addition to the need for a stronger reconnaissance force to serve the brigade, that "reconnaissance from unarmored vehicles is of doubtful value and very liable to be most costly in men and vehicles."[32] This notion supported the reasoning advanced by the men with the most experience in fully mechanized reconnaissance, and it was beginning to be borne out in Europe. In regard to the entire speech, Chaffee later commented that having completed it, "I would have the honor of being told by the President of the War College that my lecture was visionary and crazy."[33]

While the men at Fort Knox demonstrated their skills on equipment that had grown very familiar to them, their horse-mounted brethren in the 4th and 6th CavRegts were rapidly trying to expand their repertoire of skills to include those being tested at Fort Knox. The Polish disaster forced a concession within Cavalry Branch. Still adamant that large horse cavalry units were useful combat forces and not wanting to be relegated to the role of reconnaissance, Herr enacted a compromise.[34] The 4th and 6th would become horse-mechanized (H-M) regiments designed to serve the needs of a corps commander. Equipped with a horse squadron and a mechanized squadron, they were expected to operate fifty to one hundred miles in advance of a corps (usually three divisions).[35] The mechanized men moved forward of the horse units, which were to be carried on trailers until the mechanized squadron gained "contact." Troopers on horses would help maintain contact and assist the mechanized squadron by reducing roadblocks and resistance, so that the mechanized squadron's scout cars would not "BE NEEDLESSLY expended."[36]

The conversion from horse to horse-mechanized started in December 1939 and generated a considerable amount of excitement from inside the cavalry community as well as many offerings of ideas about how the new unit should be equipped and organized.[37] Colonels Grow and Crittenberger supplied Herr with ideas on organization and numbers of men. Grow's enclosure argued for standardized mechanized squadrons regardless of attachment to horse or mechanized division to be built around three similar troops or two scout troops and a single motorcycle troop. Crittenberger's model reflected Fort Knox's position, with the inclusion of a scout troop, machine gun troop, motorcycle troop and, if there were enough personnel, a rifle troop.[38] Crittenberger urged Herr to resist the War Department's solution for finding additional personnel by dismounting additional horse units. Rather, he suggested, Herr should play on public sentiment, which expressed concern that the United States was behind in mechanization. Crittenberger believed that by playing to this emotion, Cavalry Branch could acquire the men it needed through emergency measures. To him, the problems associated with the sudden mechanized expansion for these new reconnaissance units bore a striking resemblance to what had occurred in the 1920s when the air corps robbed manpower from cavalry, infantry, and artillery branch upon its creation.[39]

With the chief of staff's support, Herr focused on bringing the 6th CavRegt(H-M) up to strength. Although Cavalry Branch was less successful in filling out the 4th CavRegt(H-M), Herr expected both regiments to be "ball carriers for the entire cavalry service in the coming maneuvers."[40] But issues related to organization, training, and doctrine remained unresolved. One of the first questions to arise was who would escort the horse squadron when its mounts were on trucks and in transit behind the mechanized squadron, since the horse squadron lacked its own reconnaissance assets. Another consideration was firepower, for horse squadrons could not call on the scout cars to provide this need, and as one astute officer observed, it would go against what was being taught since scout cars were not to be "used for combat."[41] There were few good answers to the questions. Experimentation continued with the packed .30-caliber and .50-caliber machine guns. It was generally acknowledged that securing a corps' sixty-mile flank on the march would be nearly impossible.[42] Command and control posed an additional problem because the regiment's horse squadron still depended on horse-packed radios.

Fielding new equipment and training men to use it in the newly converted horse regiments posed immediate problems. There was a general lack of up-to-date equipment and few troopers trained to use what did exist. In one unit, half of the men learning radio procedures failed for lack of aptitude. Civilian driving skills did not provide men the expertise they needed for mechanized reconnaissance work; driving armored scout cars took practice as did riding a motorcycle under combat conditions. Common soldiers made the adjustment from going to the stables by substituting motor maintenance to keep their new steeds in top form.[43]

Philosophically, a much larger cross section of the cavalry community owned up to the realities of mechanized ground reconnaissance and its inherent challenges. These individuals were beginning to enter the world that the men at Fort Knox had been dealing with for years, albeit on a vastly different scale than the Germans, who were already at war. Fortunately for the army, the actual practice of war remained a young man's game, even if the often inflexible minds of the elder generation directed it. The changes presented by transitioning to mechanized reconnaissance were not lost on the junior leaders. One lieutenant summed it up well for his less fortunate peers still leading horse platoons, the same men who in a few short years would lead iron steeds into combat. His article in the *Cavalry Journal* about his experiences during the fielding and training of the 4th CavRegt(H-M) was dedicated to "those similar young heroes who, having been imbued with the doctrine of keeping their heels down and their eyes up suddenly find themselves in an atmosphere of gasoline and oil and are faced with the problem of making their decisions at 30 mph instead of six."[44] He went on to explain that training mechanized reconnaissance scouts required a considerable amount of attention to the tasks of map reading and sketching. This new way of doing business placed a higher premium on the talents of the men, because "the normal mission given a sergeant section leader and often to a motorcycle corporal will be one which if given in a horse troop would call for an officer's patrol."[45]

As the 4th and 6th CavRegts(H-M) prepared to exercise with the 7th CavBde(M), the Provisional Tank Brigade from Fort Benning, and other units in a test of the newly organized "triangular" infantry divisions that had been activated after Poland's fall, the 1st CavDiv conducted its own intensive exercises to correct the faults found in the division's fall maneuvers.[46] Seeking better performance from their mechanized reconnaissance assets, the division attempted to integrate intelligence officers into reconnaissance planning while pushing its mechanized assets farther away from the main body. This would gain them greater operational depth, thus increasing the importance of mechanized scouts that had often been excluded during exercises. Mounted attacks lost favor, and leaders learned to fight the situation on the ground, not the prepared plan, without losing sight of the need to conserve their horses.[47] COL Patton, still keeping a foot in both camps, provided his own assistance. Patton's friend, COL John S. Wood, the chief of staff for Third Army, was in charge of developing the upcoming exercise. With information obtained from Wood, Patton recommended to MG Kenyon A. Joyce, who commanded the 1st CavDiv, that he concentrate his training on radio jamming and swimming bodies of water, because a number of rivers dominated the maneuver area. Patton felt radio jamming might be an easy and effective way to disrupt the capabilities of the mechanized men.[48]

Back in Washington, Herr testified before Congress. He highlighted the inclusion of motorization and mechanization training at the Cavalry School, but also pointed out the severe shortage of vehicles at Fort Riley needed to facilitate the training. Herr even admitted that it was growing easier to obtain young men for service in the cavalry who had a fair amount of mechanical understanding and that men with mechanical knowledge outnumbered those considered good horsemen. Commenting on the demise of Poland, Herr offered that, "Judging from Spain, had Poland's cavalry possessed modern armament in every respect and been united in one big cavalry command with adequate mechanized forces included, and supported by adequate aviation, the German light and mechanized forces might have been defeated."[49] Herr also warned Congress not to put too much faith in the combat power of mechanized cavalry because, "While moving, mechanized vehicles, although heavily armed, cannot deliver any effective fire; they simply spray the landscape with unaimed bullets."[50]

Herr went on to note, "Mechanized cavalry is valuable and an important adjunct but is not the main part of the cavalry and cannot be. Our cavalry is not the medieval cavalry of popular imagination but is cavalry which is modernized and keeping pace with all developments."[51] Even the events in

Poland did not shake Herr's confidence in the force he led.

The role of mechanized ground reconnaissance was of limited importance in the overall outcome of the Third Army maneuvers conducted in Texas and Louisiana in May.[52] Most of the attention focused on the provisional armored division built from the 7th CavBde(M), the 6th Infantry Regiment (Motorized), and the Provisional Tank Brigade from Fort Benning. The horse-mechanized regiments participated, but little was said about their performance other than that there was a need to provide the trailers to move their horses.[53] The provisional armored division dominated the exercise and was a disaster for Cavalry Branch.[54] The exercise ended on the same day the Germans devastated the French by moving nine panzer divisions across the Meuse River. Armed with only twenty-eight modern tanks at the maneuvers, the United States could no longer deny that it was incapable of fighting in Europe.[55]

As the exercise ended, a small meeting of the minds took place in the basement of the Alexandria High School. Chaffee and BG Bruce Magruder, who had commanded the Provisional Tank Brigade, met with BG Frank Andrews, the assistant chief of staff of the army. The men discussed the future of armored and mechanized warfare in the United States, and they concluded that its development should continue, but outside the glare of cavalry and infantry branches respectively. Absent, although they were in the immediate vicinity, were the chiefs of cavalry and infantry.[56] COL Patton, an observer on the exercise, had seen the same 1st CavDiv he had supplied information to, beaten by the mechanized men. His participation in the session went unrecorded.[57]

Responding vigorously, Herr argued there was no need for a separate arm, citing German use of its panzer divisions as comparable to a cavalry force exploiting a penetration. Ironically, Herr reminded the assistant chief of staff that Cavalry Branch had created the mechanized brigade and was leading the way in developing corps reconnaissance regiments, thus making it appropriate for mechanized men to stay exactly where they were.[58] Herr could not overcome the forces in motion, however, and the army created the I Armored Corps on 15 July 1940.[59]

The struggle to define a role for mechanization inside Cavalry Branch was largely over and the roles were about to be reversed. The horsemen had never been interested in using mechanization to do more than assist them in performing their missions. They paid a considerable amount of lip service to the proper employment of mechanized ground reconnaissance, but their actions in numerous exercises spoke to a genuine disinterest and misuse of the assets that were available to them. Horse cavalry soldiers viewed themselves as full-fledged warriors and a valuable combat commodity. Had they not equipped themselves with heavier machine guns to counter the mechanized threat posed by other nations? The mechanized advocates, on the other hand, had always aspired to more than the limited role they were given. They advanced the use of mechanized reconnaissance because they had no alternative. Their aspirations for a combat role eclipsed their interest in the narrow field of reconnaissance, but along the way they helped define the best ways to equip such forces. After 15 July 1940, the mechanized men assumed the role they had always wanted, as an equal actor on the modern battlefield no longer constrained to those missions allowed them by the horsemen. The mechanized men were free to develop even their ground reconnaissance forces in accordance with what they had already learned and what they thought might work better in the future.

The old cavalry had always admitted a place for mechanization in the role of reconnaissance, but as world events focused new attention on its own 7th CavBde(M) in the later half of the 1930s, Cavalry Branch staunchly resisted being forced into the role of reconnaissance agency. Its interest in mechanized reconnaissance had grown, but its professional contributions lagged behind those of its mechanized peers. The blows struck against Poland and France fell with some impact on the men of the old Cavalry Branch, but it was their own army chief of staff who struck hardest.[60] By taking away the most advanced elements and thinkers on mechanization, he set them back at the very time they needed the assistance most. Even Patton, the proverbial fence sitter throughout the interwar years, could not be lured away from taking command of an armored brigade in MG Charles L. Scott's 2nd AD, when he was offered command of the 1st CavDiv.[61] Turning down command of the premier cavalry assignment in the United States Army, COL Patton wrote to Herr:

> I have kept this letter for a long time trying to make up my mind about your generous and flattering thoughts about me and the Cavalry Div. Of course it is the dream of every cavalry man who is worth a damn to command that out fit [sic] and I think I could do a job of it. However I have decided that as I was selected to come here to help make this Brigade I should in loyalty finish the job.[62]

Although it seems highly doubtful today, Patton believed that Herr was the "only person" who could understand how hard it was to decline command of the horse cavalry division, but he postulated that under the same circumstances Herr "would have done the same thing."[63] Patton tried to assuage the branch chief's ego by informing him that he believed Herr to be "dead right to push the Rec. Cav Regts." and that more would be better. Prescient, Patton predicted in 1940 that the United States would fight by corps and that single mechanized cavalry troops would be insufficient for infantry divisions. He proposed one regiment per corps, so that in theory, each maneuver division could be allotted a cavalry squadron. In this way each division would have "more men," and it would also afford "more rank" to the men who would command these yet to be formed mechanized cavalry squadrons.[64] Patton also wrote, "These Armored Divisions are pure cavalry in their functions and tactics and all the foreign writers so state."[65] How ironic that Patton, having gone with the upstart

Armored Force in the divorce of the cavalry family, clearly recognized how his adopted branch had usurped the fighting characteristics of cavalry that Herr, the letter's recipient, was so eager to preserve. Patton saw how cavalry would ultimately come to be viewed, as an agency of mechanized reconnaissance.

"Triangular" infantry divisions with increased motor mobility made it almost impossible for the horses to keep up, even with compromise combinations. Worse still, Cavalry Branch lost control of its own destiny. The chief wanted full divisions and corps of horse-mounted warriors; the chief of staff wanted nothing more than eyes for his army, and mechanized eyes at that. The cavalry had missed its chance to do something important in the last war, and it looked as if it would miss again in the next.

> It didn't take long to determine that the mechanized cavalry units were superior to the horse cavalry in mobility and fire power. (In retrospect, we probably had the most "horse power" ever put together at one time, but we didn't have it where it counted—under the hood). In other words, they ran us into the ground on many occasions.[66]—*Trooper Morrison, F Troop, 2nd Cavalry Regiment*

> If we can get by this period of ignorance and prejudice and prevent these shortsighted gumps from wiping ut [*sic*] out of the picture in their mistaken belief that the iron horse replaces one of flesh and blood, we will surely come into our own.[67]—*MG John K. Herr, Chief of Cavalry*

★   During the 1940 Louisiana maneuvers, Third Army commander, LTG Stanley D. Embick, recommended that the mechanized brigades be expanded into divisions, which they were, but he also supported the retention of the horse for reconnaissance, in combination with the further motorization and mechanization of the horse units in the army.[68] Cavalry Branch still had its horses, and during the remaining months of peace it advanced the horse-mechanized concept, thus retaining a role for the horse, even if the role concentrated on reconnaissance. The army chief of staff also required Cavalry Branch to develop an entirely new concept, organic divisional cavalry troops, to support each of the new "triangular" infantry divisions. These units developed along mechanized lines from the beginning. The 1941 General Headquarters maneuvers, involving nearly one half of the army's total manpower, tested these new units.[69] Ending only weeks before Pearl Harbor, the maneuvers led observers at all levels to conclude that reconnaissance had been poor. Evaluators criticized commanders for committing troops without adequate information and for failing to adequately secure their units.[70] These shortcomings, indicative of new units, spoke less to the organizational structures and more to the general status of the entire army in 1941. Equipment shortages plagued all reconnaissance units. Communications equipment especially affected the mechanized reconnaissance units operating at greater distances from the main body.[71] There was a general lack of training among smaller units, as the rapid expansion of the army between 1939 and 1941 injected large numbers of citizen-soldiers into the force. By the time the 1st CavDiv took the field in 1941, more than 70 percent of its personnel were replacements, for the division had expanded from 3,575 to 10,110 troopers.[72] The interwar years had seen officers stagnate in the same rank for years, denying them the opportunity to command. Now, suddenly thrust into greater positions of responsibility, they lacked the necessary experience. The problem was even truer for that part of the army, officer and enlisted alike, that had been civilians only months before.[73] If leadership and training were lacking at all levels as a result of the unexpected, rapid increase in the size of the force, the equipment shortages were far less excusable. While men trained for war with wooden rifles, mock-up tanks, and missing radios in the summer and fall of 1941, Detroit manufactured the latest, required, models.[74]

Even with the dilution of the small cadre of men who knew anything about reconnaissance, horse mounted or mechanized, and the general shortage of equipment associated with the army's expansion, the horse-mechanized experience offered valuable insight into the army's ultimate decision to discontinue the use of horses in reconnaissance. The same was true of the divisional cavalry troops' all-mechanized evolution during the same period.

The doctrinal basis of the horse-mechanized regiments changed little from their inception in late 1939 until their complete conversion to all-mechanized units in 1942.[75] As the reconnaissance agency for a corps, they operated up to 150 miles forward of the main body. After the 1940 maneuvers and the decision to equip each infantry division with its own mechanized cavalry troop, they received the additional task of coordinating with the smaller reconnaissance units now operating to their rear. Training circulars published in 1941 served as guides to the proper employment of troops, equipment, and units for the expanding army. Horse-mechanized doctrine still favored the horse and brought with it all of the old biases held against mechanized reconnaissance. If mechanized units were unable to penetrate the enemy's screen to obtain information, the *Training Circular* (of May 1941) directed leaders to bring forward horse soldiers to maximize mobility and firepower to "pass around or between the points of hostile resistance."[76] Concern for the preservation of precious mechanized assets persisted. In 1940 the commander of the 6th CavRegt(H-M) summed up the dilemma succinctly when he asked the rhetorical question, "Is it important enough to use this asset today in combat knowing it may prevent me from doing reconnaissance in the future?"[77] An observer of the army's other regular horse-mechanzied regiment, the 4th Cavalry, concluded with the same thought, remarking: "Corps reconnaissance regiments as now organized should be used primarily for reconnaissance and should be employed in combat when such employment is unavoidable."[78] Observers and participants in the horse-mechanized experiments suggested changes in equipment and organization, but their offerings were not consistent with the doctrinal definition of how they saw the corps reconnaissance agency acting. The suggestions did very little to suggest that horses were at the forefront of the commanders' minds as they had their staffs assemble the "wish lists" for future organization and equipment.

Horse-mechanized regiments still lacked organic vehicles to escort them safely while they moved behind the mechanized squadron with all of their horses and soldiers on trucks. One solution offered was the attachment of scout cars from the mechanized squadron, but this reduced the number of reconnaissance assets the regiment could use to cover the corps' advance.[79] Another option was to increase the number of scout cars in the regiment so that the horse squadron could receive its own.[80] Additional cars would improve the deficient communications assets in the horse squadron. Additional horse trailers, another solution offered, were large and heavy when loaded and thus also posed a problem. The 113th CavRegt(H-M) detailed its Pioneer and Demolitions Platoon to the horse squadron to inspect and repair bridges along the routes the trailers had to cross.[81] There was also a call for increased firepower, because commanders wanted more 37mm anti-tank guns mounted directly on the scout cars for immediate use. One commander wrote,

> During the maneuvers many opportunities for quick use of a 37mm gun occurred which did not permit time to remove the trail gun from the pintle, place it in a position and fire before the target was gone or before it was necessary for the car to pull out for safety reasons.[82]

The commander of the 106th CavRegt(H-M) even suggested that the anti-tank platoons adopt the 75mm pack howitzer.[83] Commanders also called for the inclusion of mortars to assist with the reduction of enemy anti-tank gun positions and machine-gun nests.[84] With the clamor for additional firepower, only one commander requested fewer scout cars and more jeeps in the reconnaissance platoons, so that they could "peak and sneak" with greater ability.[85] In an act of sacrilege at the end of the 1941 maneuvers, LTC Charles R. Johnson Jr., commander of the 106th CavRegt(H-M), did not recommend the retention of the horse squadron. After acknowledging the special attributes horses brought to the conduct of reconnaissance, all he wanted was another mechanized squadron.[86] He was not alone in this view.

The lack of radios hampered the horse-mechanized regiments' performance throughout their existence. Like the problem with escorts for the trailer trucks, the horse squadrons continued to borrow scout cars from the mechanized squadron, just as the horse regiments had the tendency to do in the past, to facilitate their ability to command and control the horse troops and to provide the corps timely information. The 106th CavRegt(H-M) went one step further than attaching scout cars to the horse squadron to improve their communications woes. Scouts from the 106th filed their reports directly to the corps headquarters regimental liaison officer using field telephones tapped into the existing phone system.[87] This was a creative improvisation, but not one that held real promise in the undeveloped corners of the world in which the horse cavalry was staking its future. Horse-mechanized units' after-action reviews only considered motorcycle dispatch as a temporary solution to the existing communication weakness inherent in the units' design, but the real call was for more and better radios and an increase in the number of radio mechanics.[88]

Another command and control issue arose out of the conversion of the 6th CavRegt from all horse to horse-mechanized. As the regiment expanded with the addition of troops, the commander elected to redistribute the personnel evenly throughout the regiment, attempting to have some degree of experience at every level, due to the influx of citizen-soldiers. Going into the 1940 maneuvers, COL John Millikin believed the mechanized half of his regiment performed better, perhaps reflecting the fact that, given the current composition of the nation's youth, "we [the army] may expect to develop soldiers for mechanized and motorized elements much quicker."[89] After the exercise he commented that the noncommissioned officer (NCO) leadership in the horse units had been good, but that the NCOs were not performing as well in the mechanized units. He hoped to rectify the problem with an increase in officers assigned to the mechanized portion of his squadron.[90]

In spite of all the shortcomings of the horse-mechanized concept, the dramatic increase in the number of troopers forced to embrace mechanized ground reconnaissance showed them equal to the task of developing their own techniques, tactics, and equipment. Their creativity matched that of the first mechanized reconnaissance men in the late 1920s and early 1930s. CPT Bruce Palmer Jr., a platoon leader during the May 1940 Louisiana maneuvers, used the *Cavalry Journal* to provide one of the best accounts of what was actually happening at the platoon level. Acting on an intercepted radio message, Palmer led his platoon on a night reconnaissance mission, something doctrine suggested he was ill equipped to perform. Motorcycles traveled to the rear of the formation because of their noise signature, but scout cars moved quietly through the night, a feature never attributed to them by the same doctrine. The patrol used extensive dismounted patrolling to reconnoiter the swampy and difficult terrain. When the radio failed, Palmer dispatched a motorcycle to headquarters with the important information. After all the details of how a mechanized cavalry platoon was able to accomplish its mission at night, in poor terrain, the young captain offered a conclusion that must have brought tears of joy to the eyes of the chief of cavalry.

> The author has often made one of these long night prowls afoot in rear of the enemy's lines and often has wished to high heavens for a horse. . . . Near the enemy's outpost, detruck the horses and hide the scout cars and trailer under cover. And then continue the reconnaissance from the back of the horse. Here is the mechanized, the motorized, and the horse working together in small units in their quest for information.[91]

Having proved on his own what was possible with a mechanized reconnaissance platoon, the author could not bring himself to abandon the trusty mount of the cavalry, demonstrating just how deeply the loyalty ran.

The experience of the 6th CavRegt(H-M) in the last real test of the horse-mechanized concept during the Louisiana phase of the 1941 General Headquarters maneuvers revealed increasingly creative uses of mechanization. The regiment assembled four special reconnaissance teams that came to be known as the "squealers." Each team consisted of an officer, a driver, and a radio operator. The team used a jeep that was equipped with a radio, air-ground communication panels, five days of rations, three submachine guns, two anti-tank mines, and extra fuel and oil. The regimental commander placed the teams on a five-minute recall for missions directed to get behind enemy lines, remain undetected, and report on the enemy's movement.[92] The article offered no explanation for why horses were not formed into a similar special unit, but it was clear that the leadership was starting to branch out in its thinking about the use of mechanized ground reconnaissance.

All the resourcefulness in the world could not overcome some of the glaring deficiencies of the organization, with or without the proper equipment. Horse-mechanized commanders asked for assets needed to help the horsed squadron directly or indirectly. Most of what was asked for would have led to greater mechanization in the regiment. The request for additional firepower exposed a growing gap between how reconnaissance had been collected in the past and how a reconnaissance unit had to be equipped to survive on the modern battlefield. The men who mattered most in defining the future organization for the entire army, MG Lesley McNair and GEN George C. Marshall, had seen enough.[93]

During the final phase of the 1941 General Headquarters maneuvers, held in the Carolinas, the chief of staff's office ordered the commanding general of the Second Army to form an all-mechanized cavalry regiment and an all-horse porté regiment by exchanging the respective squadrons from the 107th CavRegt(H-M) and the 6th CavRegt(H-M). McNair gave the two regiments less than a month to swap the personnel and equipment before the November exercise.[94] After the exercise concluded, the IV Corps commander recommended that the all-mechanized squadron be retained with the addition of a light tank component. Naturally this was against the wishes of the chief of cavalry.[95]

MG Herr had been led to believe by COL John Millikin that the horse-mechanized concept was succeeding. Millikin wrote Herr after the Louisiana phase of the 1941 GHQ maneuvers and praised the performance of the 4th Cavalry in its delay against Innis P. Swift's 1st CavDiv. The 4th had done well because its trailers allowed it to move 120 miles in the forty-eight hours of intense activity during the exercise. During the maneuvers, but before this particularly exemplary performance of the horse and mechanized hybrid, Millikin recounted for Herr the conversation he had had with the chief of staff, George C. Marshall. Marshall and Millikin shared lunch at the regimental command post and discussed the concept currently being tested with the 4th. Millikin "assured" Marshall that not only was he a "great believer in the Horse and Mechanized Regiment," but that he also firmly believed it was the "most capable task force for reconnaissance and security and for combat provided it was given appropriate missions." Millikin emphasized the importance of the "task force" underpinnings of the organization. It was not the horse alone that gave the 4th Cavalry its unique strengths. Rather, a combination of its mechanized troops, firepower, internal communications, combat engineering section, anti-tank section, and self-contained service section made it a good "all purpose reconnaissance, security and limited combat" outfit.[96]

Herr was livid when he heard the news that he would have to convert two of his horse-mechanized regiments for the final phase of the 1941 GHQ maneuvers. Herr's office notes reflect rage that Harry L. Twaddle, assistant chief of staff, G3, had not consulted with him before making the decision. Herr believed the test was premature and futile; he minced no words with George C. Marshall when he told him the test was a "farcical experiment, incredibly stupid" and that "it could mean nothing and . . . ought to be stopped."[97] Herr sensed a conspiracy being "plotted secretly behind my [his] back" by traitors inside his own branch.[98]

When COL John Arthur Considine, of the 6th CavRegt(H-M), crossed over in support of all-mechanized squadrons, he too earned the contempt of Herr, who said, "Considine never had much use for the horse," was behaving "like a mad dog [and] since [Charles] Gerhardt was promoted over him . . . he has [had] illusions of grandeur" and has sold out the horse-mechanized concept to "capture a star."[99] Considine had babbled "to Krueger, McNair and the G-3 representatives of the General Staff." Constantly on the defense to preserve his horse cavalry, even if it meant some compromises like wedding a horse squadron to a mechanized squadron, Herr was convinced that the General Staff had listened to Considine "only too gleefully," looking for "any recommendation to do away with the horse."[100]

Herr's attempts to make COL Considine command the all-porté regiment failed. Herr objected to the 107th being organized as such because of the poor state of training in the horse squadron of the regiment, a result of the recent large-scale increases.[101] Considine's supposed "babbling" resulted in the 107th CavRegt, drawn from the National Guard, moving to Fort Oglethorpe for the next set of GHQ maneuvers. Once at Fort Oglethorpe, the regiment broke up into two reconnaissance regiments. One, an all-horse regiment, would be commanded by a colonel from the National Guard. The other, all mechanized, was commanded by none other than COL Considine. Both of the regiments were to support the IV Corps. According to Herr it was clearly a "conspiracy of which the answer is already forecast."[102]

Herr saw at least one bright spot in the directive that ordered him to test the first all-mechanized corps cavalry reconnaissance squadron. "Fortunately" it called on the chief of cavalry to send an observer to report on "these funny doings." Herr planned to send MAJ Wesley W. Yale to "burn them up." Yale possessed all the attributes Herr required. He was considered "fearless," was a "fluent writer," and most of all was "intimately familiar with the operations of the 4th Cavalry, H&M." Yale would help expose the fact that the General Staff had "overreached themselves through sheer stupidity."[103]

MAJ Wesley W. Yale reported what Herr wanted to hear, but also detailed the very reasons the horse-mechanized concept failed. Yale went to the maneuvers operating from the premise that no one questioned the need for a light mechanized reconnaissance element for the corps' reconnaissance regiment, or the composition of the "power element" to follow up the mechanized squadron.[104] Possible solutions for the best "power element" included the existing horse squadron, another mechanized squadron, or possibly truck-mounted infantry. Yale concluded that the test was not a fair evaluation of the existing horse-mechanized regimental organization.

MAJ Yale believed that "no unit can function under the conditions which have faced present reconnaissance organizations."[105] The reasons were manifold, and they included a lack of training and a prejudice against the large unwieldy trailers, problems already well understood. Yale identified a leadership problem. The regiments' original design had been streamlined to allow corps commanders to attach additional units, such as infantry, tanks, or engineers, for specific missions, or the commanders could elect to form a number of reconnaissance detachments from the assets on hand. Yale observed that this was not happening and that by making pure regiments, one all-mechanized and the other all-porté, corps commanders lacked the necessary ingredients to form these detachments. On the other hand, if a corps commander elected to use the regiment as a unit, but was unwilling to support it with attachments, the organization would be too weak to accomplish its mission. Yale recommended that the horse-mechanized regiments pick up an additional squadron of light tanks (two companies) and a company of motorized engineers. He recognized that this was already occurring in armored reconnaissance battalions in the armored divisions.[106] Yale's report reflected his belief that the corps reconnaissance regiment's component parts should remain homogenous, even with the new inclusions "BUT ITS COMPONENT ELEMENTS ARE NOT DESIGNED TO OPERATE SEPARATELY. ONLY THE `TASK FORCE' IDEA WORKS."[107] Although he argued for the continued use of the horse, Yale's conclusions forecast the conduct of operations in Europe.

The 107th CavRegt(H-M), in what was perhaps the last feat of glory for a horse-mounted force in a large-scale training exercise, conducted a forty-mile porté movement with 150 semi-trailers and fell upon Patton's rear, where they caused considerable havoc.[108] This went unnoticed, but what had not gone unnoticed was the number of radio reports generated by the mechanized regiment versus those submitted by the porté regiment. Yale correctly pointed out that this should have come as no surprise given the small quantity of radios found in the horse squadrons compared to the mechanized units. His comment on the quality of the messages reveals much about his attitude and the outlook of those who were fighting so desperately to retain a role for the horse.[109]

> Few persons appreciate that the function of the light vehicular reconnaissance unit is to provide a large volume of radio messages giving the outline of hostile resistance; the unit has many radios for this purpose. The "power" unit on the other hand has few radios, since whether horsed or mechanized, it is designed to develop the information obtained by the light vehicular unit or to promote its advance.

Yale went on to write that it was important not to judge the units on the amount of radio reports they submitted since, "Properly used the mechanized units should obtain a large volume of relatively unimportant information; the horse unit should obtain a few items of great importance."[110] These comments and Yale's call for the retention of horses to help "filter through" enemy resistance that could not be surmounted by light vehicular reconnaissance exposed the problematic nature of retaining the horse.[111]

The earliest thoughts on the horse-mechanized concept suggested that horses should be used to go where mechanization could not because of the latter's lack of mobility. The idea of separate missions and capabilities within a single regiment had gradually moved toward a more cooperative role in which the horseman was seen as the true "power element," not the man equipped with a heavy machine gun who operated from an armored platform. While making this argument and castigating the overall merit of the information obtained by the mechanized scouts, everyone—mechanized and horse advocates alike—called for increases in mechanization, specifically the inclusion of light tanks in the organization. One can only imagine MAJ Yale's frustration as he wrote about the outstanding performance of the 8th Infantry Division's reconnaissance troop during the First Army maneuvers in the Carolinas. The division commander supported these mechanized scouts with motorized infantry, giving them the ability to overcome resistance. Yale argued that this was the very idea behind the horse-mechanized concept, and that only the 4th CavRegt (H-M) had ever employed the technique properly.[112]

COL Considine had an epiphany during the 1941 GHQ maneuvers. He realized that his "mechanized squadron performed, *by far,* most of the reconnaissance missions, both distant and close."[113] One of the most prescient observers on the eve of war, Considine wholly rejected what had been the guiding concept of stealth for the accomplishment of mechanized reconnaissance. He concluded that "a corps reconnaissance regiment which must rely on "stealty" [sic] alone" was not worth its "salt."[114] The requirement for "punch" and "power" were "habitual conditions" that required corps cavalry units to fight if they were to be "remunerative," since any other approach "emasculates cavalry, and is not true cavalry doctrine."[115] With the Armored Force having taken many of the Cavalry Branch's combat missions with it in the divorce of 1940, Considine was an early voice demanding the same combat missions for the mechanized cavalry units being specifically designed for reconnaissance. The ultimate fate of the horse-mechanized concept was only months away.

McNair started to consider the organization of the American "triangular" infantry division in 1938 but did not support the inclusion of an organic reconnaissance agency, horsed or mechanized. When the idea was finally accepted after the 1940 maneuvers, McNair's suggestion for the strength of the reconnaissance troop, 153 men, was low, and the figure climbed to 201 by the time U.S. forces went to war in 1942.[116] Equipping the streamlined divisional organization with its own reconnaissance agency gained momentum with calls for at least three reconnaissance platoons and a headquarters, so that each of the three regiments in the new "triangular" division could receive the attachment of a scouting element.[117] The War Department's decision to include a mechanized reconnaissance troop in each of the newly organized divisions immediately touched off a flurry of activity inside the chief of cavalry's office.[118]

The chief of cavalry's office recommended that each of the "triangular" infantry divisions receive an entire reconnaissance squadron that would include a mechanized cavalry troop and a porté troop.[119] The office also attempted to define an important role for the horse in the National Guard, just as it had done for the regular army. COL K. S. Bradford, who had prepared the first recommendation, cited the German use of a thirty-man horse cavalry platoon in each of its infantry regiments as the model to supply of the "square" divisions. To accomplish this, he recommended the retention of such units as the Black Horse Troop of Chicago, the City Troop of Philadelphia, the Essex Troop of New Jersey, the Cleveland Greys, the Governor's Troop of Pennsylvania, and the Atlanta Horse Guards.[120] MG Herr was not silent on the issue. In his mind, scout cars were expensive and lacked adequate protection and cross-country mobility. Stating emphatically that, "It has been my experience, not only as a regimental commander but as an observer at many maneuvers, that these scout cars are very easy to capture or destroy."[121] He added, "Almost invariably in maneuvers at least half of them are captured," and even more would be captured were there more umpires.[122] Arguing that, if used alone, scout cars were "more or less helpless," the cavalry chief urged the inclusion of motorcycles and a porté horse cavalry troop. In his view, this would provide the organization of all the assets needed to conduct reconnaissance in all types of terrain, because "this can only be done by men on their feet or by men mounted on horses."[123] Since the man on horse was superior to the foot soldier, the cavalry troop was a crucial element. The War Department offered its own views on the subject when they issued a final decision to the chief of cavalry.

> The mechanized reconnaissance troop recently approved as part of the triangular
> division was included after careful consideration of the necessity of a reconnaissance

unit and the type of that unit. The decision to make this a motorized unit rather than a partly or wholly horse was reached only after careful study and analysis, both of maneuver experience and of foreign practice.[124]

The horse cavalrymen had lost another battle in the ongoing struggle against the complete mechanization of ground reconnaissance.

Training Circular No. 10, issued on 26 November 1940, offered initial doctrinal intent for the proper role and employment of the newest reconnaissance agency. The chief of cavalry's office made it clear that it intended to control the release of doctrinal material specific to the employment and capabilities of the division reconnaissance troop. The chief of cavalry's office permitted the men at Fort Riley to distribute mimeographed handouts it had developed to support its instruction on the matter, but wanted to make sure it reinforced some important notions. These included the road-bound nature of the organization, its vulnerability to .30-caliber armor-piercing bullets, and the fact that it "cannot be expected to obtain the positive detailed information available to those units with more cross-country ability."[125] COL Bradford's comments were not all negative, and he emphasized how important it was that divisions employ the new organization in accordance with their capabilities.[126] *Training Circular No. 18* also addressed the proper use of the divisional reconnaissance troop, emphasizing many of the same points made by the chief of cavalry's office. The circular drew attention to the idea that the organization was designed for reconnaissance by stealth and should avoid combat if at all possible.[127]

The Infantry School published guidance information detailing the organization, capabilities, and limitations of the reconnaissance troops now organic to its divisions. The information closely mirrored the guidance offered by cavalry proponents and reminded infantrymen that the mechanized scouts were the division commander's asset. The scouts could be used on detachment, but only "under exceptional circumstances" should they "be divided," leading to the separate attachment of one platoon per regiment.[128] Infantry warned that such a dissipation of the troop's resources would lead to a lack of coordination between platoons.[129] With lightly equipped platoons, the infantry doctrine placed the proper emphasis on stealth. Reconnaissance troops serving infantry divisions could not only expect the corps cavalry regiment to develop the initial situation, they could also expect rapid reinforcement from the divisions they served.

Organized during the summer of 1940, the I Armored Corps lacked a dedicated reconnaissance unit, but each of the two original divisions included an armored reconnaissance battalion.[130] Each battalion contained two reconnaissance troops equipped with scout car and motorcycle platoons, a light tank company, and a company of infantrymen mounted on armored half-tracked vehicles. BG Adna R. Chaffee, father of the Armored Corps, wanted to mount the entire battalion on half-tracks, thus eliminating the scout cars and increasing the mobility of the unit.[131] The doctrinal thinking on how to employ the armored reconnaissance battalions changed little from the early days of the armored car troops in each of the mechanized regiments at Fort Knox. Performing distant and close reconnaissance, there was a greater emphasis on how the battalions integrated their efforts with their respective division commanders and intelligence officers.[132] Armored Force doctrine still emphasized stealth as the key component to successful reconnaissance, not unlike Cavalry Branch. The men at Forts Benning and Knox had already concluded that, once discovered, "the enemy will bend every effort toward" the elimination of the compromised scouts and would "institute a relentless search toward that end."[133] The armored warriors knew the enemy would do everything in its powers to destroy reconnaissance elements, not for their combat potential, but because of their ability to transmit information.[134] Commanders ordered their scouts to engage in combat only if it was absolutely essential to accomplish their mission, such as an assigned combat or security role, or to protect themselves. The Armored Force leaders also had a good understanding of the impact that time had on the conduct of reconnaissance. They were already admitting that when there was enough time, a rarity in war, stealth would prevail as the best method, but hasty reconnaissance would often lead to combat.[135] The combined-arms nature of the armored reconnaissance battalions reflected the fundamental understanding that they would often have to fight for information.

Having led the way for the mechanized ground reconnaissance development throughout the interwar years, the armored reconnaissance battalions were also afflicted by the dramatic increase in the size of the army. They too were criticized for their poor performance during the large-scale maneuvers in 1941. The 1st AD bogged down in the Kisatchie National Forest in Louisiana, a result of poor route reconnaissance. When the same mistake occurred in North Africa a year and a half later it cost the division eighteen tanks and 132 other vehicles.[136] MG Patton, who then commanded the 2nd AD, had no time for "waffle ass" in reconnaissance units. To show the absolute importance attached to good mechanized reconnaissance, and perhaps to illustrate how much learning remained to be done, Patton offered his entire division these thoughts:

> When any of you gets to a place where your experience tells you there is apt to be an anti-tank gun or mine or some other devilish contrivance of the enemy, don't ride up in your scout car or tank like a fat lady going shopping, stop your vehicle, take a walk or crawl and get a look but remember that in walking or crawling you must not go straight up the road, you must go well off to a flank probably as much as one thousand yards.

And after encouraging them to use their field glasses to get a good look at where the enemy might be, he closed the soliloquy on reconnaissance with a warning: "This walking is hard work but it is very much easier than getting killed and getting killed is what will happen to you in battle unless you use proper precautions in reconnaissance."[137]

With all their shortcomings, the armored reconnaissance battalions and their leaders demonstrated initiative and aggressiveness in the conduct of reconnaissance. Patton's own reconnaissance battalion, led by MAJ I.D. White, who had gotten his start in mechanized reconnaissance as an armored car platoon leader at Fort Knox in the early 1930s, captured GEN Hugh Drum.[138] LTC H.H.D. Heiberg, another veteran from the early days at Fort Knox, used aerial photographs to find routes for the 81st Reconnaissance Battalion that did not appear on the large-scale maneuver maps, and he attached tank platoons to reconnaissance companies during the Carolina phase. The attached armor gave the enemy the impression that it had gained contact with an entire armored regiment.[139] These actions highlighted increasing roles for the mechanized ground reconnaissance units in economy of force missions. Heiberg's decision to assign tanks to reconnaissance companies did not dramatically change the institutional norm for homogeneity in units, but it did show the possibilities of combining arms at lower levels.[140] It also demonstrated the increasing need for firepower and protection in the reconnaissance units that were organized on the principle of stealth.

The 1st CavDiv's mechanized arm of reconnaissance had gradually grown since 1939, when it was first increased to a squadron with five officers and one hundred enlisted men.[141] Increasing again in January 1941 with an injection of selective service draftees and veterans from the horse regiments within the division, the squadron consisted of two scout troops equipped with scout cars and motorcycles, one rifle troop mounted on jeeps, and a light tank company. By May 1941 the squadron was officially named the 91st Reconnaissance Squadron. The increase in relative firepower and size reflected growing recognition of the importance of mechanized units, but the squadron's organization stayed outside the debates that engulfed the horse-mechanized regiments and divisional reconnaissance troops, because it posed no immediate threat to the horse elements in the unit it served. The squadron played a vital role in the division commander's efforts to find a suitable anti-tank remedy for the 1941 maneuvers.

MG Herr was particularly concerned that the cavalry division was becoming too focused on how it was going to deal with the increasing armored threat during the upcoming General Headquarters maneuvers. He promised Swift he was doing his best to get the cavalry division its full compliment of the recently tested "bantams," some 350. These would be critical to tow the 37mm and 75mm guns that the chief of cavalry was also trying to wrangle from George C. Marshall.[142] Herr cautioned Swift not to think of "hand grenades" as a viable solution to the mounting armor threat and opined that it was "quite clear . . . they are certainly not antitank weapons, and the poor trooper is so loaded down with equipment now he can hardly mount his horse." Herr saw the best way to stop armored penetrations was to remember that "Cavalry is an offensive arm and is not designed for the purpose of blocking tanks." After all, Herr reminded Swift, "The French found this out in front of Sedan just a year ago." Instead, Herr suggested the best course of action was for the Cavalry to act "offensively in conjunction with armored elements if necessary, and make them block us, if they can."[143]

Herr further reflected on the subject: "As I see it, if we can get by this period of faddism the horse cavalry will eventually come into its own again. The armored vehicle will become as outmoded as the armored knight. Without overwhelming air to pace them they will be deprived of 75% of their power to begin with."[144] Years later the Germans, who lived in fear of Allied air superiority, would have agreed, but Herr's vision distracted him from fulfilling the task of creating the ground reconnaissance organization to help fill the gap between the air and the leading edges of ground forces. This vision was due to his undying view that the horse remained the best and most proven technology to accomplish this mission. "As soon as nations catch up with the manufacture of highly mobile antitank weapons with which to confront them the armored vehicles will begin to pass out of the picture because it costs so much less to make destroyer squadrons of unarmored vehicles toting the antitank guns around."[145] By equipping the 1st CavDiv with 37mm anti-tank guns, Herr demonstrated his conviction that the days of armored forces were numbered.

Herr also predicted that given the "parking spaces of all this horde of vehicles," the day would come when it would be "unhealthy by reason of bombers" to be part of an armored division. Acknowledging that this day was somewhat distant, since "no one is adequately prepared," he remained confident that when it arrived the horse cavalry would once again be in vogue, given its ability to disperse off roads and move cross-country before the growing tide of air power.[146] Herr also saw in air power the means to cut free of the most cumbersome aspects of the cavalry division as it existed. Air-delivered logistics combined with the restoration of pack-trains would allow the division to cut loose from its now motorized logistics tail. The promise of close air support and bombers negated the need for many of the ever growing anti-tank systems the division was now fielding to fend off the armored threat. Herr was hopeful: "If we can get by this period of ignorance and prejudice and prevent these shortsighted gumps from wiping ut [sic] out of the picture in their mistaken belief that the iron horse replaces one of flesh and blood, we will surely come into our own."[147] Since hope was never a method, MG Swift took measures to defend his division, but in the process started to arrive at a conclusion that was an anathema to Herr.

Swift concurred with Herr that the cavalry must continue to be offensive in dealing with the threat of "hostile armored elements" but "in conjunction with friendly armored units or anti-tank formations." He thought it sheer folly for the horse division as it was equipped to attack an armored force.[148] Swift had clearly arrived at what many had already concluded when he wrote Herr, with incredible candor, that horse cavalry had "become, or should become, mounted infantry." The horse was merely a means of conveyance to that place on the battlefield where the cavalrymen would be expected to dismount and "fight as well, if not better, than the best infantry."[149] Thus, the cavalry's utility still lay in its ability to do the job of the infantry, but with the continued ability to traverse difficult terrain more quickly than infantrymen mounted on trucks. With some prescience, Swift sought an immediate solution to the threat of "roving and roaming" enemy tanks with hand grenades. He hoped that a device could be made that was "light, but still powerful enough to disable a tank if it hit it in the right place." This was imperative in Swift's mind if the cavalry were to compete with the infantry and thus maintain its relevance.[150]

Herr, displeased with Swift's comment that cavalry had become little more than mounted infantry, responded with conviction:

> I cannot refrain from taking exception to our remarks that Cavalry should become mounted infantry. I can say that all true cavalrymen would regard that as heresy . . . mounted infantry is totally incapable of mounted action. The chief value of Cavalry as distinguished with mounted infantry is not only the ability to fight from the horse on occasion, particularly in small bodies, but it is also because of its much greater mobility, both because of the care taken in selecting and training superior mounts and also being so lightly equipped as to be able to transport fire power with celerity. One cannot sacrifice speed entirely to fire power.[151]

Placing superior value on mobility had long dominated cavalry thinking in the horse community and at Fort Knox. In regard to reconnaissance, the men at Fort Knox had placed a greater emphasis on protection for their scouts so that they might live long enough to perform their mission, while growing interest in the jeep in the 1st CavDiv saw the horse cavalry community moving in a different direction. As a second postscript to this particular letter, Herr had a special admonition for his friend Innis P. Swift.

> Listen Palmer, I don't know who fed you all that junk you wrote. I advise you to consult Bradford about such matters. He has been with me three years and knows the answers. You can trust him, also. He is not only capable but loyal. Use him. I am coming by for a couple of days before long.[152]

Herr now worried that the views of the commander of the most stalwart example of the relevance of horse cavalry, the 1st CavDiv, were not synchronized with his own.

Herr tasked Swift with developing an "Antitank Regiment" for the upcoming maneuvers in Louisiana. To do so, Swift reminded Herr, he would have to "deprive" the cavalry division "entirely of its reconnaissance squadron and all of its means for Antitank Defense." The proposed unit was to be composed of the existing anti-tank troop and a reconnaissance squadron organic to the division. As Swift pointed out, the division would in essence "be largely an 'old time' horse division," which should only be used if "all motor units are immobilized by mud or other adverse conditions." Swift also commented on his inability to support the ad hoc organization with portéd cavalry, because the division had some fifty-two trailers, but only eleven "prime-movers."[153]

Undeterred, Swift conducted a predeployment test of the ad hoc anti-tank squadron and pitted it against the tank troop from the division's mechanized cavalry reconnaissance squadron. With a combination of the division's anti-tank troop and the remainder of the reconnaissance squadron, 1st CavDiv tested its ability to respond to an armored threat over open terrain. The first lesson learned was the tenuous nature of air-ground communications. Planes tasked to fly aerial reconnaissance lost contact with the ground units before the first tanks even began their attack. Having lost its view from above, the provisional squadron dispatched observation posts to cover the expected approaches. This string of reconnaissance vehicles provided accurate and timely reports that allowed the towed anti-tank guns to move into good positions to engage the armored attack.[154] The foremost lesson learned was the need for redundancy in planes conducting the air reconnaissance. Perhaps as a function of gamesmanship, the attacking tanks had waited for single airplanes to depart their area before resuming their attack from a different direction.[155] Swift offered no explanation for how the division would respond if attacked during hours of limited visibility, brought about by night or inclement weather.

The division gained an appreciation for the need to have a well-rehearsed and tightly controlled "tank-tight screen," to avoid losing contact with "the enemy mechanized force." Early warning was essential since the anti-tank guns had to be towed to the correct position to make their defense.[156] In towing these guns, the division completely sacrificed the use of its mechanized reconnaissance squadron for an unintended purpose, as the machine continued to serve as the horse did in the 1st CavDiv.[157]

The jeep completely captured the attention of the army in the months leading up to 7 December 1941. With this machine the horse cavalry advocates exerted influence on the development of a major piece of equipment used for mechanized reconnaissance during World War II. Because it was vital for the towing of the unpackable 37mm anti-tank guns, MG Swift was "very enthusiastic over the performance of the Bantam car" when his division received and started testing them in January 1941. Testing pitted jeeps against both scout cars and motorcycles under a variety of conditions:

Unaided steep hill climbing

Aided "rim-rock" operations

Narrow road turn-arounds

30 miles of country road and trail

40 miles of improved road

Ability to transition to dismounted action

Ability to escape from a road block ambush

Comparative silhouette tests

Narrow versus standard gauge rough trail

With few exceptions, Swift noted that "bantam car" outperformed the other vehicles, and he became convinced "that the bantam is a grand car for such purposes as the British use as a Bren gun carrier."[158] Swift reassured Herr that, "Neither food, water, forage nor weather is going to interfere with our cutting loose to ride from hell-to-breakfast on the maneuvers but I really believe the Jeeps can accompany us anywhere."[159] He used his own jeep, equipped with an SCR-245 radio set, to travel 200 miles per day while maintaining contact with his command post. He was convinced that the "bantam can supplant the scoutcar" and looked forward to the day when all the scout cars "disappear[ed]."[160]

Given the premium value horse soldiers placed on mobility and stealth, it was no surprise that these soldiers immediately fell in love with the jeep when it was first procured by the army in large numbers in 1941. As the jeep entered service, MG Herr was still reminding everyone who would listen that in the next war roads were going to be "mighty unhealthy places" and that the cavalry needed to place more emphasis on cross-country mobility "after the manner of the hordes of Genghis Khan."[161] Herr's office told people who inquired that the "Bantam Car" was not intended to replace the horse, but primarily the motorcycle.[162] In regard to motorcycles, these same views were echoed at Fort Knox.[163] In the end, the jeep did replace the horse, but it certainly did not share all the characteristics long advocated by the leaders in mechanized reconnaissance. The horse cavalrymen could take credit for integrating the use of the jeep into mechanized reconnaissance against the strongest objections presented by men like COL Charles L. Scott, who had consistently called for more armored protection and firepower for those operating at the fore of all friendly forces.

The initial announcement heralding the quartermaster general's purchase of "several hundred" jeeps projected their probable employment in the role previously allotted to the motor tricycle in the motorcycle platoons.[164] The 6th CavRegt(H-M) received its first eight jeeps in December 1940. With the same zeal displayed by earlier generations of mechanized scouts, the troopers immediately sought to gain familiarity with their new "iron ponies."[165] They wasted no time modifying their new steeds to meet their own specific reconnaissance needs. They added skid plates to the undercarriage to protect the drive train and make it easier to slide the vehicle off of obstructions, they attached gas-can brackets, and they fabricated a new trailer hitch to speed the employment of the 37mm anti-tank gun that the new vehicle was expected to tow into battle.

Radio mounts constructed by the men enabled them to install SCR-245 radios, providing the vehicle with the same capability as the scout car.[166]

One of the constant criticisms of mechanized reconnaissance had been its inability to swim across water obstacles. Over the years this task had become more difficult for horse cavalry with its host of packed weapons and motorized supply systems. In time, a number of articles appeared detailing the best ways to train horses who had never been forced to swim, as well as how to prepare equipment for water crossings. The jeep offered alternatives that were never applicable to the heavier scout and armored cars. The troopers at Fort Oglethorpe built makeshift rafts with thirty-two empty ten-gallon cans. Having thus floated the jeep, they improvised two different means of propelling the vehicle across the water under its own power. One version used paddles attached to the rear wheels. The other involved the use of a vehicle axle converted into a scaled-down version of a riverboat paddle wheel.[167] The men of the 9th Reconnaissance Troop stationed at Fort Bragg were no less enamored with the newest addition to their reconnaissance organization. They chose to use the technique described in training circulars and in the *Infantry Mailing List* that involved wrapping the jeeps in the tarps taken off of large trucks. Once ensconced in green canvas, soldiers floated the jeep across the river or towed it across using the winch on a larger truck.[168] Having solved the problem of how to cross water obstacles, the men turned their attention to integrating the jeeps into the tactical operation of platoons.

Relying on the jeep's low silhouette and high mobility to avoid casualties, emerging doctrine placed jeep-mounted men on "point" to conduct reconnaissance for the platoon's scout cars. Horse-platoon leaders started to accompany mechanized platoons in their own jeeps in order to conduct personal reconnaissance "very rapidly" before returning to the porté detrucking point to rapidly commit their platoons.[169] With horse-platoon leaders now riding in jeeps, there emerged yet another slow, but sure encroachment on the horse's continued utility.

One week before the attack on Pearl Harbor, LTG McNair reflected on the state of training in the rapidly expanding army. He singled out the lack of good reconnaissance and security measures as a major training shortfall.[170] Events in Europe led to the eight-fold expansion of the United States Army that was felt across Cavalry Branch.[171] The creation of the Armored Corps and loss of the 7th CavBde(M) forced Herr to seek compromises that he hoped would secure a future for the horse. The horse-mechanized concept was his premier vehicle to demonstrate a contemporary response to the twin disasters in Poland and France. Mechanized reconnaissance troops in each of the new "triangular" infantry divisions signaled the direction which George C. Marshall and Lesley McNair were inclined.

Herr remained convinced that the maneuvers conducted in 1940 and 1941 were "rigged to limit the activities of the cavalry, for the pressure was on from certain quarters to eliminate the mounted service."[172] Having relegated mechanized forces to the flanks of divisional maneuvers in the past, Herr now experienced a dose of his own medicine as the speed of mechanization forced horses to the flanks and limited their ability to deploy and fight.[173] The Armored Force and Cavalry Branch exchanged roles, as the former took the latter's combat functions. Herr and men like him did not accept this willingly and continued to defend their horses throughout the coming war and beyond.

> To them we must have seemed a vision from another century, wild-eyed horses pounding headlong; cheering, whooping men firing from the saddles.[174]—*LT Edwin Price Ramsey, 16 January 1942, Luzon*

★   After the attack on Pearl Harbor, the descendent of the first mechanized reconnaissance unit in the United States Army, the 91st Cavalry Reconnaissance Squadron (CavReconSqdn), moved out to patrol the Mexican border.[175] One could have viewed its deployment with some irony, or as a prescient act indicative of what was to soon follow on a larger scale. Much of the debate about the role of mechanization centered on the issue of where the country might expect to fight, and Mexico dominated the interwar possibilities. Those who had seen limited potential for the mechanized scouts viewed the terrain along this border as being too rough and broken for vehicles, but it was here that the first "iron-mounted" men saw service when war found the nation. The men of the 91st CavReconSqdn were also joined by more traditionally mounted units within their division, among them the 2nd CavRegt that moved to Papago Park, between Phoenix and Tempe, Arizona, to conduct mounted border patrol, and to guard airfields and mines.[176] Much was still to be decided, but there was little time.

The period between Pearl Harbor, at the end of 1941, and the first cavalrymen splashing ashore in North Africa under cover of darkness on the morning of 8 November 1942 was extremely important in the development of mechanized ground reconnaissance units. Horse and machine continued to coexist within Cavalry Branch. After years of interwar development and experimentation, there was little time left for organizational and doctrinal changes, and the changes made and not made during this period had a direct bearing on the conduct of World War II in the European theater. Thus, during the interlude between the attack on Pearl Harbor and the introduction of American combat forces in North Africa, the War Department established or retained a baseline of organization and doctrine that guided mechanized cavalrymen to war to perform the missions demanded of the latter within the limitations of the former.

With one major exception, there were in fact few changes to the organization of the mechanized reconnaissance units that had been evolving since 1928. The exception, of course, was the corps cavalry regiment. Born from the crisis in Poland in 1939, it was suited to the chief of cavalry's efforts to find some accommodation between horse and machine. The horse-and-mechanized regiment that emerged had undergone extensive testing prior to Pearl Harbor. The chief of staff of the army decided the organization's fate in the early months of 1942, when the regiment underwent the first and most dramatic change of any of the mechanized cavalry organizations during World War II: it suffered the removal of its horses.

Large-scale maneuvers conducted days before Pearl Harbor had convinced senior army leaders of the need for a reconnaissance agency dedicated to the collection of information for army and corps-sized units. One observer, LTC John A. Hettinger, the assistant chief of staff, G2, for the 8th ID, explained what he had seen to one of his colleagues working in the office of the chief of cavalry. Given the extended frontages "occupied by divisions, corps and the army," only a "bountiful use of cavalry" could keep "all of the gaps . . . patrolled properly."[177] After "filling the gaps," Hettinger noted that the corps had "absorb[ed]" the mechanized reconnaissance squadrons and that army-level commands would be left "groping about for some type of trained reconnaissance elements that can do their job."[178] He further believed that this "immense job" had been given all too little "consideration to date." Hettinger saw some hope in the fact that the army was "rapidly becoming reconnaissance conscious."[179]

LTC Hettinger offered some critical advice to his friend, based on his observations during the recent large-scale maneuvers. Hettinger saw "the greatest opportunity for expansion that the cavalry has had since this trouble began."[180] What Hettinger then said may have affronted the last chief of cavalry, who was far more interested in the fighting characteristics of the cavalry than the lesser mission of reconnaissance.

> Another thing that should not be lost sight of is the fact that we need a cooperative type of cavalryman with these units and not one who wishes to build up a small army of his own and fight a private war on the side without furthering the division, corps or army commander's plans. The army needs a division of cavalry. The corps, at least a reinforced brigade and all of the divisions need a regiment.[181]

Undersized mechanized reconnaissance units forced commanders to borrow valued assets from the very units they expected to do the fighting. Hettinger estimated that the current arrangement only provided the army with a total of 2 to 3 percent of its force, with elements dedicated to reconnaissance. His interpretation of the existing field service regulation saw a need for a level of reconnaissance troops approaching 30 percent.[182] Certainly, Hettinger's call for a far-larger percentage of troops to collect information through reconnaissance reflected his current job as a division G2.[183] Other observers sent thoughts to Washington.

Ordered by the War Department to test "Cavalry Reconnaissance Units," MG O.W. Griswold prepared a report outlining his observations during the recent Carolina maneuvers. He concluded that the corps needed a dedicated agency for the "protection of the flanks of the corps, for distant reconnaissance and long range counter-reconnaissance action." In his view, anything less than a regiment would be too small to accomplish these tasks.[184] His experience in Louisiana and in the Carolinas led him to conclude that "the corps reconnaissance regiment should be wholly mechanized."[185] He cited the mechanized squadrons' ability to carry out distant reconnaissance and noted the inability of "horse cavalry" to "advance rapidly enough to effectively reconnoiter in front of, or protect, the advance of motorized infantry elements."[186] He damned the porté trailers for their size and weight, both of which contributed to their lack of "cross-country mobility" and the inability to conceal them.[187] Griswold further recommended that the corps reconnaissance regiment be equipped with enough organic firepower so that it could conduct "limited attacks to secure information," since that would often, "if not usually," be required.[188] The same assets Griswold called for in the organization of the corps cavalry regiment, such as light tanks, antitank guns, and reconnaissance troops, would also allow it to effectively perform counter-reconnaissance and delaying actions. Even if interwar opponents of mechanized reconnaissance continued to insist that stealth was more fundamental than fighting ability, Griswold, a critical observer above the fray with a high degree of prescience, saw the need to be able to fight for information.

Herr took exception to Griswold's recommendations. He believed that Griswold had mismanaged his cavalry assets. Worse still, Herr was convinced that the faulty conclusions being drawn about the utility of horse cavalry were "based on the unrealities" built into the exercise's design. Specifically, Herr cited the "accelerated" and "fantastic" speed that characterized the initial movements with opposing forces arrayed "75 to 150 miles apart."[189] This, Herr concluded, rewarded the commander who raced as many men forward as fast as possible on every available vehicle. In essence, the design of the recent maneuvers denied reconnaissance units, particularly slower-moving horse elements, the opportunity to develop the situation before the commitment of larger formations of armor and infantry. Furthermore Herr argued, "vehicles simply cannot perform complete reconnaissance." Only horses were capable of "filtering through almost anywhere between the roads."[190] Herr provided his own recommendation and argued that every unit, from army to regiment, needed a dedicated reconnaissance unit that at a minimum "moves faster than the main body."[191] Herr conceded the role mechanization would play in terms of reconnaissance during the war confronting the country, but he insisted on a combination of horse and mechanized units. Acknowledging the superior ability of mechanized forces to move on established road networks, he was equally quick to point out, in the language of the interwar critics, that this was really only good for providing "negative reconnaissance."[192] Herr also articulated why "many high commanders" arrived at faulty conclusions about how reconnaissance forces should be organized, returning to the theme that war games rewarded a race to contact since each side was confident that nothing was operating to their immediate front.[193] There was little doubt that Herr sought to satisfy the army's reconnaissance needs, even with considerable mechanized elements. It was also very clear that he sought to retain horses in large numbers with a plan calling for at least one horse cavalry division per field army. He envisioned even larger organizations of mounted men.

In late January 1942, Herr sought permission to take command of "large groups of cavalry, such as cavalry divisions and separate brigades," since they were "as much GHQ troops as [were] armored divisions." He saw his rightful place of duty in command of these forces in the American southwest with "similar authority to that exercised by the Chief of the Armored Force."[194] Very much in synchronization with the contemporary doctrine, Herr continued to view his branch as a combat arm and sought the training opportunities to prepare large horse-mounted formations for war. Only days after he requested to move at the head a large body of horse cavalrymen on the familiar terrain of his not so distant past, Herr received two telling blows that assured he would neither realize his personal request nor see his branch retain the identity he had fought so hard for it to maintain.

On 3 February 1942, Herr wrote his friend, MG Innis P. Swift, who commanded the 1st CavDiv, that "the Chief of Staff has decided to mechanize all these corps cavalry regiments on the advice of Leslie McNair and his G-3, and without any consultation with me on the subject."[195] West Point ties being of such a nature that they followed officers throughout their careers, Herr could not resist a jibe at Swift, remarking: "Frankly, this fellow McNair, this classmate of yours, is the most ignorant person in the world with respect to the horse and has been able, due to his position, to exercise a very baneful influence." Two days later, Herr made his feelings known to the chief of staff, George C. Marshall.[196] It was one of his last opportunities to lobby for the horse, while he wore the uniform of a general officer.

Pearl Harbor set into motion a chain of events that reordered the War Department. Marshall had lost confidence in the department's ability to process information and serve as an effective command post for the direction of the war. His unhappiness with the organization of the War Department pre-dated Pearl Harbor, but the department's inability to issue a "follow-up" warning to the Hawaiian Department provided the catalyst for change.[197] President Franklin D. Roosevelt signed Executive Order 9082 on 26 February 1942, and the order took effect on 9 March 1942. Implementation of the order drastically reduced the General Staff and gave Marshall "paramount authority . . . under the President in the broad sphere of strategy, tactics, and operations, the most important functions of command."[198] The reorganization also eliminated the General Headquarters and all branch chiefs and replaced them with the Army Ground Forces organization. Army Ground Forces was responsible for every aspect of organizing, training, equipping, and preparing the doctrine for the American ground forces contribution to World War II.[199] Marshall and McNair were now free to build the army any way they wanted.

One of the casualties of the reordering of the War Department was MG John K. Herr, who would go down in the history books as the last chief of cavalry. As the last chief, Herr's term expired in March 1942, and neither he nor any of the other branch chiefs were replaced.[200] With Herr went the last hope for some accommodation to be reached between the horse and the vehicle. Herr may have lost his ability to directly influence the role Cavalry Branch was to play in World War II, but he certainly did not lose his passion for what had been his life's service. As president of the Cavalry Association he wielded his influence in the *Cavalry Journal,* supplying readers with a steady diet of horses in action, even if the horses in question were more often than not those of the Soviets or Chinese.[201] The valiant action of the 26th Cavalry's Philippine Scouts also gained acclaim in accounts of their valiant but futile effort in the Philippines in the early days of 1942.[202] Perhaps hungry for coverage of their own horse-mounted warriors, the suggestion was put forward in the pages of the journal that the United States should form units of "cavalry commandos." These specially trained men were to carry out operations behind enemy lines using horses for mobility.[203] A month after this call to arms appeared in the September–October edition, the branch had its own heroes, but rarely men on horses.

With the last impediment to wholesale mechanization removed, the process went forward rapidly. The army's two cavalry divisions, the 1st and the 2nd, did not ship a single horse overseas during World War II. The 1st CavDiv dismounted before shipping to Australia. Retaining its "square" organization to compensate for its smaller numbers, members of the division fought with distinction under GEN Douglas MacArthur throughout the remainder of World War II, but they fought as infantrymen. The next time the old horse division remounted, it would see the sons of these men in helicopters in another Asian theater, Vietnam. The 2nd CavDiv, with its African-American brigade composed of the 9th and 10th CavRegts, was completely broken up to create other units. The 2nd and 14th CavRegts that made up the other brigade of the 2nd CavDiv, along with other army's non-divisional regiments also supplied a large number of personnel to form the newly created 2nd, 3rd, 11th and 14th Armored Regiments of the newly created 9th and 10th ADs. None of the National Guard's horse cavalry divisions were federalized, and many of the division's members went on to see service in field artillery, coastal artillery, and tank destroyer units.[204] Only later, after the personnel had been distributed to new units, did Army Ground Forces resurrect the regimental headquarters of the 2nd, 3rd, 6th, and 11th CavRegts, around which they formed mechanized cavalry regiments.

Once the chief of staff "decided to mechanize all . . . corps cavalry regiments on the advice of McNair and his G-3" and "without any consultation with" with John K. Herr, the process moved swiftly. This led to the full mechanization of the 4th and 6th CavRegts of the regular army as well as the National Guard regiments operating in the horse and mechanized configuration.[205] The result was an all-mechanized regiment consisting of a regimental headquarters and two mechanized cavalry squadrons. Each squadron consisted of two reconnaissance troops and a support troop that included light tanks and assault guns.[206] Changes were not limited to the cavalry.

On 1 March 1942, the Army Ground Forces reorganized the existing armored divisions, establishing the organization the United States would take to war in North Africa before the year was over. Each division lost one of its three tank regiments, the remaining regiments consisting of two medium tank battalions and one light tank battalion. The restructuring of the division also eliminated the brigade headquarters and replaced it with what came to be known as Combat Commands. The 1942 armored division contained only two Combat Commands, A and B. Combat Commands allowed for the flexible combination of the division's resources on a mission-by-mission basis. The division retained a regiment of armored infantry and saw an increase to three battalions of armored artillery. The division also retained its armored reconnaissance battalion while the remaining regiments retained their organic reconnaissance companies. By the end of 1942, the army stood up fourteen of the sixteen armored divisions that saw combat in World War II, a remarkable accomplishment given that only two armored divisions existed in 1941.[207]

The events in Europe that accelerated and ultimately brought about the victory of mechanization over the horse for reconnaissance taught the losers valuable lessons. The summer of 1940 convinced the British that light tanks were good for nothing but reconnaissance. The British also abandoned all of their open scout cars and simple armored cars, finding them insufficient. Believing "the need for an overhead armor is absolutely imperative," they sought it in the form of wheeled reconnaissance vehicles with tank-like bodies, or light tanks.[208] The Cavalry School at Fort Riley was not oblivious to what was happening in North Africa and Europe, stating, "the scout car is not considered a suitable reconnaissance vehicle for mechanized reconnaissance," but that it had to serve as a substitute until something better was available.[209] American scouts went to North Africa in open scout cars and were also equipped with jeeps, because the scout car replacement was not expected to be available until September 1942. It was in fact not fielded until 1943.[210] Motorcycles remained in all the reconnaissance organizations as well but were also viewed as having limited utility for carrying messages and doing traffic control, and they continued to be replaced by jeeps.[211] Weapons were another matter.

The largest weapon found in any reconnaissance organization was the 75mm howitzer, organic to the platoons of the armored reconnaissance battalions and found in the support troop of the 1942 mechanized cavalry squadron.[212] This particular weapon system was not controversial, but the 37mm cannon gained more attention. All existing army doctrine in 1942 viewed the 37mm cannons on the light tanks in the mechanized cavalry squadrons and armored reconnaissance battalions as an effective anti-mechanized weapons systems. Doctrinally, so were .50- and .30-caliber machine guns with armor-piercing bullets at close range against light armored vehicles.[213] Reports from abroad changed this view and led to the modification of existing doctrine in staff studies conducted at the Cavalry School at Fort Riley.[214]

While the organization of the Cavalry Branch had undergone some modifications since Pearl Harbor, there was little time for the army's and Cavalry Branch's doctrine to reflect these changes. Therefore, as the nation went to war, the horse remained the centerpiece of cavalry doctrine and army doctrine related to cavalry. All doctrine addressed cavalry mechanization but relegated it to the role of performing reconnaissance missions for units still equipped with horses. The doctrine the nation took to war also clearly indicated how the Armored Force replicated the combat missions still allotted to horse cavalry units.

The army published its premier how-to-fight manual, *FM 100-5, Operations,* in May 1941, the first official update of army doctrine since 1923. It was an extremely important manual because it established the doctrinal foundation that guided the efforts of a rapidly expanding army of citizens who were taking up the tasks of soldiers.[215] To accomplish all the tasks that soldiers had performed in the past, the 1941 edition of *FM 100-5* anticipated a continued mission for cavalry. The army defined cavalry as a force composed of "highly mobile ground units, horse, motor, and mechanized" characterized "by a high degree of battlefield mobility" that allowed it to rapidly and easily move firepower from "one position or locality to another."[216] With these attributes, the doctrine considered:

> Cavalry is capable of offensive combat; exploitation and pursuit; seizing and holding important terrain until the arrival of the main forces; ground reconnaissance; ground counterreconnaissance (screening), both moving and stationary; security for the front, flanks, and rear of other forces on the march, at the halt, and in battle; delaying action; covering the retrograde movements of other forces; combat liaison between large units; acting as a mobile reserve for other forces; harassing action; and surprise action against designated objectives deep in hostile rear areas.[217]

In 1941, Cavalry Branch retained all of its former missions including combat. This was the same doctrine the nation took to battle in 1942.

In the chapter dedicated to intelligence and reconnaissance matters, *FM 100-5, Operations* acknowledged the role of aviation, pointing out that "ground reconnaissance elements . . . cannot obtain a complete picture of the enemy situation to any great depth in rear of the hostile screen."[218] The portion of the manual describing the horse cavalry division echoed this sentiment.[219] Ground reconnaissance was effective for maintaining "continuous contact . . . [and able to] operate under weather conditions which preclude air reconnaissance."[220] The manual then praised horse units for their ability to conduct detailed ground reconnaissance "within an appropriate area," whereas mechanized reconnaissance units were "of great value" on more distant missions or along "an extensive front."[221] This too summarized interwar thinking.

The army anticipated the need with its 1941 doctrine for an organization that did not even exist for those occasions:

> When cavalry divisions or adequate mechanized reconnaissance forces are not available, and the reconnaissance mission indicates the probability of serious combat or necessitates operations at a considerable distance from the main forces, a composite force consisting of available mechanized reconnaissance elements and either porté horse cavalry or motorized infantry may be desirable. Such a force may be reinforced by other arms.[222]

This broad mandate provided the army's leaders the doctrinal license to improvise organizations to meet its reconnaissance needs. It also provided the same ability for broad interpretation that allowed for the task organization of mechanized reconnaissance units to fulfill non-reconnaissance missions. Doctrine called for concentration, but allowed that under some circumstances, "reconnaissance elements may be held in reserve" either to support ongoing missions or provide commanders with the flexibility to direct their efforts in new directions.[223] In regard to the actual conduct of reconnaissance missions, *FM 100-5, Operations* stated:

> Ground reconnaissance elements gain and maintain contact with the enemy, and by working through gaps and around the flanks and rear, endeavor to ascertain the strength, movements, composition, and disposition of the enemy's main force, and the approach of enemy reinforcements.[224]

This recognized in a doctrinal way the expectation that reconnaissance units would have the room to maneuver on the battlefield to find the flanks and gaps, reinforcing the interwar notion and expectation that the next war would be one of mobility.

Gaining and maintaining contact with the enemy was sought "chiefly through the use of patrols."[225] The doctrine placed a premium on stealth, for "weak reconnaissance elements" wanted "to avoid combat unless it [was] necessary for gaining essential information."[226] To achieve success, commanders needed to contain the enemy's counter-reconnaissance force, so that ground reconnaissance force could push "reconnaissance around the flanks."[227] Again, the dedicated reconnaissance forces were not to become involved in penetrating the enemy's counter-reconnaissance screen; rather, they were to maneuver to find the gaps and flanks. Even so, army doctrine did anticipate the need to fight for information, stating, "essential information can frequently be obtained only through attack."[228] In doing so, *FM 100-5, Operations* anticipated the need to develop "limited objective[s]" when it was impossible to "brush aside or envelop" the enemy.[229] The doctrine warned commanders that "such reconnaissance may bring on a general engagement."[230] Doctrinally speaking, more passive reconnaissance remained the standard, with the assumption that the battlefield would allow for enough maneuver to find the gaps and flanks. The reconnaissance organizations then in the army reflected this idea in their basic organization and were generally unable to fight for information unless reinforced.

Because the Armored Force was not recognized as a branch unto itself, it was not described in the chapter detailing the basic characteristics of the arms and services. Rather, *FM 100-5, Operations* placed the description of the armored division in the same chapter as the description of the horse cavalry division. There the armored division was described as "a powerfully armed and armored, highly mobile force" chiefly characterized as possessing a high degree of "protected fire power" and "battlefield mobility."[231] These characteristics were remarkably similar to the description afforded cavalry, and they afforded the armored division the ability to operate at extended distances with its self-contained logistical support assets. Unlike the cavalry, however, with its horses, the doctrine still described the armored division as being sensitive to "obstacles, unfavorable terrain, darkness and weather."[232] Also unlike the cavalry, the armored division's list of possible missions did not include reconnaissance. It was doctrinally considered "capable of engaging in all forms of combat, but its *primary role is in offensive operations against hostile rear areas.*"[233] One could argue that by 1942 the Armored Force had usurped the prized missions of the horse cavalry, leaving Cavalry Branch the reconnaissance mission.

The cavalry field manual, *FM 2-15, Employment of Cavalry,* also published in 1941, superseded *Cavalry Field Manual,* volume 3 (3 January 1938). The new manual echoed *FM 100-5, Operations* in regard to the branch's identity as a combat arm "organized primarily to perform missions requiring great mobility and fire power."[234] More emphatically, the doctrine stated that, "the primary mission of Cavalry is combat."[235] A key limitation was the inescapable fact that animals and machines all required care and maintenance. The "effectiveness of Cavalry over extend periods" depended on "the efficiency of the supply and maintenance system."[236] Even so, horses still only need food, water, and rest to "renew their strength wherever they happened to be," while vehicles were "dependent upon supplies, such as fuel, oil, and spare parts."[237] Logistical requirements were not the only differences.

Horse cavalry rightfully continued to claim the "ability to operate on almost any type of terrain and under practically all types of weather conditions."[238] The doctrine anticipated that in large horse units "well-conditioned men and animals" could march thirty-five miles per day for an extended period of time and "in an emergency, 125 miles in 48 hours."

Although mechanized units had far greater ranges, 180 miles per day—or under extreme conditions up to 300 miles in twenty-four hours—and up to 480 miles in forty-eight hours, they were still limited by "mud, deep snow, poor visibility, and adverse weather conditions."[239] In regard to firepower, large horse cavalry units retained their relevance with the inclusion of large amounts of firepower. This included their artillery, mortars, and anti-tank guns, in addition to the small arms and machine guns found in the horse cavalry troops as well as each cavalry division's mechanized squadron.[240] Ironically, the preponderance of the horse division's firepower resided with seemingly road-bound mechanized or motorized assets. The horse cavalry was trying to claim both mobility and firepower, yet the preponderance of its firepower rode in the very vehicles it claimed that for lack of mobility assured a continued role for the horse.

In regard to organization, the 1941 *FM 2-15* was particularly out of date by the time the United States entered World War II, reflecting Herr's greatest vision, a cavalry corps composed of "two or more cavalry divisions" and assorted "supporting or reinforcing units."[241] Descriptions of horse divisions, brigades, regiments, and horse-and-mechanized units were now irrelevant, but used the same definitions of reconnaissance as the 1941 edition of *FM 100-5, Operations*. The "mobility of its mechanized elements on roads and of its horses on practically any type of terrain" made cavalry "well-suited to perform reconnaissance missions using horse elements, mechanized reconnaissance elements, or a combination of the two."[242] Cavalry doctrine in 1941 called for the reconnaissance detachments to operate "from 1 hour to 2 days" in advance of the main body of forces with the normal distance being "1 day's march." This was to allow enough time for the supported commander to make decisions with the information gained by the reconnaissance detachments. While the doctrine clearly recognized the large disparity in the distance and speed at which mechanized units could travel, it still bound these units to the march rate of the main body. In another interesting way, the doctrine got at the disparity by addressing the organization of reconnaissance detachments. Squadrons of horse cavalry could expect the attachment of scout cars, heavy machine guns, anti-tank guns, pack radios, motorcycles, and "transportation facilities."[243] Mechanized elements could expect few attachments because they already had the necessary heavy weapons and communications equipment to forge ahead on their own and assumed they could expect the striking power of horse cavalry to follow rapidly in porté trailers. In either case, horsemen or mechanized scouts could also expect the attachment to engineers and possibly artillery to help them in their mission.[244]

The other side of the reconnaissance coin was security, not being surprised by the enemy. Sound reconnaissance carried out by the "regular reconnaissance agencies of the command as a whole" prevented unwelcome surprises.[245] Other means of providing security included the use of advanced guards and covering detachments. Because of "their vulnerability to ambush and their unsuitability for sustained defense," the doctrine saw an advanced guard role for mechanized elements limited to the service of "mechanized and motorized units."[246] When paired with horses, the mechanized elements provided assistance by operating to the front and flanks and on "distant security missions primarily along roads" or by following the horsed elements and helping maintain communication with the main body. Moving only one to two hours in advance of the main body, the mechanized elements could have expected to provide minimal warning of the threat of a mechanized or air attack, since they never would have been more than eight to ten miles away. The reduction of any resistance along the route of march remained the responsibility of the horse element of the advanced guard.[247] *FM 2-15* also cautioned against using mechanized elements in covering detachments, the units used to secure the main body's movement in the face of the enemy, because of "their characteristic vulnerability" once their freedom of maneuver was restricted by being in contact with the enemy. Furthermore, the manual directed that when used in covering detachments, mechanized elements "are best used to conduct reconnaissance" in support of the covering detachment.[248]

FM 17-22, Armored Force Field Manual, Reconnaissance Battalion appeared in August 1942.[249] It followed the doctrinal precepts established in *FM 100-5, Operations,* which clearly considered "reconnaissance" one of the five major functional components of the armored division, the others including command, striking, support, and services.[250] This echelon was to perform "ground reconnaissance," and in the very next sentence the doctrine unequivocally stated that "if necessary, it fights for information."[251] Whether it was fighting for information or obtaining it passively, the armored division still accomplished the function of reconnaissance with the armored reconnaissance battalion, at the time three reconnaissance companies supported by a single company of light tanks and attached aviation assets. Doctrinally, it provided the timely information that allowed the division to exploit "the results of the reconnaissance."[252] To accomplish its missions, *FM 17-22* emphasized the need for the battalion to effectively liaison with the regimental reconnaissance companies and with the division's artillery headquarters and engineer battalion.[253] In this respect the outlook of the manual thoroughly reflected the Armored Force's legacy of combined-arms thinking. Like much of the language describing the use of mechanized reconnaissance units inside the 7th CavBde(M) as a whole, during the interwar years, *FM 100-5, Operations* and *FM 17-22* expected the reconnaissance element of the armored division to "seize key terrain" in advance of the division.[254] The notion of getting out in front of the armored division was directly linked to another very important theme, distance.

Instruction at Fort Knox emphasized that armored reconnaissance battalions performed reconnaissance "up to 150 miles in advance of the division," but that one hundred miles was considered the maximum distance under normal circumstances.[255] The battalions were to gather information about the route, enemy, and the terrain. The only way the division could benefit from the capability the armored reconnaissance battalion brought back to it was to allow the battalion to move "out several hours ahead of the division," so that it would have sufficient time to perform its important task.[256] Characteristic of the cavalry doctrine, the Armored Force instruction called for the battalion to move to the flanks once the division committed itself to combat.[257] The Armored Force clearly recognized the need for ground reconnaissance units to operate at some distance from the main body, given the rate at which friendly armored divisions were capable of moving, as well as the speed at which the enemy could be expected to move. The division's reconnaissance agency operated at these extended distances while reconnaissance units subordinate to the armored regiments covered the terrain between the armored reconnaissance battalion and their own units.[258] For this purpose, each of the division's two armored regiments retained a reconnaissance company.[259] The Armored Force, true to its interwar development, continued to value the impact that reconnaissance had on its ability to freely maneuver the striking elements found in the armored regiments. It gained this freedom by emphasizing reconnaissance at great depth and in echelon with the armored reconnaissance battalion and the regimental reconnaissance companies, each playing important roles, but in a mutually supporting manner. At the Cavalry School at Fort Riley, in 1942, they were also confronting the issue of depth on the battlefield with new limitations.

The Department of Tactics wasted little time examining the doctrinal implications of the War Department's decision to take the horse out of the horse-and-mechanized regiment. In its "Report on the Organization of a Mechanized Reconnaissance Regiment as the Cavalry Component of an Army Corps," published at the Cavalry School on 14 February 1942, the Department of Tactics signaled a shift in Cavalry Branch doctrine in regard to mechanization and its capabilities. In regard to depth, the report expected that mechanized regiments in support of corps could be expected to move as far as 125 miles in advance of the corps they supported.[260] As they moved forward, mechanized reconnaissance units were still to use "stealth," but now there was a greater emphasis that they be "provided with the strength to penetrate the opposing screen" rather than seeking the gaps or finding the flanks.[261] Having gained contact with the enemy at extended range, it was now recognized that contact had to be maintained. Subsequent review of tactical doctrine also called for a force with the strength to "pierce the enemy screen" and continue until the "enemy contour" could be discerned. Likewise, this action was meant to prevent the enemy from gaining the same kind of information.[262] The changes in expectations for what was expected of a mechanized cavalry regiment were subtle, but represented the first acknowledgment that Cavalry Branch had to find a way to perform its long-standing doctrinal mission of reconnaissance without the horse.

One of the first iron ponies to see active service during the interwar years, 1932. Courtesy
of the National Archives

Proven technology, George
S. Patton Jr. sits a horse
during the interwar years.

Courtesy of the Patton Museum
of Calvary and Armor, Fort Knox,
Kentucky

Armored cars conducting reconnaissance for the 1st Cavalry Regiment (Mecz) during the 1930s. Courtesy of Special Collections, Norwich University, Kreitzberg Library, Northfield, Vermont.

Iron ponies at the halt in
1936. Courtesy of the Patton
Museum of Calvary and Armor,
Fort Knox, Kentucky

Old mounts meet new mounts during the Louisiana Maneuvers of 1941.

Charge! With drawn .45 caliber pistols, men of the 104th Cavalry Regiment move into action less than one month before the attack at Pearl Harbor. Courtesy of the National Archives

Open scout car on maneuvers in 1941. Courtesy of the National Archives

Horses could swim, but heavier weapons needed to maintain relevance on the modern battlefield required additional procedures and equipment; notice the small canvas boats. Courtesy of the National Archives

Machines serve horses; the horse and mechanized attempt to maintain traditional cavalry's relevance on the modern battlefield, September 1941. Courtesy of the National Archives

An iron pony—jeep—
takes on a water obstacle.
The army developed
techniques to wrap the
entire jeep in a truck
canvas and then winch it
across an un-fordable body
of water. Courtesy of
the National Archives

Veterinarians give way to mechanics. Ironically, during the early 1930s, the regimental veterinarian often used a car to move to the head of horse-mounted formations at the end of the day's march to inspect the horses. Courtesy of the National Archives

The tactical doctrine reviews also captured the very essence of what had taken place with the removal of the horse from the horse-and-mechanized regiment. What had been given up was 530 riflemen who had ridden on horses to the assistance of the mechanized men. These riflemen were replaced with a combination of thirty-four light tanks and twelve 75mm howitzers in the support troops of the newly organized regiment.[263] The reorganization earned the remark that "the traditional cavalry shock element is in the support troops."[264] Granted, the regiment gained a substantial amount of pure firepower at the expense of small arms capability, but it also lost the troops doctrinally deemed essential to support the same light tanks offered in lieu of rifle troops. At the same time, unofficial doctrine was beginning to recognize that without the horses and the men who rode them, the mechanized elements had to act alone, in some cases, to fight for the information the corps needed. Unlike their reconnaissance brothers in the armored reconnaissance battalions, they could not necessarily depend on the swift arrival of supporting arms directly under the control of a division commander. Only time would tell if the new organization directly supported the increasing burden being shifted to mechanized reconnaissance units.

Patton, no stranger to the subject of mechanization, believed the Desert Training Center served to teach the army the finer points of living and fighting in the desert. In the very first two paragraphs of his instructions on tactics and techniques, Patton pointed out that the desert, with its lack of "terrestrial cover," challenged the ability of friendly forces to remain hidden. The natural conditions also lent themselves to "inevitable air attack" and opposition by enemy armored forces. Patton concluded the only way to properly secure friendly forces from ground attack was through "distant reconnaissance both by ground units and by air observation units." Air units had to "HAVE A RADIO WHICH IS CAPABLE OF COVERING TWO WAYS WITH THE RECONNAISSANCE ON THE GROUND," if they were to effectively synchronize their efforts with those performing the same mission on the ground.[265] Unfortunately, even though this reflected every level of army doctrine pertaining to mechanized ground reconnaissance and mechanized operations in general, it was a notion that remained unsupportable for lack of the proper radios.

Patton anticipated armored reconnaissance battalions or the occasional "corps reconnaissance battalion" preceding the main body of friendly forces by "three to five hours." It was imperative that once in contact with the enemy, the battalion "*MUST RETAIN CONTACT* and work around the hostile flanks to discover what is in the rear."[266] As a point of departure from what the Armored Force School was teaching at Fort Knox, Patton explained that at this point in the battle the reconnaissance battalion's "primary mission is information not fighting." Aside from the seemingly difficult task of maintaining contact with the enemy without fighting, Patton further distorted whatever message he was really trying to get across. Later in the document, in the same kind of clipped prose that he loved to dispense advice with or record simple but salient ideas, Patton remarked on how best to arm the men performing the reconnaissance mission. They, he believed, should have "a 37[mm] mounted co-axially with a .30 caliber machine gun." This he allowed would "permit them to fight."[267] Patton's vagueness on the proper employment of mechanized reconnaissance elements mirrored the ongoing confusion about how things would really work.

While Patton directed maneuvers, MG Charles L. Scott went to North Africa to observe the British army at war, before he took over the leadership of the Armored Force Replacement Training Center at Fort Knox. He was there to see how desert war really worked. When he returned he published an indictment of the current American approach to mechanized reconnaissance.[268] Scott, the long-time advocate of armored reconnaissance vehicles, who had suggested in the early 1930s that one day tanks would have to accompany the leading elements of the reconnaissance effort, saw his reasoning validated in North Africa. Field Marshal Erwin Rommel changed the nature of the reconnaissance game by raising the stakes in North Africa. Units organized to conduct "pure reconnaissance," going, seeing, and reporting, were unable to survive long enough to complete their missions and, in some cases, were unable to pause long enough to send their messages of warning. Because of what he had seen on the ground, Scott warned of the inherent risks assumed by placing a weak reconnaissance unit on a flank or in the rear. Scott, with the concurrence of the British officers he served with, felt the armored reconnaissance battalion organization, the most robust ground reconnaissance agency in the United States Army as it entered World War II, was the best on hand to face the new threat. He warned that mechanized reconnaissance units had to do more than passively observe, they had to act offensively to fulfill their assigned tasks of long-range reconnaissance and security for the flanks and rear. Scott concluded, "We'll never win the war by observation—but by fighting units that can both observe and do some killing."[269]

Since the end of World War I, Cavalry Branch had acknowledged that there would be a place for mechanization in the role of reconnaissance. World events of the late 1930s focused increasing attention on the branch's own 7th CavBde(M), yet as a branch it resisted being forced into agency pure reconnaissance role. Their interest in mechanized reconnaissance increased as a function of survival and relevance, but their professional contributions lagged behind their mechanized peers. The fall of Poland and France reverberated across the Atlantic resulting in George C. Marshall making a decision that had a telling impact on the old cavalry arm.[270] By creating the Armored Force, Marshall took away the most advanced elements and thinkers on mechanization. Even Patton chose an armored brigade in MG Scott's 2nd AD rather than the prestigious 1st CavDiv.[271] Herr's inflexibility prevented him from responding to the impact of world events in 1939 and 1940 in a manner that assured Marshall that Cavalry Branch was ready for growing challenges appearing on the horizon.

Mechanized, horse-mechanized, and horse cavalry units had all felt the impact of the eight-fold expansion that occurred in the United States Army between the invasion of Poland and attack on Pearl Harbor.[272] In that relatively short period the struggle over the proper role for mechanized ground reconnaissance accelerated and changed. The birth of the Armored Force validated the thoughts and arguments of the pioneers who developed specialized reconnaissance units at Fort Knox during the interwar years. Having lost the 7th CavBde(M), Herr sought a compromise that would secure a future for the horse, namely corps cavalry regiments employing the horse-mechanized concept that was an old idea that gained greater traction after the twin disasters in Poland and France. Each of the new "triangular" infantry divisions received a fully mechanized reconnaissance troop, signaling another important advance in the development of mechanized reconnaissance. Marshall and McNair had no interest in Herr's horses, and the newest, and smallest, reconnaissance agency reflected the changing attitudes.

Herr continued to believe that the large-scale maneuvers were "rigged to limit the activities of the cavalry, for the pressure was on from certain quarters to eliminate the mounted service."[273] In his opinion, exercise restrictions forced the horse units to the flanks while mechanized forces raced to contact before the horse soldiers could deploy and fight delaying actions from ambuscades.[274] The long struggle tore the branch apart as the Armored Force and Cavalry Branch exchanged roles, the former taking over the latter's combat functions. Forced to accept what was in their eyes a lesser role focused on reconnaissance only, Herr and men like him continued to resist what had become apparent at the end of World War I. In the end, mechanization and the men who refused to believe they were incapable of accomplishing every aspect of reconnaissance emerged with the leading role as the United States made its final preparations for the war that was rapidly engulfing the world. Official recognition of this victory was only months away while the debate surrounding the horse and mechanization would echo throughout the war and beyond.

As the United States prepared for war it made some final adjustments to the organization of those units serving as the eyes of units tasked to carry out combat functions. There were few actual changes, and most of the interwar ideas gained from the mechanized reconnaissance debates remained intact. One major adjustment, however, did occur: the complete removal of the horse from the equation that underpinned all mechanized reconnaissance units in Cavalry Branch, except for those serving infantry divisions. John K. Herr disappeared from the uniformed debate but carried on the fight for his beloved horse outside the army. The premier doctrine for training the rapidly expanding army, *FM 100-5, Operations,* predated the major organizational change in mechanized reconnaissance units, but subsequent Cavalry Branch doctrinal publications were not radically different. The Armored Force doctrine and organization of mechanized reconnaissance units continued to reflect the more aggressive interwar approach to the problem of gaining information that had taken place at Fort Knox, independent of the horse. All of these doctrines and theories were about to be tested in North Africa and the Mediterranean theater of operations, but only on a limited scale, because none of the corps cavalry regiments, now all mechanized, saw service there. How well would the armored reconnaissance battalion serve the armored division? How well would the mechanized reconnaissance troop serve the infantry division? How well would the cavalry division's mechanized reconnaissance squadron perform without the support of its horse cavalry division, which was destined for the Pacific? In any case, the first mechanized cavalrymen were about to enter combat in a most unusual manner, astride neither horse nor "iron pony."

4

War in the Mediterranean

★   Under the cover of darkness five enlisted men and one officer slipped into a small rubber raft and paddled away from the submarine USS *Barb*, at 2225 hours on 7 November 1942. Seven miles to the east lay their objective, the entrance to the harbor of Safi on the coast of French Morocco. Their mission was to mark the entrance, so that in the morning American destroyers could range close enough to support the Allied invasion of North Africa, Operation Torch. For the men in the rubber raft, seven miles of open sea posed a challenging route, reduced with each stroke. For the United States, their journey represented barely perceptible steps toward the nation's ultimate objective, Berlin. Riding neither horse nor jeep, two of the men, trained as mechanized reconnaissance scouts, were out front leading the way.[1]

The invasion of North Africa provided the United States Army an opportunity to gather lessons on the employment of ground reconnaissance before even larger land operations took place in Western Europe. The continued campaigning in the Mediterranean contributed little to the ongoing development of mechanized ground reconnaissance, but remained important to those who opposed the wholesale replacement of horses with vehicles. To an extent, the remainder of the Mediterranean campaign validated some of the prewar beliefs of these men and gave them hope for the return of horses. In all fairness to the die-hard horse advocates, even the army was not convinced that horses were no longer necessary. One month prior to the invasion of Sicily, Army Ground Forces ordered the Cavalry Replacement Training Center to have 1,100 "horses and horse specialists" trained by August and September 1943.[2]

With North Africa and Sicily liberated, even though fighting continued in Italy, the army instituted additional changes that affected the conduct of mechanized reconnaissance. These changes, some minor and others more complex, touched on the organizations serving or expected to serve in front of every corps, armored division, and infantry division. The changes were important because of the impact they would have on the fighting in Europe, and for Cavalry Branch's ongoing fight to maintain its identity. The conversion of armored reconnaissance battalions, ARBs, into mechanized cavalry reconnaissance squadrons was a major step toward reconciling the divorce that had taken place in 1940, but not one that helped Cavalry Branch regain any ground in its quest to continue being recognized as a major contributor on the battlefield. The organization of corps reconnaissance regiments, yet to see combat, changed again. Subtler than the decision to remove all of the horses, the regiments' conversion to mechanized cavalry groups was significant. Part of McNair's pooling concept, it indirectly led to Cavalry Branch recapturing some of its combat arms identity even if the manner in which this happened was largely inconsistent with the army's and Cavalry Branch's doctrine. The long-held doctrinal belief that relegated mechanized cavalry units to the task of reconnaissance remained largely unchanged even in the light of what had been learned already. The army's most over-arching doctrine clearly maintained a role for the horse while cavalry doctrine reflected the sweeping changes that had taken place in 1942. With the removal of the horse from all reconnaissance organizations, cavalry doctrine started to reflect a more aggressive attitude, akin to the beliefs long held at Fort Knox and in the Armored Force. The imperceptible beginnings of a reunion of Cavalry Branch and the Armored Force had begun.

The action in North Africa was important to the continuing evolution of ground reconnaissance equipment, doctrine, and organization, because it provided the first feedback based on combat experience. Relatively few mechanized ground reconnaissance units saw action in North Africa, but their performance revealed what had and had not been accomplished between the wars. Infantry divisions used their organic cavalry reconnaissance troops with mixed results from the beaches of the Torch landings to the ultimate capture of Tunis. Only the 1st AD's 81st ARB saw extensive service with an armored division, but with decidedly mixed results, especially during the events in the early days of the Battle of Kasserine Pass. The 91st CavReconSqdn, detached from the 1st CavDiv as it deployed to the Pacific theater without its horses, also went to North Africa. It was fitting that the squadron, descended from the very first mechanized reconnaissance unit in the United States, saw service as a II Corps asset, but was generally employed in support of infantry and armor divisions while operating in North Africa and the Mediterranean theater. The experiences of the squadron typified the mismatch between the intended purpose of mechanized reconnaissance units and their actual employment. What remained untested was the corps reconnaissance regiment, a formation that had been evolving since 1939. Mediterranean combat also provided material for the undying debate about the future of the horse. As in years past, the debate largely centered on the traditional combat role of horse cavalry, the role that had been assumed by the Armored Force, but at times still touched on ground reconnaissance.

With few exceptions, most of the units slated for Operation Torch had no desert experience. The men of the 1st AD trained in Ireland; the 9th ID started its training for amphibious assaults in 1941 and moved to North Africa from the east coast with the Western Task Force under Patton's command. The 1st and 34th IDs, like the 1st AD, traveled to North Africa as part of the Mediterranean Task Force that originated in England and thus missed the opportunity to hone their skills in the desert.[3]

The 1st AD's contribution to the initial phase of the Torch landings consisted of the detachment of the division's Combat Command B (CC B) led by BG Lunsford E. Oliver. The Combat Command was further divided into two task forces with COL Paul M. Robinett commanding Task Force Red and BG Oliver's CC B headquarters retaining control of Task Force Green; each had landing objectives east and west of Oran, but neither task force included elements from the division's ARB.[4] The task forces did rely on regimental reconnaissance companies, a remaining vestige of what had been learned at Fort Knox during the interwar years.[5] Oliver's Task Force Red used its mechanized reconnaissance company to secure a key road junction that facilitated the rapid inland advance of follow-on forces and prevented Vichy units from counterattacking into the flanks of the advancing task force.[6] COL Robinett's Task Force Green projected its mechanized reconnaissance elements in a similar fashion. LT Richard Van Nostrand's reconnaissance platoon, first across the beach, provided security. Once relieved, the platoon advanced to secure key terrain and establish enough distance between itself and the main body of the task force to provide early warning. This reflected a sound use of the characteristics and capabilities of the lowest echelon of mechanized reconnaissance found in the division. Equipped with vehicles and radios, the scouts moved rapidly inland to provide early warning of any possible counterattacks and to seize key terrain, in this case against no resistance. Just as the doctrine called for, they were quickly relieved of their initial terrain objective and pushed back out on reconnaissance to fulfill their intended purpose.[7]

The mechanized cavalry reconnaissance troops found in the assaulting infantry divisions made varying contributions. Other than the heroic efforts of the scouts turned amphibians, the 9th Reconnaissance Troop contributed little to the landings on the west coast of Africa. Only a single platoon from the divisional reconnaissance troop saw action during the initial landings at Port Lyautey on the Moroccan coast while attached to the 60th Combat Team, and then only as assault troops, not as a mechanized reconnaissance platoon.[8] 1st Reconnaissance Troop, 1st ID, contributed more outside Oran, Algeria, because it was able to come ashore relatively intact on 8 November 1942. With 149 men assembled by noon on the first day of the invasion, only the lack of vehicles hampered the troop; just four scout cars and four jeeps were available for use the next day. As contact developed between the division's infantry regiments and the enemy, the troop often found itself on the division's flank, securing it from envelopment, coordinating with the adjacent armored task force operating to the south, or reconnoitering new routes.[9]

Vichy resistance in North Africa was as limited as were the opportunities to evaluate the utility of mechanized reconnaissance units. The Allies fared less well as they thrust east to Tunisia from their secure beachheads. Having moved east they found a more worthy opponent whose initial rebuke, when combined with poor weather that limited mobility, insured that the first real test of American reconnaissance units would not take place until February.

The 81st ARB's first test as a nearly complete unit came at the end of January 1943 in western Tunisia.[10] There the division became increasingly fractured, no different from the equally dispersed 1st and 34th IDs.[11] The 81st ARB, organizationally best suited to act as a single unit, was incapable of supporting its now dispersed parent unit spread out into three separate combat commands. Against the wishes of MG Orlando Ward, commander 1st AD, the II Corps commander, MG Lloyd Fredendall, ordered a raid on the small Axis outpost at Sened Station.[12] With the attachment of B Company, 81st ARB to CC C, the reconnaissance battalion entered the fight on a small scale. COL Robert Stack, the commander of CC C, departed Gafsa for Sened Station at 0400 hours, 24 January 1943. He used C Company of the reconnaissance battalion with an attached tank platoon from D Company in a manner reminiscent of war games conducted during the interwar years. Stack ordered the reconnaissance company to move beyond Sened Station and establish a position that would prevent reinforcements from coming to the aid of the beleaguered garrison. The plan worked well with Stack's small combined-arms force capturing 96 prisoners in three hours of fighting before returning to Gafsa by 1800 hours the same day. The men of C Company, 81st ARB, found an American P-40 fighter plane in perfect condition and added it to the growing menagerie of spare vehicles, which with the addition of a few Ford trucks, had been in the service of the German Army.[13] Piecemeal employment of the 81st ARB in the vicinity of Maknassy and Sened Station remained the norm.

While 1st AD attempted to capture Sened Station in preparation for a more general advance to Maknassy, the Germans seized the initiative. Having recovered from collapse in the Western Desert, Field Marshal Erwin Rommel's Afrika Corps was ready for action, first taking control of a key exit from the eastern dorsal of the Atlas Mountains from ill-equipped French forces holding Faid Pass on 30 January 1943.[14] This was a prelude to a larger offensive that came to be known as the Battle of Kasserine Pass. Relative to the minor combat operations against Vichy forces and small raids aside, this was the beginning of the first real test of the concept of American mechanized reconnaissance.

The American response to the seizure of the Faid Pass was a poorly executed counterattack on 31 January. Although it was a combined-arms affair built around elements of CC A, the armor and infantry battalions attacked into a curtain of enemy tank, mortar, heavy machine gun, and anti-tank fire.[15] With the division's reconnaissance battalion already scattered among two other combat commands south of Faid Pass, American effort was unable to support the counterattack. Thus, the battalion continued on its respective missions with CC D at Sened Station and CC C attempting to open Meloussi Pass. Another failed counterattack on 1 February left CC A in defensive positions waiting for a new plan.

While infantrymen and other supporting forces of the units of LTC John K. Waters (Patton's son-in-law) and COL Thomas D. Drake dug in on the hills of Djebel Lessouda and Djebel Ksaira that overlooked Faid Pass to the east, the 81st ARB, less B Company, which was still attached to CC C, returned to Sbeitla for much-needed maintenance.[16] 1st AD detached B Company from CC C and reassigned it to CC A on 2 February 1943, and it then took up patrolling toward Meloussi Pass, providing some security to CC A's southern flank. Having refitted, the remainder of the 81st ARB moved out in support of CC A, with A Company taking over responsibility for B Company's sector on 13 February. C Company took up a position west of A Company, but also oriented toward the southeast.[17] The battalion's distant patrols operated no more than fifteen miles from CC A's southernmost position and sent no patrols east between CC A and the German-held Faid Pass.

The 81st ARB's summary of operations, written after the disaster at Kasserine Pass, which actually started just west of Faid Pass, indicates that the battalion believed "the enemy was assembling for a stiff defense or for an attack."[18] The first real indication of what was about to happen occurred during the night of 13–14 February. Company A reported hearing tanks to the east, while C Company relayed reports from the French that the Germans were massing infantry at Sened Station and Maknassy.[19] In the morning, C Company detected the southern arm of the German double envelopment at 1020, as "30 tanks in V-formation" moved northwest from the vicinity of Maknassy with "100 trucks of enemy infantry" identified in addition to the tanks at 1320.[20] With an assortment of light tanks, scout cars, and jeeps, the reconnaissance company fought a delaying action and withdrew to the northwest, pressed by the onslaught of German armor.[21] Company A was less fortunate.

Like C Company, A Company had generally been oriented to the southeast. The men of A Company also detected the German envelopment that swung well west of their position, a position that C Company was able to move away from during the afternoon of 14 February. What A Company could not detect from its position, but B Company could, was the progress of the northern enveloping force that had boiled out of the Faid Pass that morning, cutting off the line of retreat. Robert Marsh, serving with an assault gun in B Company, recalled that men serving on an "outpost" reported sixty-three advancing enemy tanks. An officer in charge apparently told these men they must be liars, since there were not sixty-three tanks in all of Africa.[22] For the next two days, A Company fought against what must have seemed "like all the tanks in Africa." Although they had some assistance from cut-off elements of CC A who were cut off from their unit, by 16 February the men of Company A had no option but to destroy their equipment and attempt to walk out on foot. Of the more than 100 members of the company, only two officers and fourteen men escaped.[23] Before the stranded members of CC A succumbed to the Germans, 1st AD mounted one last effort to reach them.

After the disaster at Faid Pass, 2nd Battalion, 1st Armored Regiment, moved during the night of 14 February and the early morning hours of 15 February to an assembly area close to the American forces that were cut off on Lessouda Mountain. At 0330 lieutenants from the regiment's own reconnaissance company provided a limited amount of information about the enemy disposition and strength, and the terrain north of the marooned members of the regiment. Absent from their report were any terrain maps or sketches. Wasting little time, the battalion issued its order at 0400. With only two 1/100,000-scale maps for the entire battalion, it prepared to attack what was estimated to be between fifty-five and sixty tanks in the area of Sidi-Bou-Zid and the two hills occupied by survivors of 1st AD's CC A.[24]

LTC J. D. Alger's battalion moved blindly as it sought the enemy at Sidi-Bou-Zid. The tank battalion led its supporting artillery and infantry in columns of tank companies with other battalion assets interspersed. On the flanks, platoons from the 701st Tank Destroyer Battalion moved along with Alger's own battalion reconnaissance platoon. In his typed report prepared long after the disaster Alger noted, *"No Reconnaissance was provided from the jump-off position East."* With D Company's tanks in the lead, the battalion fought its way into Sidi-Bou-Zid. The last report the battalion heard from its own reconnaissance platoon was that it was having a hard time negotiating the wadis on the battalion's southern flank as it moved to the east. Once established in the small town, the battalion was trapped between the pincers of a double envelopment.[25]

Alger need not have led CC C into the attack at Sidi-Bou-Zid with minimal reconnaissance to the front and flanks. B Company of the division's reconnaissance battalion remained intact and patrolled to the northwest of the cut-off elements of CC A. C Company patrolled to the southeast of Sbietla, west of Sidi-Bou-Zid.[26] In fact, a single platoon from B Company reported to Alger at 0850, 15 February, prior to the ill-fated counterattack. Although the platoon leader was unable to tune his radios to CC C's frequency, he was prepared to take up a position on the right (southern) flank of Alger's attack.[27] Before the attack commenced, the B Company commander recalled his platoon leader, leaving Alger with a decidedly smaller reconnaissance unit to place on his vulnerable flank.[28] That day, Robert Marsh, who reported on the enveloping tanks from the north, witnessed columns of smoke climbing into the desert sky as testament to the piecemeal nature of the failed counterattack.[29]

After the calamities of 14 and 15 February, 3rd Platoon of B Company, 81st ARB, ventured onto the battlefield to destroy the abandoned vehicles of the 1st Armored Regiment's reconnaissance company left near Lessouda. Among the vehicles, they encountered German soldiers who elected to run rather than fight. A German tank approached, but the assault gun found in each reconnaissance platoon scored a hit at an extended range.[30] The battalion commander even had a brief encounter with a German tank during the retreat from Faid Pass. Operating from Sbeitla, LTC Charles Hoy decided to venture east, in the direction of the oncoming Germans, for a personal reconnaissance in his own "peep." "Tooling down the road," Hoy saw something that brought out the old horse cavalryman in him when he calmly instructed his drivers to "Whoa! Turn around." The driver complied, but Hoy, unable to apply spurs to his jeep, now shouted, "GET GOING!!" as the German tank oriented on Hoy's iron pony.[31] And so ended the 81st ARB's contribution to the Battle of Kasserine Pass.

The 81st Reconnaissance Battalion reconstituted itself while performing limited missions for the division and II Corps. The battalion drew light tanks from the division's armored battalions when the headquarters and headquarters company transferred men into the line companies to make up losses.[32] At the highwater mark of Rommel's advance, what remained of the reconnaissance battalion served south of Tebessa, securing the Allied southern flank. As the German tide receded, the reconnaissance companies advanced east across the littered battlefield, with B Company ending near Faid Pass and C Company operating west of Gafsa while the corps refitted.[33] During this time the companies operated seventy miles apart, well beyond a mutually supporting distance and thereby stretching the logistics capacity of the battalion to the limit.[34]

Throughout the remainder of the campaign in North Africa, the 81st ARB, now with a modified organization after the loss of an entire reconnaissance company, fought intact while refining its patrolling skills. It established a number of important observation posts that allowed it to report on the activities of the Germans operating in the area and became intimately familiar with German mines, minefields, and booby traps.[35] For all its reconnaissance work during March, there was no breakthrough on the Maknassy front. The battalion achieved greater success at El Guettar.

Now assigned to COL C.C. Bensons's task force, which bore his name, the 81st ARB led the effort to penetrate the eastern dorsal of the Atlas Mountains on 30 March 1943. While the infantry and armor of the task force attempted to make a breakthrough, the scouts conducted extensive dismounted patrolling, manned observation posts, and acted as forward observers for artillery fires. When "Benson Force" finally broke out on 7 April, after Field Marshal Bernard Montgomery's Eighth Army penetrated the Chott Position southeast of the El Guettar and made the German position untenable, the 81st Reconnaissance Battalion found a gap in an existing minefield that allowed for the restoration of mobility. Once through the gap, the battalion led the task force until it established contact with Montgomery's 12th Lancers.[36]

Having established contact with the British, 1st AD and the rest of II Corps moved to northern Tunisia for the final assault on Tunis, completing the liberation of North Africa. An untested reconnaissance unit joined II Corps for the final effort.

The 91st CavReconSqdn arrived in Casablanca, Morocco, on 24 December 1942, after completing preparations at the Desert Training Center in August.[37] The manner in which the corps and division commanders employed the 91st CavReconSqdn illustrated how an organization built and intended for mechanized reconnaissance could be used for a host of other tasks for which it was often ill suited. As the future revealed, the 91st CavReconSqdn's employment in North Africa was indicative of how it would be used throughout the rest of the war and how other similarly organized units could expect to be used in the future.

After debarking at Casablanca, the descendants of the very first mechanized reconnaissance unit took their turns at a variety of tasks. Between 11 and 28 January they helped provide security for the meeting between President Franklin D. Roosevelt and Prime Minister Winston Churchill. February found them, like other mechanized reconnaissance units such as the 1st Reconnaissance Troop of the 1st ID, monitoring the Spanish-Moroccan border and patrolling critical rail lines of communication that ran to the east, where the fight was beginning to develop again. The squadron was finally called to the front on 5 April 1943, traveling to Tebessa, Algeria, where it joined II Corps. II Corps attached the squadron to the 9th ID, then operating as the northernmost U.S. division in the II Corps sector that had shifted from central to northern Tunisia. After traveling more than 1,000 miles in five and a half days, a feat unimaginable for any horse cavalry unit, the 91st CavReconSqdn took up its first combat mission just east of Abiod.[38]

The 9th ID assigned the 91st CavReconSqdn a ten-mile section of the front that had formerly been occupied by an infantry regiment. It also placed the squadron on the division's boundary, requiring it to maintain contact with the 1st and later the 34th IDs. Here the men of the squadron remained in a relatively static position facing the Germans' Barantrin Regiment, while they gained an appreciation not only for their current organization but also for the inadequacy of their equipment in handling the infantry tasks they were now called on to perform. Lacking entrenching tools to prepare defensive positions—they were expected to operate mounted in most cases—the troopers learned to improvise using the hand tools from their vehicles or individual knives.[39] Between 23 April and 4 May, the squadron learned even more during II Corps' general offensive.

Advancing in their own ten-mile zone, the troopers did not move in the van of the division. Rather, in most cases, they attacked on foot in a traditional cavalry-as-a-combat-arm role, complete with modern horse holders in the form of vehicle drivers.[40] They also learned the futility of trying to bring along the plentiful heavy machine guns in their organization when attacking dismounted. The weight of the guns, tripods, and ammunition made them too cumbersome to be of much assistance while maneuvering against the enemy, and too heavy to lead to anything but exhaustion for the men carrying them.[41] Under constant small-arms fire, the squadron used artillery assistance to slowly force the Germans back toward Mateur, despite an inability to move mounted and to bring to bear the few heavy weapons at its disposal, like mounted 37mm guns. The squadron's light tanks were unavailable for support, because they had been detached to support the Free French attack taking place seven miles north of Mateur.[42] The squadron accomplished its missions, despite the mismatch of equipment, organization, and doctrine. As far as the squadron's early days in combat, a unique example occurred that illustrated the still unresolved issue of how to use mechanized reconnaissance troops.

During the general advance, C Troop was unable to advance against a defended hill. Intense machine gun fire ripped through the "ditty bags" carried on the backs of troopers as they crawled forward.[43] The squadron commander, LTC Harry W. Candler, moved to the stalled sector and ordered a new attack. The company commander refused, citing the exhausted condition of his men. The colonel drew a line in the sand and ordered all who would attack with him to cross the line. Most of the men did so, yet all of the regular (prewar) noncommissioned officers filling the platoon's sergeant positions refused. So did the commander, who was subsequently arrested. Candler led what reminded one trooper of the "mock cavalry charges" held at Fort Riley in 1942 as they advanced up the hill on foot complete with the bellowed word "CHARGE!"[44] Under the direction of new-army noncommissioned officers the attack carried the hill, and the colonel earned the respect of the troopers who accompanied him.

Had the old-army sergeants spent any time with the 91st CavReconSqdn during the interwar years, they would have recognized the task of attacking dismounted on foot as something they did not do. They rode to the front, they rode to the flanks, they dismounted when necessary, but they left the fighting to the cavalrymen (mounted men of the 1st CavDiv) under circumstances like this. The new troopers attacked, perhaps not knowing this was not the intended purpose of their unit, but certain that disobeying orders would get them a court martial for mutiny.[45] These troopers represented the future of all mechanized reconnaissance units. With a rapidly growing army, no unit could expect to have more than a handful of officers or noncommissioned officers versed in the interwar debate about the employment of a mechanized reconnaissance unit. The troopers did what they were told with what was on hand and were little interested in what the doctrine had to say. LTC Candler was of the old army and the old "fighting" cavalry. To attack dismounted was nothing new, and in fact, would have been what he expected to do under a variety of circumstances. Fighting dismounted was something his squadron continued to do in North Africa, Sicily, and from one end of Italy to the Swiss border in May 1945.

Soon after the fighting ended in North Africa, the army attempted to capture the important lessons learned. The process of obtaining the information took on a variety of forms including unit-after-action reports (AAR), observer-board reports, and the formation of specialized boards to examine specific issues. In some cases, the veterans of the fighting in North

Africa returned to the United States to write new doctrine and publish articles in the *Cavalry Journal*. Some of this led to changes in the doctrine, organization, and equipment of the mechanized reconnaissance units, and experience gained in North Africa sustained the debate about how best to accomplish the task of ground reconnaissance.

Almost exactly one month after the cessation of hostilities, the 81st ARB prepared its "Report on Combat Experience and Battle Lessons for Training Purposes." The report deemed the doctrine used in North Africa sound, and stated that the missions the battalion had been ordered to carry out were generally "suitable."[46] The report attributed missions run afoul to: unclear orders, misinterpretation of orders, terrain, lack of drive, not keeping the mission foremost in mind, and abandoning the mission when light resistance was encountered. The report drew attention to specific issues that plagued the battalion and contributed to its poor performance at times.

Of topics related to training, the ability to read maps had certainly been a critical skill since the first mechanized scouts rode forward of their parent units with radios, and skill in map reading remained an area for serious improvement.[47] One commentator noted that once it had been simple to send any man to the front of a marching body of troops to signal "enemy in sight," but in the modern world of mechanization, long-range weapons, and radios, the task could no longer be left to just anyone. Ground reconnaissance had become a "complicated and vital phase of the military art."[48] The rapidly expanding American army probably contributed to the after-action-report's call for more training and development for the all-important noncommissioned officers who were so critical in accomplishing all missions, but this was certainly not a problem unique to mechanized reconnaissance units. The experiences of LT Wendell C. Sharp, a man who had seen extensive action in North Africa, captured the essence of what it meant to operate at the leading edge of battle and the importance of training the needed noncommissioned officers required to support him. He wrote his wife that he had grown accustomed to "a steady diet of reconnoitering, attacking, patrolling, and playing clay pigeon by day—and more patrolling, mine-sweeping, etc. at night." The stress made him feel "like a cat—always tense and ready."[49] Clearly the army, and specifically the schoolhouses at Fort Knox and Fort Riley, had to do their part in producing more men capable of carrying out the tasks of leaders like Sharp.

Other topics for improved training included night reconnaissance, radio security, use of key terrain, use of aerial photographs, and the sending of timely and accurate reconnaissance reports. One proposed solution was the introduction of a *"TEN COMMANDMENTS"* for inclusion in the training of all officers and noncommissioned officers. Examples for inclusion on this list included: know where I am, always provide for local security, dig in, make sure all subordinates know the mission, know what I will do if I am attacked right now.[50]

Unique to reconnaissance units were the few items that addressed organization and equipment specifically. Of particular importance were the weapons on hand. The 75mm howitzer, organic to the scout platoons of the ARB, was the only weapon capable of killing an enemy tank. The report emphasized that "no Reconnaissance should be without the 75mm," calling for the inclusion of this weapon system in each platoon mounted on a tank chassis.[51] The report recommended retaining the second officer then serving in each platoon, because he assisted with command and control and enabled the necessary flexibility to split the platoon into two sections capable of performing independent missions.[52] Two other organizational issues and one equipment issue came to the fore. There was a recognized need for a closer relationship with the division's artillery units, even though this relationship was already spelled out doctrinally.[53] The solution proposed was that artillery forward observers travel with the reconnaissance battalion.[54] This assumed that the battalion was actually moving forward of the division, which was rarely the case in North Africa. The AAR also called for greater oversight of regimental reconnaissance companies, citing the ability of the ARB to properly train them, something the report suggested was lacking without specific evidence.[55] The army implemented this idea in all armored divisions. And under the heading of "Miscellaneous," MAJ Michael Popowski, the battalion executive officer, called for the use of horse cavalry to patrol the flanks. He emphasized how this would have assisted the battalion in the rough terrain "during the Kasserine affair."[56]

MG Ernest N. Harmon, who had begun the North African campaign in command of the 2nd AD and assumed command of the 1st AD in April 1943, prepared his own thoughts about what had been learned. In regard to reconnaissance, he was extremely direct. He attributed an over-emphasis on speed, learned in peacetime maneuvers where logistics, casualties, and reconnaissance were not portrayed accurately, as one of the reasons forces in North Africa did not take the time to carry out quality reconnaissance. Capturing the essence of the interwar vision of what mechanized reconnaissance units should have been doing, Harmon noted that division commanders "must always be thinking in terms of what is going to happen six to twelve hours ahead of the present."[57] Mechanized reconnaissance units had rarely been deployed with enough depth to provide some of the early warning required to help commanders think that far in advance. One reason the reconnaissance men had been unable to achieve depth was their inability to overcome enemy resistance. Harmon cited the futility of trying to destroy German heavy tanks with anything less than the guns found in tank destroyer units, guns certainly not organic to reconnaissance units. He was particularly critical of the idea that the half-track mounted 75mm was an acceptable "antidote" for German armor.[58] Harmon's comments were primarily focused on the infantry and armor operation and only tangentially touched on the mission of mechanized reconnaissance units, but divisional intelligence officers addressed the issue directly.

Their report included input from the infantry divisions, whose reconnaissance troops saw extensive action.[59] The divisional officers thought an
entire squadron or battalion of mechanized reconnaissance assets would have
better served their divisions, the same-sized element called for by Herr before
the war. Individual troops lacked the manpower or equipment to "effectively
combat German reconnaissance elements."[60] German reconnaissance cars
carried 75mm guns, and thus divisional troops needed attached tank destroyer
platoons to conduct reconnaissance by any other means than stealth.[61] The
81st ARB and the 91st CavReconSqdn were marginally better equipped with
their light tank companies and assault guns. Individual troops also lacked
the mass for extended operations, and the organic firepower to fight for
information. The board disparaged the concept that stealth was an effective
means for conducting reconnaissance in anything larger than "a small patrol
of two men."[62] LTC Hoy, commander of the 81st ARB, supported these views;
he placed a premium on the ability of officers with "cunning and daring" to
penetrate the enemy defenses as deep as five thousand yards. Hoy emphasized
that the technique that had been derided for being "OK for maneuvers, but
not in war" had been extremely successful.[63] The report mentioned the lack
of corps cavalry, but concluded the divisions would have benefited little since
these formations would have been performing corps missions.[64]

The army sent its own observers to North Africa to collect information from
the battlefield. These observer reports passed through the Army Ground Forces
Headquarters and directly to the organizations capable of effecting the required
changes.[65] Among the observers, LTG McNair visited North Africa himself to
check on the status of the army he built for George C. Marshall. It was here that
he first demonstrated his propensity to move a little too far forward in an effort
to get close to the action, and was wounded by a shell fragment.[66] McNair's
observation teams provided an excellent perspective on the performance and
limitations of the mechanized reconnaissance units in North Africa.

With maps that depicted roads where in reality only camel tracks existed,
operations in North Africa once again reminded everyone that there was still
no substitute for a "personal looksee."[67] Observer teams confirmed the need for
mechanized reconnaissance units to fill the gap between frontline units and the
capabilities of aerial reconnaissance. It often took too long to develop the film
and disseminate the information down the chain of command, resulting in the
situation changing before the pictures were developed.[68] This further reinforced
the need for ground reconnaissance units capable of sustained operations in
direct support of ground commanders.

Observers also cited inadequate training and equipment as factors
contributing to the inability to get as much information about the enemy
as possible. They singled out the light tanks and tracked vehicles for praise,
but also called for more firepower, with a minimum benchmark established
as 57mm cannon.[69] Comments on jeeps were mixed. Some deemed them
better than the motorcycle, but one observer noted that anyone still riding a

motorcycle was trying to get a jeep.[70] Even so, to make it practical for combat, the jeep had required a number of modifications, such as a pipe on the front bumper to protect the occupants from wire strung across the road by the Germans, and sandbags on the floors to provide some protection from mine blasts.[71] Another observer recognized the need for armor capable of stopping a .30- to .50-caliber bullet, long the standard during the interwar debates on what a reconnaissance car should be able to stand up to.[72] Early practitioners and advocates of mechanized reconnaissance had warned that crews must be able to survive first contact with an enemy rifleman. They had also recognized that mass-produced vehicles, like the jeep, would ultimately need a variety of specialized components were they to survive at the leading edge of battle. Combat in North Africa validated their beliefs.

Another point made by the observers, but absent in the after-action-reports of the 81st ARB and the articles prepared for the *Cavalry Journal,* dealt with the composition of Axis reconnaissance forces. Combat in North Africa had not played out as interwar maneuvers or war games. Rather than racing toward similarly or lesser-equipped mechanized reconnaissance units, American troopers confronted a far more sophisticated enemy. German units habitually employed Mk III and Mk IV tanks, and their static outpost lines included strongpoints equipped with 75mm guns and in some cases towed 88mm guns.[73] In the offense, the Germans preceded the movement of their main body with a highly mobile reconnaissance unit operating up to three hours in advance.[74] Out-gunned, American units were ill equipped to fight an effective counter-reconnaissance fight or penetrate Axis positions. Even in the vast spaces of North Africa, other than during the first raid at Sened Station, going around or bypassing the enemy was not an option. Clearly, the calls for more firepower and armor protection provided early indications that all was not right in the organizations that had been developed during the interwar years, especially with respect to equipment.

As cavalry veterans returned to the Cavalry School at Fort Riley, they wrote articles for the *Cavalry Journal* and documents for instructing cavalrymen destined for combat. The commanders of the 81st ARB, LTC Hoy, and the 91st CavReconSqdn, LTC Candler, were among those who returned to the schoolhouse to write. The *Cavalry Reconnaissance* series was a unique blend of combat narrative and footnotes that directed the reader to the appropriate army doctrine justifying the actions of a unit in a given action. *Cavalry Journal* articles were similar and provided a forum to continue the debate about the utility of the horse and the deficiencies of mechanized reconnaissance.[75]

Hoy's article, "The Last Days in Tunisia," prompted editors of the *Cavalry Journal* to insert italicized comments favorable to the retention of the horse. Hoy recounted the extensive movement made by II Corps from southern Tunisia north to their new sector for the final offensive. The editors noted that had II Corps had a *"horse cavalry division,"* it could have made the *"march with ease"* citing the 1940 Louisiana maneuvers as evidence of the horse cavalry's ability to fight and move over extended distances.[76] Hoy noted that the withdrawing German forces continued to delay Allied pursuit by sowing mines and blowing up bridges. In one instance it took the engineers a full twenty-four hours to bridge a *wadi* to allow the advance to continue. The *Cavalry Journal* editors noted that had a horse cavalry division been available, it *"would not have had to change gait in crossing that wadi any place along the 3-mile front."* And only two paragraphs later, in response to Hoy's account of the his ARB conducting "mop up" operations in the wake of the armored division's advance, the editorial comment that *"horse cavalry can follow tanks in a breakthrough and perform this mission"* appeared.[77]

In his conclusion, Hoy, a combat veteran and commander of a mechanized reconnaissance battalion, demonstrated his unwillingness to seek improvement in the arena of mechanized reconnaissance by improving the existing doctrine and organizations. Rather, Hoy, in a manner that must have pleased Herr, sought the solution to the problems encountered in North Africa by returning to that which had been abandoned, the horse. Hoy cited the unnamed "doughboy" as "being sure that if he had horse cavalry, the tank-cavalry-air team really would hit 'pay dirt' in pursuit and exploitation."[78] Hoy missed the point that the real purpose of his unit was to serve the needs of his division and that what he was suggesting, the return of a traditional horse cavalry division, sought to replicate what it was that the armored divisions were expected to do on the modern battlefield. In his last paragraph he offered that, "Reconnaissance by horse cavalry units is *unquestionably mandatory* in certain terrain, but *there is a bigger job to be done by horse cavalry."*[79] How soon Hoy had forgotten the fields filled with the burning carcasses of American tanks that had charged out blindly against superior German panzers.

Hoy was not alone in his belief of the continued need for horse cavalry. MG Manton Eddy, commanding general 9th ID, unabashedly remarked, "We could have used horse cavalry. We had some 280 native animals which I used for supply a gread [*sic*] deal and on occasion for operations." COL C.C. Benson, a frequent contributor to the *Cavalry Journal* during the interwar years and who had often taken a progressive stance on the subject of mechanization, commanded the 13th Armored Regiment during the North African campaign. He "wished" many times that he had had a "squadron of horse cavalry." Even the commanding officer of the 34th Reconnaissance Troop believed, "Horse cavalry would have done wonders" in North Africa.[80] All of these remarks, collected by a board of observers, lacked context. One set of remarks included in the board's finding, and probably the most important, did provide context.

GEN Dwight D. Eisenhower, commander of all Allied forces operating in North Africa, acknowledged the request of many of his subordinate commanders for horse cavalry. He even acknowledged that a number of "horses and donkeys" had been requisitioned and served a role in moving supplies. However, Eisenhower reminded all that:

. . . the advantages of horse cavalry did not outweigh the need for shipping space which was and still is critical and vital. To have been of value during the North African campaign, it would have been necessary not only to transport the animals, but also the grain, for only green forage is available. Grazing might suffice in an emergency, but horses in combat must have grain and they cannot graze where the bullets are flying.[81]

LTC Candler of the 91st CavReconSqdn also provided "SPECIFIC OBSERVATIONS AND CONCLUSIONS" that reflected the wide variety of techniques he had employed in his unit to accomplish the missions assigned him by the divisions he served. As an old horse soldier, the first thing Candler noted was that his troopers had performed the "dual role of cavalry and infantry."[82] Having acknowledged the reality of how his unit had been employed, he then focused on what had worked well and what needed improving. The jeep served as his squadron's "principal reconnaissance vehicle," while scout cars were better suited to moving radios. He believed the squadron was "over-weaponed and under-manned." This probably reflected the frustration associated with the extensive dismounted fighting the squadron conducted, which exceeded the capabilities of its doctrinal organization in regard to rifle strength and the inability to call on the rifle strength of the horse cavalry division his unit was accustomed to supporting. The large number of crew-served machine guns was great as long as they were mounted on vehicles. Unfortunately, the vehicles proved unable to accompany the troopers when they advanced on foot. What had worked well was using the heavier crew-served weapons, especially the 81mm mortar and 37mm gun to watch over and support by fire the actions of those dismounted.[83]

Candler also seized on the more critical equipment issue in regard to how his squadron was to doctrinally fulfill its mission of reconnaissance, the radio. In his opinion the radios had been adequate, but he still used messengers to maintain contact with higher headquarters, an important matter that would revisit the mechanized cavalrymen awaiting their turn to fight in Europe. Candler sought the addition of an officer in each of his scout platoons similar to the organization of the reconnaissance platoons in the ARB. He believed this would provide the necessary leadership to command the scout cars that often watched over and reported on the actions of the jeep-mounted reconnaissance sections. With an additional officer, the reconnaissance platoon leader would be able to focus all of his attention on his primary mission, reconnaissance.[84]

The manner in which the 1st AD employed the 81st ARB throughout the campaign in Africa all but ensured the reconnaissance men would fail to perform well at anything beyond the small-unit and individual level. The operational distances envisioned for the use of mechanized reconnaissance units had been measured in hundreds of miles. North Africa, with its expansive terrain should have been the ideal place to let these ideas take wing. To some extent, as John K. Herr had predicted, once the distance between friend and foe closed, mechanized units found it difficult to maneuver. Just as MG Charles Scott had predicted after his visit to the British Eighth Army before the United States launched its campaign into North Africa, reconnaissance units had to be robust enough to fight for information and live long enough to report what they discovered. The 91st CavReconSqdn, thrust into a different role than it had prepared for, demonstrated the ability of mechanized cavalry units to execute missions their doctrine warned against. That they could do so reflected the courage and determination of the unit, but not necessarily the wisdom of the commanders who chose to use them in this manner.

The fighting in North Africa provided the first real test of the majority of the mechanized reconnaissance formations and doctrine. What remained untested was the ability of a mechanized cavalry regiment to operate forward of a corps in the capacity of corps cavalry, even though the commanders in North Africa thought these regiments would have been very useful except during the final stages of the campaign outside of Tunis. In contrast, in late 1942, about the same time Americans were entering the fight in Africa, which had been ongoing since 1940, the Germans operating there elected to consolidate their reconnaissance battalions. The 15th and 21st Panzer Divisions with the 580th Reconnaissance Battalion, which had been serving as a general headquarters unit, were combined into a single reconnaissance brigade. The brigade remained under the control of the headquarters of the Panzer Army of Africa.[85] Thus formed, the Germans massed their mechanized reconnaissance units and, by assigning them to the army-level headquarters, were using them in a manner somewhat similar, in regard to command and control, to the still evolving American doctrine pertaining to "corps reconnaissance" regiments. Perhaps providing an early indication of how American corps cavalry units would ultimately be used in the future, the first mission mentioned for the absent mechanized cavalry regiments was that of "gap filling," and then reconnaissance.[86] Future operations in Sicily and Italy would not provide the opportunity to test the value of this critical component of what remained of the Cavalry Branch. Continued operations in the Mediterranean did provide more opportunities to evaluate the existing organizations and doctrines that had already been tested in North Africa.

Wars should be fought in better country than this.[87]—*Major General John P. Lucas*

★   While troopers of the 91st CavReconSqdn provided security during the Casablanca Conference in January 1943, little did they know that the decisions being made there would affect them for the remainder of World War II. The Combined Chiefs of Staff concluded that Sicily should become the Allies' next objective. By securing Sicily they could consolidate their growing control over the Mediterranean Sea's lines of communication and possibly knock Italy out of the war. At a minimum, the Allies hoped the capture of Sicily, placing them only two miles from the tip of the Italian peninsula, would assist the Soviets by diverting even more Axis resources to defend Italy and the southern edge of the Third Reich.[88]

The fighting in Sicily led to Italy in September 1943. COL Hamilton H. Howze, a veteran of the fighting in North Africa and a cavalryman, succinctly captured the action there from the perspective of a mounted warrior.

> Of my service in Italy I'll omit much. There were many weeks on that peninsula during which the fighting occurred in such tightly forested mountain terrain that there was no possibility for the proper employment of large armored forces. Often we were consigned to the role of general reserve . . . tank forces as compared to the infantry, had it easy. The close terrain of Italy kept us often unemployed while the infantry divisions spent long, uninterrupted, and often punishing months in the line.[89]

The nature of the terrain seriously impacted the ability to conduct mounted operations. Small towns and cities established in classical and medieval times often occupied the high ground for defensive purposes. The mounted avenues of approach into these cities and villages were often steep and winding, with chariots and mules in mind, but decidedly not military vehicles.[90] While COL Howze's tankers may have had it easy relative to the infantrymen because of the terrain, the mechanized reconnaissance men were afforded no such break and saw continuous service in Sicily and Italy, even if they were not mounted or conducting missions focused on reconnaissance.

The campaign in Sicily lasted from 10 July–17 August 1943, and was disproportionately led by former cavalrymen in the general officer ranks. Patton commanded the U.S. Seventh Army. MG Lucian K. Truscott commanded 3rd ID, and MG Terry de la Mesa Allen continued to command 1st ID. MG Hugh Gaffey, who had served as Patton's chief of staff during his brief stint as II Corps commander in North Africa, took command of 2nd AD in May 1943. The non-cavalry officers rounding out the American contribution to Operation Husky included LTG Omar Bradley as the II Corps commander, MG Troy Middleton commanding 45th ID, and MG Eddy Manton, who still commanded 9th ID.[91] With the addition of the 91st CavReconSqdn as a corps asset, the total contribution of mechanized reconnaissance forces was limited to the divisional cavalry troops in each of the infantry divisions and the ARB organic to the 2nd AD. Opposing Patton's Seventh Army and Montgomery's Eighth Army was General Alfredo Guzzoni's Italian Sixth Army, a collection of 200,000 Italian soldiers backed by 50,000 Germans.[92]

Like North Africa, mechanized reconnaissance troopers saw action on 10 July 1943 as American forces came ashore, in some cases landing only minutes after the initial infantry assault waves. The troopers secured bridges and roads while fighting sharp encounters with German tanks and the dreaded 88mm cannons.[93] Early days in Sicily mirrored what had been seen in North Africa, but something new was about to happen once Patton gained some room for his Seventh Army to maneuver. Patton's drive to Palermo on the western tip of Sicily started in earnest on 13 July 1943 after General Harold Alexander, the 15th Army Group commander, gave him permission to conduct a "reconnaissance in force" toward Agrigento, twenty-five miles west of Licata.[94] The port of Agrigento, if captured, would enable Patton to discontinue bringing his logistics over the beaches and thus increase the efficiency of the army he had ashore. Moreover, Montgomery's Eighth Army, driving for the island's only strategic objective, Messina, provided Patton with an outlet to employ his lone armored division in a lightning dash.[95]

Lacking a corps cavalry regiment, Truscott's 3rd ID carried out all of the preliminary reconnaissance in the direction of Agrigento with assets organic to his division in support of the newly created "provisional corps" under the command of LTG Geoffrey Keyes.[96] The 3rd Reconnaissance Troop played an instrumental role in the early days of the operation, notably, LT David C. Waybur's platoon holding off the attack of four Italian tanks with nothing more than a three-jeep patrol.[97] Truscott saw to it that the troop's third platoon, which had not received its own jeeps and scout cars, was mounted on vehicles for the drive to the northwest. Each platoon then served one of the division's three infantry regiments as they probed toward Palermo.[98] Company B of the 82nd ARB lent assistance to this effort by continuing to support BG Maurice Rose's CC A of 2nd AD as it drove out of the beachhead.[99] These early ad hoc efforts, in theory, gave way to a more organized reconnaissance effort.

The 82nd ARB consolidated on 18 July 1943 at the same time the 2nd AD pulled together for future use as an intact armored division. The reconnaissance battalion was to lead the way for the "provisional corps," but from the beginning it was at a disadvantage. Company A did not deploy to Sicily, leaving only two reconnaissance troops and the light tank company to cover the entire corps' advance as it started north.[100] Although the battalion commander later wrote about his unit's service as a corps reconnaissance battalion, an examination of the manner in which he employed it lends less credibility to this notion. While the 2nd AD waited in reserve until Keyes committed it on 22 July 1943, the two reconnaissance companies worked for the 82nd Airborne Division and the 3rd ID on 19 and 20 July. Each of the companies was further divided down to the platoon level in the service of their divisions. The reconnaissance battalion commander's account of each of the platoons' actions during those two days provides no indication that their efforts were in any way controlled by the reconnaissance battalion headquarters, nor were the actions of the Third Reconnaissance Troop integrated into the overall scheme of maneuver. Nonetheless, the platoons were able to overcome relatively light resistance with the weapons organic to their platoons in what the official history has described as "little more than a road march."[101]

Released from the 82nd Airborne Division, Company C led 2nd AD's final advance into Palermo on 22 July, identifying minefields and pockets of resistance that were easily overcome by CC A.[102] In this way the battalion contributed to the division's success much in the manner intended. To the east, Company B, Eighty-Second ARB assisted COL Sidney R. Hinds in his attempt to capture an Italian battleship believed to be anchored in the harbor at Palermo. The cavalrymen's opportunity to write a new page in the history of the American cavalry tradition was dashed once they learned the ship had sailed only the day before.[103] Another patrol from the 82nd ARB captured General Giovanni Marciani, commander of the Italian defense force stationed in Palermo. With the subsequent capture of GEN Giuseppe Molinero, commander of the port defenses, the city fell to Patton's rapid thrust that evening.[104] Marching one hundred miles in four days, the drive to Palermo validated the "indispensable role" of the armored division and recaptured the essence of the 2nd AD's participation in the prewar maneuvers under Patton's command.[105] Patton wrote in his diary that the drive to Palermo would be studied by "future Command and General Staff School" students as "a classic example of the use of tanks."[106] He credited his success to a willingness to hold back his tank units until the infantry found the holes in the enemy line through which to send the tanks "in large numbers and fast."[107] But the success occurred against anything but stiff resistance. The reconnaissance effort remained focused at the platoon and troop level and the concept of corps-level reconnaissance was only nominally tested.

The 91st CavReconSqdn also saw action during the Sicilian campaign. Less the small contingent that participated in the 10 July assault landings, the balance of the squadron assembled for combat in the last days of July.[108] Proceeding to the front, it "filled the gap" between Bradley's II Corps and the British 30 Corps and secured the flank of Terry Allen's 1st ID in the process.[109] It supported the 1st ID during its drive on Troina.[110] The squadron's service in Sicily was characterized by dismounted action, overcoming roadblock after roadblock, and often fighting as individual troops and platoons rather than as a complete unit. Just as there were few opportunities for the 82nd ARB to employ its light tanks during its brief drive on Palermo, terrain and the nature of the fighting precluded any use of the 91st CavReconSqdn's light tank company. The closest manner in which the squadron was employed as a corps asset, even when it was directly attached to the 1st Infantry and the 9th IDs, was in its capacity of maintaining contact with the British Eighth Army's 30 Corps.[111] Like the divisional cavalry troops and the 82nd ARB, the 91st CavReconSqdn experienced nothing new in its employment during the Sicilian campaign.

The relatively short campaign in Sicily generated a few ideas specific to mechanized ground reconnaissance. Based on what he had seen, Patton called for doubling the size of each of the infantry division's reconnaissance forces and called for the addition of a reconnaissance company to each of the GHQ tank battalions, which had become so important for "maintaining the integrity of the armored divisions" by providing infantry commanders with the support they needed, but not at the expense of the mass of the armor division.[112] Patton saw in the horse yet another solution to the constant delays presented to the mechanized reconnaissance men in the defiles by the destroyed bridges and other obstacles to mounted movement.

Having commanded an army-sized organization, Patton put forth not only his ideas in regard to the proper organization of corps and armies, but also hope for those who saw in Sicily the proof that they had been right in regard to the continued role of the horse on the modern battlefield: "In almost any conceivable theater of operations, situations arise where the presence of Horse Cavalry in a ration of a division to an army will be of vital moment. Had we possessed Horse Cavalry in Tunisia or Sicily, not a German would have escaped."[113] This was clearly a comment on the traditional role of cavalry as a fighting force and not an indictment of the utility of mechanized reconnaissance units on the whole, especially given the nature of the rest of his comments about how armies and corps should be organized. Patton's comments resonated with those who had opposed the wholesale elimination of the traditional cavalryman mounted on a horse. One wonders how men on horses would have fared any better than men mounted on jeeps, if they were regularly forced to dismount when confronted with artillery, machine guns, and other aspects of the modern battlefield. Perhaps Patton's views were shaped by the action of the Fourth Tabor shipped to Sicily with 678 Berber soldiers, 117 horses, and 126 mules. These French colonials fought well, but gained a greater reputation for their service as superb mountain infantrymen, not for any mounted exploits.[114] Even the Goumiers had undergone a transformation from their earlier days of being a fully mounted unit to one that relied on their animals more for logistics and local reconnaissance. Their locally procured animals suffered the realities of modern war. Many were lost to shellfire, enemy action, and even accidents on the treacherous mountain trails where they operated.[115] The Goumiers used their animals in the same manner the Americans had in Sicily and would in Italy, to move supplies, and certainly not to conduct the kinds of reconnaissance missions envisioned for the mechanized formations then in the field.

The Sicilian campaign saw the employment of new techniques to gain the operational depth required for cavalry to provide commanders early warning. MG Lucian K. Truscott, commander 3rd ID, enhanced his organic ground reconnaissance through the creative use of observation aircraft. Mechanized ground reconnaissance remained important as long as the air reconnaissance assets were controlled above the division level. In the case of the Seventh Army during the Sicilian campaign, a division could expect only eighteen tactical reconnaissance flights per day with each flight lasting approximately thirty minutes.[116] The time taken to disseminate the information from the army level, through the corps down to division level limited much of its tactical value. On a chaotic beachhead, likely to be defended, Truscott wanted to avoid as many unexpected surprises as possible. The answer came from within his division artillery section.

By 1943, the army's Table of Organization and Equipment authorized each division artillery (DIVARTY) to add an Air Observation Post (Air OP) for each of its firing battalions.[117] Truscott approved the plan that led to the construction of runways on two Landing Ships Tank (LST) destined to support his amphibious assault at Licata. Not only did the light aircraft organic to the division help adjust rounds, their intended purpose, they also assisted in guiding landing craft to the correct beaches and provided timely reports on both the friendly and enemy situation. Before the campaign was complete, they were flying route reconnaissance missions and even helped resupply an infantry unit cut off during the fighting around Troina.[118] For men like Truscott this was a boon. Even though the idea of using aircraft for route reconnaissance had been pioneered at Fort Knox during the 1930s, the mechanized reconnaissance men were not directly linked to the valuable resource right over their heads. Rather, the mechanized reconnaissance men found themselves fighting on foot for every turn in the road, minefield, and blown bridge. It would be no different in Italy, now in full view across the Straits of Messina. Even with his creative use of air observers to augment his reconnaissance capability, Truscott reached back to his roots to enhance his ability to operate in Sicily and in the process probably helped to provide the basis for Patton's argument that horse cavalry would have been effective.

During the drive to Palermo, Truscott placed MAJ Robert W. Crandall, a cavalry officer who served in the 5th CavRegt during the interwar years under Truscott command, in charge of what became the Provisional Pack Train and the Provisional Mounted Troop of four hundred mules and over one hundred horses. Lacking specialized equipment and trained personnel, the provisional units were not efficient, but Truscott credited them with speeding his coastal drive to Messina. Cited primarily for their contribution to the logistics effort that kept the infantrymen supplied with food, water, and ammunition, the animals also provided some ability to move across country in Truscott's continued attempts to cut off the retreating Germans. Unfortunately, since the Germans were fighting a delaying action, they often waited until the last minute before retreating to the rear on trucks behind the protection of yet another roadblock or blown bridge.[119] Committed in such small numbers, there was little the mounted men could hope to achieve, but their actions attracted the attention of the horse advocates. Their actions also convinced Truscott it was worth the shipping space to take this provisional unit to Italy. After the conclusion of the Sicilian campaign, the 3rd ID reorganized its provisional mounted units for service in Italy. The organization grew to include a pack train for logistics, a pack battery for fire support, and a troop of "mounted infantrymen" for reconnaissance.[120] But even Truscott disbanded the Provisional Mounted Squadron, celebrated in a later *Cavalry Journal* article, once the 3rd Infantry Division became bogged down in defensive operations after the failed breakout from the Anzio beachhead during the Italian campaign.[121]

The Allies invaded the Italian peninsula in September 1943, but bogged down in the rugged mountains along the Gustav Line in October 1943 where Monte Cassino dominated Highway No. 6, the most direct avenue to Rome. Planners and commanders conceived the Anzio landings, which took place on 22 January 1944, as an ambitious means of restoring mobility to the battlefield by enveloping the German positions and appearing in their rear only thirty miles south of Rome. The envelopment failed to achieve the desired result, and other than a brief period of maneuver after the breakout from Anzio to Rome and beyond, and the last month of the war, there were few opportunities for the larger mechanized reconnaissance organizations to perform their doctrinal missions. For the most part, the scouts served as infantrymen in the mountains, while the light tankers assisted with their logistical needs by leading pack trains, and the assault-gun crews continued to provide fire support.[122]

Just as the fighting in North Africa and Sicily provided the horse advocates what they believed was rich material for their personal campaigns to reestablish the rightful place of the horse on the battlefield in reconnaissance and combat roles, Italy provided the same inspiration. John K. Herr's son-in-law, COL Willard A. Holbrook, Jr., himself the son of the first chief of cavalry, wrote him the following observation even as he trained with the 11th AD in California: "I am tickled to death with the turn combat is taking to show up some of our chiefs in their opinion of cavalry. I would not be surprised if some well trained horsemen were allowed to get into the coming spring fuss."[123] One can only speculate from where he thought these well-trained horsemen would come, since the 1st CavDiv was already deployed to the Pacific and the 2nd CavDiv was on the verge of being disbanded.

Herr also heard from another former cavalryman inclined to address his former chief's interest in all things related to horse cavalry. After expressing his gratitude for the role Herr had played in getting him promoted, MG C.H. Gerhardt observed that

> Both Truscott and General Lucas felt that at least a regiment of our U.S. Cavalry would have made a tremendous difference in that campaign [Italy]. All the divisions had pack trains of captured mules and equipment, and Truscott has a Provisional Mounted Troop for scouting and patrolling.[124]

Ironically, at the time the letter was written in January 1944, Lucas and Truscott were just coming to grips with the Anzio beachhead where Truscott ultimately disbanded his mounted provisional troop. One of the most interesting letters about the action in Italy came from yet another old cavalry family and was directed to a most vocal proponent of the horse at the *Cavalry Journal*, Hamilton Hawkins.

Hamilton H. Howze, like Willard A. Holbrook, Jr., was from an old cavalry family and had also married the daughter of a former chief of cavalry. His uncle happened to be Hamilton Hawkins. In a hand-written letter drafted in February 1945, Howze offered his "opinions on the possibilities of cavalry in the pursuit phase" that had followed the breakout from Anzio.[125]

> It is my opinion that a corps of two cavalry divisions would have utterly destroyed the German *14th Army,* in toto; we could have marched a [unintelligible text], practically, to the Brenner Pass. Lesser cavalry forces should have doubled the 5th Army's take of prisoners, which in actuality was about 40,000. That is provided the high command should have been able to hold that cavalry as cavalry, should have resisted the temptation to send the cavalry, dismounted, into the mountains during the winter to relieve the very weary doughboys.[126]

Years later in an interview conducted at the Army War College, Howze was careful to qualify his comment on the utility of horses in Italy. He rather said that two divisions of horse cavalry only would have been useful had "it been kept in a training status" and "out of the line, until the propitious moment" arrived. Howze reiterated that only a "*bonus* of two well trained horse cavalry divisions would have been a tremendous benefit." What he also offered the interviewer that he probably did not offer the die-hard horse advocates, was his impression of what had happened in Italy when his tanks encountered Turkomen cavalry in the service of the Germans. He noted that the outcome of these encounters was "not a great endorsement of horse cavalry."[127]

Although Howze's letter to "Uncle Ham" never specifically addressed the mechanized reconnaissance units in 1st AD, he left little doubt about his frustration with the inability to clear routes to the front in a timely manner. Frustrated with the rate of reconnaissance afforded him with dismounted infantry operating to either side of any given road, Howze ordered his tanks to "barrel down the road" until "bang and I'd have a flamer."[128] Recognizing that the "lead tanker" knew he was the "next one to die" and did not appreciate this, Howze conceded that it was "difficult . . . to make speed."[129] Nothing found in Howze's divisional or regimental reconnaissance agencies had the equivalent firepower and protection as the tanks he was using to lead the way. Howze did offer a solution.

To overcome and bypass the endless defended roadblocks, Howze called for the use of a squadron of horse cavalry. Riding to the flanks, where he insisted they "would have encountered NOTHING," armed with tank killing bazookas, the horse cavalrymen could have enveloped and reduced the successive roadblocks and rapidly increased the armored division's rate of advance. In closing his letter, Howze begged the questions on John K. Herr's mind: "I wonder if we'll be smart enough to send mounted cavalry to the Pacific for operations in China, and whether we'll reform divisions after the war. I doubt it, and it will cost us."[130] Howze was not alone in his sentiments of what had plagued mechanized forces in Italy. The army, as it had in North Africa and Sicily, dispatched observers to Italy. The army also recalled, in July

1944, MG Ernest N. Harmon, commander of 1st AD, a man unequalled in the tactical employment of armored forces. Harmon, the external observers, and those assigned to the mechanized reconnaissance units all recorded their views. Equipment and organization continued to dominate their remarks.

If one could consider the horse a piece of equipment, the observers were no different from horse advocates in that they continued to record the desires of some units for the return of horse cavalry given the difficulty of the terrain in Italy. In the horse, they also saw the ability to get off the roads to pursue and cut off retreating enemy forces, or to exploit breakthroughs.[131] The horse's replacements, the jeep and the M8 Armored Car, garnered far more attention. Troopers and observers liked the M8, but like its predecessor the scout car, the M8 came with a number of serious limitations. As long as the weather was dry, it proved capable of limited cross-country mobility in "moderate terrain." Equipped with enough sandbags on the floor, it provided limited protection from the ubiquitous land mines found in Italy, but at the expense of space in the crew compartment.[132] Harmon classified the 37mm cannon it carried as "another weapon that has been built in large quantities and which has no practical use except in small quantities in the infantry to operate against machine gun nests."[133] Harmon's comment was not directed at the M8 Armored Car, but not even the light tanks in the ARBs and squadrons carried anything bigger. The high hopes held out for the arrival of the M8 since the days of campaigning in North Africa and Sicily were dashed in Italy. The jeep received scant attention.

Mechanized reconnaissance units took matters into their own hands to solve the many mobility problems that confronted them in Italy and sought solutions that blended equipment needs with ad hoc organization. The 91st CavReconSqdn identified the impact that the constricted terrain in Sicily had on its operations. Before it arrived in Italy, its members collected a number of pieces of engineering equipment and the 15-percent-over-strength in personnel to man their provisional "reconnaissance engineer platoon."[134] The squadron commander believed it was particularly important for the 91st CavReconSqdn to have its own engineers, since as a corps asset it often moved to different sectors of the front and could not depend on the support of higher headquarters. The army rewarded the squadron's initiative in April 1944 by stripping it of its personnel and equipment above what was authorized for it in the Table of Organization and Equipment.[135] Harmon cited the need for more armored bulldozers in his report as well.[136] Cavalrymen in the field, as they had through the formative years of mechanized reconnaissance, continued to apply creative solutions to overcome their problems, but their creativity was dashed by the army's overarching manpower and equipment shortages.

The organization that received the most attention from the observers in Italy was the infantry division's reconnaissance troop. The terrain in Italy forced the reconnaissance troops to operate dismounted most of the time, yet they still patrolled, manned observation posts, and maintained contact between friendly adjacent units. Troops were also subject to having their platoons dispatched on separate missions that placed them beyond mutually supporting distances.[137] Calls for an increase from the troop-sized units serving the infantry divisions to become squadron-sized elements were common, but the reasoning for the increase in size was not always consistent with why the mechanized reconnaissance troops had been formed in the first place. Citing the inability of the existing organization "to deal with the German reconnaissance battalion" was a very reasonable call for the expansion of the troop to squadron-sized strength, given the implied assets that came with the squadron organization.[138] Calls from the field for larger reconnaissance organizations for each of the infantry divisions lent substance, even if in retrospect, to John K. Herr's rabid defense of a more sizeable cavalry force as the nation entered World War II.

Neither a function of equipment, nor a function of any existing organization, air-ground cooperation in the field of reconnaissance continued to attract the attention of the observers. Not all of them viewed the inclusion of light reconnaissance aircraft as necessarily the solution to the ongoing ground reconnaissance problems found in Sicily and Italy, but everyone seemed to agree that there had to be better and more direct communication between the air and ground reconnaissance assets.[139] Commanders valued the planes for their ability to direct accurate indirect fire, and in some cases their mere presence was enough to silence German artillery units fearful of receiving counter battery fire.[140] Even as units continued to discover new uses for the light observation aircraft, or in many cases rediscover uses long established at Fort Knox during the interwar years, the planes continued to remain beyond the reach of the mechanized reconnaissance units intended from the earliest days of mechanization to work directly with them. Harmon, a real fan of light airplanes for reconnaissance work, commented that "contrary to expectation [the light observation aircraft] has been able to survive on the battlefield in spite of hostile air."[141] The issue of air-ground cooperation remained solved only on an ad hoc basis. Long identified as a critical component of successful mechanized ground reconnaissance, the air-ground link remained broken.

Not surprisingly, as Howze's division commander, Harmon's account of how the division conducted ground reconnaissance in the pursuit north of Rome after the Anzio breakout echoed the techniques Howze had adopted. Infantrymen riding on tanks dismounted to reconnoiter, since for the most part Harmon's ARB operated to the division's flank, maintaining contact with adjacent units.[142] He adopted this technique because

> During a pursuit the use of reconnaissance elements, as the leading elements in a force, merely because contact has been lost locally, is not justified in most circumstances. Where the nature of the expected resistence [sic] can be in any way deduced—and it usually can—appropriate composition of the column should be made ahead of time to overcome it; usually time is saved in the long run if medium tanks lead.[143]

Harmon recognized the inability of his ARB to overcome anything but the lightest resistance and thus used them on his flanks. By putting the reconnaissance battalion on the flanks, albeit for good reasons, Harmon placed himself on the horns of a dilemma he acknowledged earlier in his report. It was an observation that would have made John K. Herr proud.

Writing about speed on the battlefield, Harmon concluded that "all movement . . . is relatively slow and deliberate" and that interwar maneuvers had lent a "false picture of speed" because of the failure to accurately portray casualties and logistics.[144] Harmon remarked that his division had lost "many tanks . . . in the early days under the false training idea" that rewarded "boiling down the road." Speed could only be achieved by "going rapidly from one reconnoitered place to another" and "by thinking ahead and being prepared with the solution for [an] emergency when it arises." He cautioned that division commanders should be "thinking in terms of what is going to happen from six to twelve hours ahead of the present."[145] Only good ground reconnaissance coupled with thinking ahead could save countless lives. Herr had recognized the false sense of speed, and American doctrine had long recognized the need to conduct reconnaissance well in advance of the main body of forces. What Harmon pointed out with the use of his reconnaissance battalion to the flank was the inability of the contemporary organizations, dedicated to the task of getting that information well in advance, to perform as desired. Designed with one expectation of what war would be like in the future and confronted with a different reality, the American army was learning as it went, and it was finding solutions. The horse advocates continued to point to what they viewed as a proven technology and organization.

The fighting in Sicily and Italy presented few opportunities for the employment of mechanized reconnaissance units in the ways envisioned during the interwar years. The campaigns in Sicily and Italy did present plenty of opportunities for mechanized reconnaissance units to make numerous contributions, more often dismounted than they might have expected. A major contributing factor to the manner in which the mechanized cavalry units saw service was the terrain. The mountainous and built-up terrain limited the mobility of all units, but combined with an enemy skilled in the art of defense it became nearly impossible for the lightly equipped mechanized reconnaissance units to precede the larger organizations they served. Another factor that led to their employment in ways never intended simply boiled down to the need to rotate haggard infantrymen in and out to the line. The mechanized cavalrymen may not have been organized, trained, and equipped for extended dismounted service, but they served in this capacity often. Mechanized reconnaissance units in the service of corps remained absent from the battlefield still waiting to make their debut. Only on rare occasions were battalion- and squadron-sized elements able to operate mounted, although on these occasions they were still often found on the flanks and not necessarily in front leading the way. Citing the inability of a mechanized cavalry troop to cover an entire infantry division's frontage was inconsistent with the basis of its creation, namely, the ability to reconnoiter a few specific routes for the division's advance.[146] Continued calls for increases in the divisional cavalry troops did reinforce the last chief of cavalry's belief that each infantry division required an entire squadron to carry out all of the division's reconnaissance and security chores. The last chief of cavalry also saw himself vindicated in other ways.

Harmon, more experienced in the employment of armored divisions in combat than any other commander at the time, commented specifically on the unrealistic speed of movement that had characterized the prewar maneuvers. It was this speed that had largely been the undoing of horsed units. It was the same quest for speed that had long characterized the development of the lightly armed mechanized reconnaissance units that were expected to range out to find the enemy. The fighting in Sicily and Italy only presented opportunities for fast-paced operations when Axis forces were in full retreat, as had been the case on the drive to Palermo and during the final push into Rome and beyond. Terrain, combined with the enemy's determined defense, limited the ability of mechanized units to operate as they were intended. Herr was somewhat vindicated by the initial reports coming from Sicily and Italy. The horse had returned to the battlefield.

Provisional and Allied horsed units did see limited service in Sicily and Italy. The majority of their service fell under the realm of logistics, with the requirement for donkeys and burros to carry supplies to mountain positions. Truscott's own mounted reconnaissance troop experiment was short-lived once confronted with the artillery-swept conditions of the Anzio beachhead. Most calls for the employment of horse cavalry divisions at the time did not factor in the cost of what other types of units might have been left out in order to pay the penalty in shipping space to bring over a horse cavalry division. None of the commentators suggested that they would have willingly given up an infantry or an armored division to have a horse cavalry division. Perhaps they did not realize it, but German success off the east coast of the United States had a dramatic impact on the American ability to deploy forces to Europe while the fighting raged in North Africa and in Sicily and Italy. The United States did not deploy any divisions to Europe from October 1942 to March 1943, and only eleven divisions crossed the Atlantic between March and November 1943.[147] General Howze, with time to reflect, further captured the dilemma that forced mechanized cavalrymen into the line as infantrymen. When he wrote his uncle, Hamilton Hawkins, Howze wondered aloud if the commanders could have resisted the urge to use the horse cavalry divisions as infantrymen, or if they would have been able to keep them in the rear where they could train and remain capable of exploiting a breakthrough. Given the limited shipping and the growing emphasis placed on the build-up of forces for the cross-channel attack in 1944, it seems unlikely. Before this final test, the army visited more changes on the organization and doctrine of the mechanized cavalry units expected to lead those units marshalling in England.

> I cannot refrain from taking exception to our remarks that Cavalry should become mounted infantry. I can say that all true cavalrymen would regard that as heresy . . . mounted infantry is totally incapable of mounted action.[148]—*MG John K. Herr to MG Innis P. Swift*

★   During the summer of 1943, LTG Lesley McNair and army chief of staff, George C. Marshall, concluded that eighty-eight divisions would be enough to win the war rather than the 200 divisions they had originally planned for in 1941. McNair and Marshall were depending on the pooling concept to deliver victory. Corps would be the army's premier tactical fighting unit, tailored for each mission. Thus, McNair concluded by 1943 that the division would be the largest organization with a fixed table of organization and equipment. He did not expect the streamlined divisions he created in 1943 to fight on alone; McNair expected corps commanders with the support of their equally streamlined staffs to be "the key headquarters for employing all combat elements in proper tactical combinations."[149] The twenty-two corps that saw action in WWII had a pool of corps-controlled artillery, engineers, tank and tank destroyer units, and cavalry groups, and a host of other supporting troops to help them accomplish their missions.[150] With this, McNair hoped to enable corps commanders to have the utmost flexibility to blend together the proper combinations of forces to achieve the mission, while not wasting precious assets by permanently assigning them to combat divisions that would not always be committed to the fight.[151] The result was hoped to be just enough infantry and armored divisions to win the war with the support of the pooled assets in the proper combination at the proper time and place. Pooling also allowed McNair to save on precious shipping space and other natural resources, since each division now had fewer vehicles (fewer tires and lesser requirement for fuel, oil, etc.) and no horses.[152] The pooling concept affected units performing mechanized ground reconnaissance directly and indirectly. Indirectly, only the test of combat would reveal the impact of pooling corps cavalry reconnaissance assets. Directly, the realignment of the army in 1943 impacted ARBs serving in armored divisions and the all-mechanized corps cavalry regiments.

McNair reorganized American armored divisions in 1943, still fully wedded to envisioning them as modern cavalry for pursuit and exploitation missions, not for creating penetrations, killing tanks, or assisting the advance of infantry divisions. Troops drawn from the corps pool, such as tank destroyers and independent armored battalions, when properly employed could fulfill these tasks. The reorganization took place during the last quarter of 1943 with two exceptions. The 1st AD retained its 1942 configuration until after it had liberated Rome in June 1944 while 2nd and 3rd ADs retained their 1942 organization throughout World War II.[153] Fundamentally, armored divisions became smaller and more balanced.

While the division retained its ARB, it did receive a new name, Cavalry Reconnaissance Squadron (CavReconSqdn). Other than the additional reconnaissance troop assigned to the cavalry reconnaissance squadrons of the armored divisions, all cavalry squadrons now looked alike, ending the slight differences that had distinguished the mechanized reconnaissance squadrons and battalions to date. Moreover, now every single unit, troop sized and larger, that was to perform the task of ground reconnaissance, bore the word "cavalry" in its official title.

McNair's elimination of the regimental headquarters echelon was not limited to the armored divisions. He also eliminated the regimental headquarters of the all-mechanized corps cavalry regiments he created in 1942. In doing so he applied the same logic that resulted in the concept of pooling. In place of the regimental headquarters, McNair inserted the group headquarters. Cavalry groups were nothing more than a headquarters and a headquarters troop capable of commanding and controlling any given collection of assets assigned to the group by the corps for any given mission.[154] Each group habitually contained the two mechanized reconnaissance squadrons that had previously been the regiment's 1st and 2nd squadrons.[155] This reorganization had immediate and lasting impacts.

Often lamented in official and unofficial correspondence of mechanized cavalry officers after the war, the reorganization of the cavalry regiments destroyed the regimental identity.[156] Because McNair sought maximum flexibility, he demanded that these squadron-sized elements be logistically and administratively independent. As regimental headquarters became more austere group headquarters, squadrons gained the needed assets to operate independently of the group.[157] This flexibility was two-edged. It provided for the flexible assignment of other pooled assets to the cavalry group, such as additional engineers, artillery, tank destroyers, and infantry, to accomplish missions the group might receive from the corps to which it was assigned. The organization of independent mechanized cavalry reconnaissance squadrons loosely organized into cavalry groups presented the possibility that commanders might be tempted to commit these specialized corps reconnaissance assets in piecemeal fashion. This second condition led to later charges of misuse of mechanized cavalry reconnaissance squadrons.[158] Based on the experiences of the 91st CavReconSqdn in North Africa, Sicily, and Italy, the signposts marking the way to this misuse were perfectly evident, even if the subtle misuse of mechanized reconnaissance units during the interwar years was not. The establishment of cavalry groups in place of cavalry regiments in combination with the pooling system built on the premise of having the leanest possible fighting force created the conditions for a potential disaster.

Published in June 1944, nine days after D-Day, the War Department issued World War II's last version of the *Field Service Regulations,* which provided a broad doctrinal overview of how the army intended to fight the campaign in Europe. Little changed between the 1941 and 1944 editions, with the notable exception that the horse-mechanized concept and organization was absent from the 1944 doctrine. The army still provided this description of cavalry: "Cavalry consists of highly mobile ground units of two type: horse units and mechanized reconnaissance units."[159] The doctrine reminded readers that the "efficiency of cavalry" still depended on the "condition of its mounts and vehicles," and that horses could be "transported in trucks or semi-trailers in order to increase their mobility or to conserve animals."[160] Even if the horse-mechanized concept was officially absent, yet still alluded to, there was no question about the role of the horse.

Retaining the premier mission of offensive combat, the 1944 doctrine continued to celebrate the horse cavalry's ability to carry out its missions "over almost any terrain and under all conditions of weather." But not completely blind to the changes that had occurred in war and the advent of motorized and mechanized forces, the horse cavalry's doctrinal niche for exploitation remained in "terrain not suited for armored or motorized units."[161] The doctrine further acknowledged the long-held understanding that: "Horse cavalry habitually maneuvers mounted, but ordinarily fights on foot. As a rule, mounted maneuver is combined with dismounted action."[162]

The 1944 doctrine did provide minimally more coverage than the 1941 version on the topic of *mechanized cavalry*. Having achieved dual subject-heading billing with *horse cavalry*, the very term *mechanized cavalry* reflected a subtle evolution in the doctrine over the 1941 term *mechanized reconnaissance units*.[163] Even so, one would not find the words "horse cavalry" italicized in either edition; thus there was still a subtle subordination even after a year and half of war. Ironically, thus far the only organized horse cavalry units that had seen action (rather than the provisional units seen in Sicily and Italy), had done so in a dismounted capacity, and the only cavalry units to have fought mounted did so in a mechanized capacity. Further, the doctrine made it perfectly clear that *mechanized cavalry units* were "organized, equipped, and trained to perform reconnaissance missions employing infiltration tactics, fire, and maneuver . . . [and] engage in combat only to the extent necessary to accomplish the assigned mission."[164] The doctrine again clearly defined the subordinate relationship that mechanized reconnaissance units performed in regard to the larger units they were reporting to on "the location of enemy forces . . . [and for whom they were] providing timely warning of ground and air attacks."[165]

On a final note about mechanized ground reconnaissance, the doctrine spoke directly to one of the chief reasons so much effort had been invested in the idea between the world wars, namely, aviation. The "operations of mechanized cavalry and aviation" were viewed as "complementary," since aviation was expected to provide the information to facilitate the better employment of ground reconnaissance units.[166] The space between opposing forces remained the domain of mechanized ground reconnaissance forces until the main bodies of forces became engaged and the reconnaissance men withdrew to the flanks, liaison roles, or the rear.[167] This did not reflect any change in the optimism that had prevailed between the wars that the next war would be one of maneuver, not fighting from trenches. The role of aviation had changed dramatically between the wars, but not really in regard to reconnaissance. Even the hidebound "horsey" types saw the important role the airplane would play in the realm of "strategic reconnaissance" at the end of World War I.[168] The real tragedy was that the doctrine still did not provide a doctrinal link between those in the air and those on the ground, both of whom were in search of the same thing, information. The coming campaign in Europe would continue to provide colorful ad hoc examples on par with MG Lucian Truscott's use of artillery observer planes in the Mediterranean, but a doctrinal solution still escaped the army.

As the United States made its final preparations for the drive on Berlin, it implemented a number of organizational and doctrinal changes that directly affected the agencies of mechanized ground reconnaissance. The decision to convert all but two of the armored division's ARBs into cavalry reconnaissance squadrons reinforced Cavalry Branch's growing ascendancy in all matters related to mechanized reconnaissance. There were minor changes associated with this conversion, but the combined-arms approach of the organization remained intact, and some of the more aggressive views of the ARBs began to appear in the mechanized cavalry doctrine. The decision to convert the non-divisional mechanized cavalry regiments into cavalry groups was perhaps the most important organizational change. The corps cavalry regiment concept had not been tested under battlefield conditions. What had been learned already, particularly in North Africa, was that the dissipation of reconnaissance assets did not work. McNair's effort to create the maximum amount of flexibility in the application of pooled resources came at the expense of the mechanized cavalry regimental commander by doctrinally allowing for the employment of individual squadrons. Beyond the control of the regimental, now group, commander, it would be up to corps and division commanders to properly employ the mechanized cavalry reconnaissance assets at their disposal. Although the doctrine continued to make it clear how they were to be used in accordance with their organization and prewar expectations, the economy of combat divisions arrived at by pooling made these heavily armed, lightly protected, specialized troops vulnerable for other uses. Even the army's anachronistic belief in the combat characteristics of horse cavalry, in contrast to the purely reconnaissance characteristics of mechanized cavalry, could not guarantee that the corps cavalry regiment concept would be carried out correctly. Ironically, the doctrine's open-endedness for other uses, when commanders deemed it necessary, provided the very means for mechanized cavalrymen to fully reclaim all the attributes associated with their branch, including the ability to contribute directly to the fight. Only the test in Europe would reveal the impact that their organization and equipment, including the lack of horses, had on their ability to perform not only reconnaissance, but the other missions that might fall their way.

5

D-Day to VE-Day

CAVALRY GROUPS

ACROSS EUROPE

During the war in Europe, the 4th Cavalry
[Reconnaissance Squadron] was repeatedly committed
in every type of role except one—we were not
dropped by parachute or glider.

—*LTC John R. Rhoades, Commander 4th Cavalry*
Reconnaissance Squadron

★   In the early morning hours of 6 June 1944, a group of four men, armed only with knives, slipped over the side of a boat. Against a curtain of darkness, periodically punctuated by flashes and claps of artillery, they made their way to the small islands of Iles St. Marcouf, some six-thousand yards off the coast of Normandy near what the world would come to know as Utah Beach. In their wake followed a small task force composed of men from Troops A of the 4th and 24th Cavalry Reconnaissance Squadrons (CavReconSqdn). Under the command of LTC Edward C. Dunn, the troopers seized the small mine-infested islands at 0430 on that morning of 6 June 1944, two hours in advance of the American divisions assaulting across Omaha and Utah beaches. In doing so, the cavalrymen joined the paratroopers in leading the way for the ground forces bent on invading France and conquering Hitler's Third Reich.[1]

After years of development, the United States had finally committed mechanized reconnaissance units intended to serve with corps-sized units to the fighting in Europe, albeit in a role no one would have predicted during the interwar years. Like so much of what was to take place during the eleven ensuing months, elements of the 4th CavGrp and other cavalry groups made important contributions to the success of D-Day, but not in the manner long envisioned for them. Like the cavalry reconnaissance troops and squadrons assigned to infantry and armored divisions, they were organized, equipped, and doctrinally expected to perform reconnaissance. Like the smaller mechanized reconnaissance units that preceded them into combat, cavalry groups confronted realities of war that did not square with the expectations of the interwar years. In this environment, in the service of generals with barely enough soldiers to accomplish the mission, the cavalry groups began to reclaim the combat functions long ascribed to their branch, functions that had been usurped by the Armored Force in 1940. It was in Europe, due to changes already accomplished in what remained of Cavalry Branch's most dominant organization, the cavalry group, as well as the manner in which commanders employed these changes, that the reconciliation between Cavalry Branch and the Armored Force made its greatest headway. By the time the European crusade was complete, the men wearing crossed sabers, the symbol of Cavalry Branch, no longer accepted a subordinate role to horse cavalry, still doctrinally recognized and celebrated on the cover of the May–June 1944 edition of *Cavalry Journal* with a picture of a cavalry trooper assigned to the 3rd Provisional Cavalry Troop saddling a horse. Whether riding on rubber rafts, horses, jeeps, armored cars, or tanks, cavalry was cavalry. Mechanized cavalry, by campaign's end, no longer meant just reconnaissance; it often meant fighting. The first six months of campaigning in Europe presented the cavalry groups with a variety of experiences and challenges. They confronted the *bocage,* the thrill of breakout and pursuit, and crashed into Hitler's *Siegfried Line,* the vaunted West Wall. All the action of the first six months on the Continent provided grist for the unending debate on the continued role of the horse. Information and lessons learned from combat contributed to finding ways to improve the future performance of the all-mechanized ground reconnaissance units. The last six months of World War II in Europe provided lasting impressions that shaped final conclusions about the utility and future of mechanized ground reconnaissance units, because they gave additional opportunities to evaluate these specialized units in combat. The officers of the U.S. Army's General Board who studied mechanized reconnaissance units after the war drew the majority of the historical vignettes used to illustrate their conclusions from this period.[2]

The German offensive that came to be known as the Battle of the Bulge shattered the front of the U.S. First Army in December 1944. In the opening hours of the offensive, the German drive nearly destroyed the 14th CavGrp, revealing critical limitations of mechanized cavalry organizations in static positions. Rather than focusing on the demise of the 14th CavGrp, the General Board focused on successful contributions of other mechanized ground reconnaissance units during the same battle. Mechanized reconnaissance participated in the counterattacks that reduced the German salient by using mobility, firepower, and dismounted fighting.

With the "bulge" reduced, a month of limited activity ensued until another exhilarating breakout occurred. As mechanized cavalry units led the advance or protected the flanks of eastward-probing columns, conditions again proved ideal for showcasing the capabilities of mechanized ground reconnaissance units on a fast-moving battlefield. This time the pursuit carried the cavalrymen into the arms of Soviet allies, who in some instances happened to be riding horses. The final pursuit, more than any other phase of the European campaign, captured the General Board's attention when it went looking for examples to illustrate its findings, while calls continued for the restoration of horses for reconnaissance and combat assignments. Regardless of their shortcomings, mechanized cavalry units contributed to the victory achieved in Europe.

★   PART ONE   ★

Just as the invasion of North Africa featured mechanized reconnaissance men blazing the way to the landing sites near Safi, modern cavalrymen again rode in from the sea, this time to secure islands off the coast of France. The 4th ID's reconnaissance troop arrived on the Continent at 0930 on D-Day, but with only two operational vehicles it failed to link up with the paratroopers at Ste Mere Eglise. The entire troop assembled by 9 June and throughout the remainder of the next two months supported the 4th ID's fight in the *bocage* with security patrols and reconnaissance. The troop also experienced frequent attachment to the 24th CavReconSqdn.[3] The 1st Cavalry Troop, which had already seen action in North Africa and Sicily, crossed the beaches of Normandy without its equipment one day after its parent division stormed ashore.[4] The same was true of the 82nd ARB, which followed immediately in the wake of the 2nd AD. Permanent fixtures in every division and well established by 1944, these units ultimately received less attention in the final analysis of what had been good and bad about ground reconnaissance after World War II ended. Far more numerous than the cavalry groups, their contributions to the first six months in Europe were considerable, but they are not the primary focus of this chapter.

Like previous missions in which commanders committed mechanized reconnaissance units in piecemeal fashion, the same disastrous results awaited the untested cavalry formations. CPT William Larned's Troop B, 4th CavReconSqdn found itself attached to the 82nd Airborne Division from D-Day until 3 July 1944. As if scripted in Hollywood and taking cues from the Wild West genre, Larned's mission was to link up with isolated paratroopers to provide the additional firepower needed to repulse German counterattacks, not Native Americans. LT Gerald H. Penley led his platoon across Utah Beach at 0930 on D-Day and fought through to elements of the 82nd Airborne Division at Ste Mere Eglise. Joined a few days later by the remainder of Troop B, the 82nd Airborne Division assigned the cavalry troop to do combat patrols in the hedgerow country and enjoyed the additional firepower the troop brought to the division with its 37mm cannons and machine guns. It was on one of these platoon-sized patrols that the troop lost an entire platoon of men and vehicles, less one jeep and two scouts who managed to escape.[5] Lacking the support of the troop's other platoons, and beyond the support of yet other assets found in every mechanized reconnaissance squadron, the lightly armed platoon suffered heavy casualties.

Early fighting in Normandy provided one fascinating example of the value of stealth, one of the salient features of the interwar debate on the desired characteristics of a mechanized reconnaissance unit. In helping take Auderville, in support of the 9th ID, Troop B, 4th CavReconSqdn, proved under combat conditions the value of wheeled vehicles for mechanized reconnaissance. Confronted with a continuous line of German defenders, the cavalrymen found a hill behind friendly lines that allowed them to gain enough speed on the descent into enemy lines to gain the momentum needed to coast undetected into and beyond the German positions. Completely surprised, the Germans retreated under pressure from the attacking American infantrymen as the cavalrymen dashed on to Auderville, where they surprised the garrison and took control of the village by daylight.[6] This was not characteristic of most of the fighting in Normandy. The squadron learned to concentrate the fire of Troop E's assault guns with Troop F's light tanks to create the firepower needed to support the dismounted attacks generated by the reconnaissance troops. Reflecting on the squadron's first thirty-nine days in combat, it was not lost in the unit's official history that although the men had been trained and equipped for mounted service, their "missions had been almost consistently dismounted." The fighting resulted in 138 casualties, of which twenty-four were enlisted men and officers killed in action.[7] Not fighting was not an option.

As fighting continued, the U.S. VII Corps drove north to capture Cherbourg and clear the Cotentin Peninsula, and the British Second Army found itself fighting toward Caen, where it was opposed by heavy concentrations of German armor. Between these two forces, *bocage* country limited mobility and observation and slowed the American efforts to move off the beaches. The same was true of the sluggish streams found near Carentan that created marshes devoid of cover and concealment, leaving the few roads as the only avenues for mounted advance. Although the action ultimately embraced four American corps of General Bradley's First Army as they fought their way free of the constricting terrain, the XIX Corps carried the action that was oriented toward the vital road junction at St-Lô. The drive on St-Lô also presented the first opportunity for the wholesale commitment of an intact cavalry group, the 113th.

Astride the Vire River, St-Lô controlled all the roads in the immediate area, but especially the lateral routes that allowed the Germans to shuttle forces along their front depending on where the Americans attacked. The zone of attack for the XIX Corps straddled the Vire River and was fifteen miles wide at the line of departure.[8] First Army assigned the 113th CavGrp to the XIX Corps for its drive on St-Lô, and XIX Corps ordered the cavalrymen to support the 30th ID during the opening phase of the offensive, by maintaining contact between the 30th ID and the 83rd ID of the adjacent VII Corps.[9]

When the offensive began on 7 July 1944, the 30th ID led the way by gaining a crossing of the Vire-Taute Canal.[10] While the infantrymen advanced and gained the far shore, the cavalrymen in the 113th CavGrp, led by COL William S. Biddle, awaited their turn to cross into the ever expanding bridgehead created by the infantrymen and combat engineers. It took the cavalry group six hours to cross on the single-lane bridge, and once across they immediately encountered mud as deep as four feet along the trail that led to their sector near the village of Goucherie. Leading with a light tank, an assault gun, followed by another light tank, and then a reconnaissance platoon, the squadron interspersed heavier firepower throughout the column.[11] When the squadron reached its objective in the early morning hours of 8 July 1944, it was able to overcome the initial resistance, but it took the effort of two troops to dislodge a single enemy platoon. Only demolition charges set in the hedgerows allowed the light tanks and assault guns to maneuver into positions to support the dismounted attacks of the reconnaissance troopers.[12]

As morning arrived, the 125th CavReconSqdn moved in on the left of the 113th CavReconSqdn, which had established a generally north/south line. In doing so the 125th linked the 113th CavGrp with the rightmost element of the advancing 30th ID. The flank of the XIX Corps was secured, but the terrain and stiff enemy resistance nullified any hope of the cavalry advancing mounted to the west. The cavalry had already begun to advance dismounted without the benefit of its light armored protection and vehicular-mounted weapons. Others braved "the gauntlet of heavy flanking fire from the hedgerows." COL Biddle decided at 1600 on 8 July 1944 that his cavalry group would take up defensive positions, placing a higher premium on maintaining contact with the advancing 30th ID. By doing so, Biddle made sure the Germans could not counterattack into the flank and rear of the division his unit covered.[13] Although the cavalry group maintained contact between two separate corps, a traditional mission in many respects, terrain and enemy opposition prevented the group from advancing in the manner envisioned during the interwar years. In the most traditional sense of American cavalry, the specialized mechanized reconnaissance men found themselves fighting on foot. The next day, 9 July 1944, the commander of the XIX Corps attached the 113th CavGrp to CC B, 3rd AD.[14] Rather than reinforcing success, the situation forced MG Charles H. "Cowboy Pete" Corlett to divert armor assets to hold open the shoulder of a penetration because the cavalry group lacked the strength to do so.

The 9th ID attacked through the 113th CavGrp on 10 July 1944, but it too met stiff resistance and made little headway, which allowed a gap to develop between the division and the continued advance of the 30th ID. XIX Corps was fortunate that the Germans had been unable to mount anything more than localized counterattacks. The 113th CavGrp had avoided a serious test, but suffered fifty-three casualties, lost three jeeps, three light tanks, and one armored car for the gain of a few thousand meters.[15] Unable to achieve any depth between the enemy and the corps it served, the 113th CavGrp failed to provide early warning when the German Panzer Lehr Division attacked the 9th ID in the early morning hours of 11 July 1944. The attack along the boundary between the 9th and 30th divisions disrupted communications, but the 9th Cavalry Reconnaissance Troop helped restore the division's situational awareness. This allowed the 9th ID to seal the penetration by committing ground and air forces, but it cost them a day of planned advance.[16] XIX Corps placed the 113th CavGrp in reserve beginning on 11 July 1944.

The *bocage* of Normandy stymied the efforts of all mechanized ground reconnaissance units in June and July 1944. Troops assigned to divisions continued to contribute to their divisions' success within the limits of their capabilities. In the case of the 4th CavGrp, squadrons continued to see their troops employed on independent missions. The 113th CavGrp, with an early opportunity to demonstrate the full might of an intact unit, could not overcome the Germans or the terrain that lent the defenders such an advantage. With their host of supporting weapons from mortars at the platoon level to the assault guns and light tanks found in each squadron, the mechanized reconnaissance troopers quickly adapted to the dismounted techniques required of them. Even if the mechanized cavalrymen were largely unable to carry out their primary doctrinal mission of distant mounted reconnaissance, they contributed to the campaign. There was no call for the reintroduction of horses as the solution for the static nature of the fighting. For all the frustration presented by the Norman hedgerows, the opportunity to perform a "real cavalry mission" was about to take place. This would be the opportunity for the interwar faith in the mobility of the future battlefield to come to the fore.

With a well-established beachhead and with the preponderance of German armored forces confronting Field Marshal Bernard L. Montgomery at Caen, it was time to get on with liberating France. The Twelfth Army Group commander, GEN Omar N. Bradley believed that "only a *breakout* would enable us to crash into the enemy's rear where we could fight a war of movement on our own best terms."[17] The interwar faith in the "war of movement" was about to be realized, if ever so briefly. With the restoration of movement, the offspring of the Cavalry and Infantry Branches—the Armored Force with its armored divisions—dazzled the world as they raced across France and ultimately Belgium under the air cover of MG Elwood A. "Pete" Quesada's XIX Tactical Air Command.[18] While units like the 4th AD captured the headlines, the cavalry groups finally came into their own and made important contributions.

Operation Cobra, with its carpet of bombs, created the penetration that allowed mobile units into the rear of the German Army. Having joined a frontline battalion to witness the army he built in action, LTG Leslie McNair died during the carpet bombing that preceded the breakout when a short bomb found him and hundreds of other American soldiers in their foxholes.[19] McNair did not live to realize the full capabilities of the fully mechanized corps cavalry regiments, now groups, as they burst forth from the rupture created by Cobra and performed in the manner long anticipated during the interwar years. The loss of McNair was felt by the pro-mechanization faction of the cavalry community. MG Charles L. Scott, one of the pioneers of mechanized ground reconnaissance at Fort Knox during the interwar years, wrote, "Our armored units have lost a great mentor, director and leader who had the forsight [sic] and judgement [sic] to make provisions in our army for a greater proportion of armor to other arms than that of any other army in the world, including Germany."[20] With the way cleared, there was little time for maintenance as extended daylight during August found units driving off their maps and being forced onto tourist maps to find the way across France and Belgium while still having sharp encounters with the wounded but undefeated German units. Fuel shortages at the beginning of September 1944 and the respite they brought to the retreating Germans marked the end of the brief period of war of movement.[21]

Waiting for the opportunity to redeem himself, Patton stood ready with his Third Army. The Allies had long planned to liberate the Brittany Peninsula's ports, St. Malo, Brest, Lorient, and St. Nazaire, and this task fell to MG Troy Middleton and his VIII Corps. In addition to clearing Brittany and reducing hold-out German garrisons, Middleton secured Third Army's ever expanding southern flank. The 2nd CavGrp covered the area between Angers and Nantes along the Loire River, a particularly important task, because VIII Corps became overextended after the loss of the 4th AD, as the pursuit continued east and north of the Seine.[22] Throughout early August, the 2nd CavGrp had worked with or was attached to the 4th AD as it raced south and east. Now as VIII Corps settled down to reduce the channel ports, Third Army reassigned the 2nd CavGrp to MG Manton Eddy's XII Corps on 22 August 1944.[23] Although they had not fought in the manner experienced by the other cavalry groups in Normandy, the 2nd CavGrp experienced regular contact on the Third Army flank. Marching more than three hundred miles during the first weeks of August, although not moving forward of a corps or division in a reconnaissance role, the 2nd CavGrp lent security to the breakout by screening the exposed southern flank of Third Army.

The 2nd CavGrp had not been alone on the Brittany Peninsula during the early days of August 1944. The 15th CavGrp provided the principal maneuver units of Task Force A, which joined the 4th and 6th ADs in their initial thrust south from Avranches into the Brittany Peninsula.[24] BG Herbert L. Earnest commanded the collection of cavalry reconnaissance squadrons, a tank destroyer battalion, and combat engineers assembled at La Repas, twenty kilometers north of Avranches, on 1 August 1944. Task Force A followed the 6th AD, and believing the route clear, the mechanized cavalrymen moved at speeds as high as forty miles per hour as they slashed west from Avranches, as if on mounts "with stripped saddles."[25] The race ended abruptly when the lead platoon drove into an ambush at Baguer-Pican in the early morning hours of 3 August 1944. Out in front, as if leading a horse cavalry charge at one of the 1st CavDiv's interwar maneuvers, rode the cavalry group commander at forty miles per hour with the lead reconnaissance platoon in tow. Rounding a bend that concealed a well-constructed road block, the commander's car burst into flames at the first report of a German anti-tank gun. Surrounded on both sides of the road by ditches and hedgerows, it was impossible for the fast-moving column to avoid becoming easy targets. After the exertions of the remainder of the 15th CavReconSqdn, three survivors of the lead platoon escaped. Those not killed spent the remainder of the war in a German prisoner of war camp on the island of Jersey, until it was liberated in May 1945. Among the internees was the overzealous group commander.[26] Although its initial action was reminiscent of the mad dashes of the Louisiana maneuvers, the 15th CavGrp soon found itself conducting the same kind of operations that were carried out by the other cavalry groups. With the support of the assault guns and light tanks, they too performed extensive dismounted operations to clear pockets of German resistance and remained stationed in Brittany until February 1945.[27]

As Third Army poured through the breach at Avranches and VIII Corps drove south into Brittany, Patton's XV Corps drove southeast keeping pace with the First Army, which exploited the breakout. Early on, a thirty-five mile gap developed between VIII Corps and XV Corps, which Patton filled with the 106th CavGrp between Louvigne and Rennes.[28] When VIII Corps' advance pushed further into Brittany, the 106th continued to support XV Corps' exploitation by conducting reconnaissance east from Fougères to Mayenne and Laval, then on to Le Mans. Against light enemy resistance, the 106th participated in the XV Corps' single-week eighty-five mile dash, often leading or protecting the open flanks.[29] As XV Corps started north toward Alençon after it captured Le Mans on 8 August, the 106th CavGrp continued to range to the corps' right, where it met minimal resistance.[30] As the noose tightened around the German Seventh Army in the Falaise pocket, the XV Corps turned north and headed for the Seine. The 106th led this advance, covering in excess of fifty miles on 15 August and arriving at Dreux west of Paris by nightfall. The group remained west of Paris protecting XV Corps' flank until 27 August, when it was briefly attached to XII Corps for twelve days, carrying out reconnaissance as that corps advanced east. The group concluded its two-month, four-hundred-mile journey across France by returning to the XV Corps as it moved to take up position on the Third Army's southern flank.[31]

The 3rd CavGrp burst forth from the Norman hedgerows in August to accompany MG Walton H. Walker's XX Corps' drive to the east. Like the 2nd CavGrp, it too had seen extensive service, initially along the Loire River screening the advance of Patton's Third Army. At times it had moved in advance of XX Corps in a sector as wide as seventy miles, and by 20 August it reached the Seine and crossed over at Tilly on 25 August. With one squadron following the 7th AD and the other screening the corps' right flank, the group moved across the World War I battlefield at Chateau Thiery and crossed the Marne on bridges captured by the 7th AD. When XX Corps reached Reims, it reoriented east in a quest for a bridge over the Meuse at Verdun, where its four-hundred-mile foray culminated.[32] Third Army was out of gas.

> "But we halted at the frontier for a reason unforeseen,
> Not because of hostile action but for lack of gasoline."[33]

For the next five days, patrols from the 3rd CavGrp ranged between the Meuse and Moselle Rivers, operating platoon-sized units on captured fuel while pushing east to the limit of their capability. One platoon made it as far as Thionville, near the Luxembourg border, where for a few hours it managed to defuse an intact bridge over the Moselle that the Germans had rigged for demolition. Increasing German opposition forced the small cavalry force, equipped with only six jeeps and three armored cars, to give up the bridge. A former French marine who guided them to the bridge was equally helpful in leading the platoon to safety.[34]

Other patrols had similar experiences, but by 3 September even the 3rd Cavalry suffered the constraint of limited gasoline. The mechanized scouts made it to the Moselle's edge and in many places reported no German resistance. They obtained this information by fighting, not by "sneaking and peeking," but lacked the power to seize and hold critical crossings, a mission long envisioned in the early development of mechanized reconnaissance doctrine. By the time the flow of fuel resumed, the Germans had consolidated their defenses and forced a long campaign to reduce the fortress of Metz on the Moselle.[35]

Farther south, in XII Corps' sector, "Tiger Jack" Wood led the 4th AD over the Meuse River on 31 August. On 1 September, 4th AD was equally immobilized by the fuel shortages plaguing the entire Third Army. MG Wood siphoned fuel from the vehicles in his division, so that his 25th CavReconSqdn might continue its advance and at the same time maintain some pressure on the retreating Germans. Like the 3rd and 43rd CavReconSqdns to the north, the 25th CavReconSqdn moved against limited resistance to report that there was little to bar the advance of the division into Lorraine.[36] With enough gasoline to resume the offensive, the 25th CavReconSqdn again saw its assets allotted to the combat commands, where it served with distinction during the crossing of the Moselle on 11 and 12 September and in the subsequent encirclement of Nancy. Scouts accompanied the penetration, reaching as far as Arracourt by 14 September 1944.[37] Here, BG Bruce Clarke pushed Troop D of the 25th CavReconSqdn, which was attached to his combat command, even further to the east. Prepared to continue moving east before the Germans could re-form, the XII Corps commander, MG Manton Eddy, decided to use the force to help the infantry divisions consolidating the corps' gains on the east bank of the Moselle. The Germans took advantage of the reprieve to organize their reeling forces for a counterattack intended to envelop the exposed Third Army penetration.[38] The first wave of GEN Hasso von Manteuffel's Fifth Panzer Army's counterattack fell on the 2nd CavGrp.

The 2nd CavGrp, having been attached to the XII Corps since 20 August, moved more than three-hundred miles to resume its role protecting the Third Army's exposed southern flank, which now belonged to Manton's corps. By 29 August the group had shifted from the corps' flank to its front, conducting reconnaissance as XII Corps advanced first on the Meuse and then on the Moselle Rivers. Like the 3rd CavGrp in a sister corps to the north, the 2nd CavGrp continued to push toward the Moselle when the rest of the corps' units ground to a halt on 1 September. Using captured German fuel in their jeeps, the mechanized cavalrymen reached the Moselle on 2 September and continued to patrol in the vicinity where the Madon River joined the Moselle until the corps could resume offensive operations.[39]

The 2nd CavGrp's position on the XII's flank was particularly important. The Allies landed in southern France on 15 August, and LTG Alexander Patch's Seventh Army was moving north. The 2nd CavGrp sat astride the German route of escape and on 6 September, the 43rd CavReconSqdn, in conjunction with the 696th Field Artillery Battalion, blocked a retreating German column, killing 151, destroying 30 vehicles, and capturing an additional 178 enemy soldiers.[40] In conjunction with CC B, 4th AD, the group gained control of Lunéville, but based on the reports of prisoners captured on 17 September, 2nd CavGrp commander COL Charles Reed became convinced that the Germans were preparing to launch a major counterattack and requested tank destroyer assets from the XII Corps headquarters. His request was denied.[41]

The next morning, the advanced guard of the 111th Panzer Brigade struck the screen line established by the 42nd CavReconSqdn. This was part of Field Marshal Hasso von Manteuffel's Fifth Panzer Army's counter-offensive across the salient created by the 4th AD's penetration. COL Reed was wounded when rounds fired from his assault guns bounced off the armor of the advancing German tanks. The 42nd CavReconSqdn fought a spirited delay that allowed the remainder of the 2nd CavGrp to retreat to Lunéville, and by 1200 that day, elements of the 4th AD advanced with the added support of tank destroyers to maintain the defense at Lunéville. Preserving XII Corps' flank had been costly. The 42nd CavReconSqdn's commander died in the heavy artillery fire that characterized the fighting, and the squadron lost twenty-seven of its vehicles.[42] One journalist, following the rapid advance of Patton's Third Army, observed that the "2nd CavGrp, the unit which made a story-book dash across France and always moved so fast it never had to dig foxholes," finally had to slow its pace when confronted with an onslaught of German tanks at Lunéville.[43] The attack at Lunéville should have served as a warning about the inherent limitations of cavalry groups, but other factors set the conditions for an even bigger disaster a few months later.

Patton enhanced his ability to command and control his forces through the creation of his own organization, the Army Information Service. Also known as Patton's "household cavalry," this helped Patton keep track of forces on a dispersed front. His household cavalry may have mirrored Field Marshal Montgomery's "phantom" network of reporting, as suggested by historian Martin Blumenson, but it also reflected the interwar use of early mechanized reconnaissance units serving traditional horse cavalry units.[44] More in line with the argument that the Army Information Service was not necessarily an idea borrowed from the British, at least in terms of word choice, Hasso von Manteuffel cited Patton's use of "cowboy-aides" and "saddle-orders," something he was familiar with as a cavalryman himself.[45]

While still in England, the 6th CavGrp acquired extra radios to fulfill its role as the Army Information Service, and it conducted intensive training to become the eyes and ears of an army that operated from Brest, on the Atlantic coast, to the Moselle River, on the doorstep of Germany.[46] Patton activated his household cavalry when the Third Army arrived on the Continent. As the detachments joined their units, roughly a reconnaissance platoon per division, they carried a letter from Patton explaining that they were not there to comment unfavorably on the unit's performance, but to ensure a secure line of communication between the supported unit and army headquarters.[47]

By 15 August, Patton committed fifteen detachments, which consumed an entire cavalry squadron, leaving one squadron from the 6th CavGrp to perform missions for Third Army. Army Information Service planners had not anticipated the need for a higher headquarters, the cavalry group, to provide centralized command and control for the many detachments. Initially there were problems with the radios, which remained untested while the units operated under a veil of radio silence. Motorcycle scouts and jeep couriers proved the most reliable means of communication until wire could be strung.[48] The sheer volume of radio traffic forced the detachments to shift their efforts from monitoring radio nets to having scout platoons obtain the latest information directly from the senior commanders or from the front. The cavalry group headquarters took these reports directly from the field to conduct extensive battle tracking with situation maps and copies of orders to better direct their reconnaissance detachments to the action so as to gather the most up-to-date information.[49] This allowed Patton to have nearly a complete picture of an exceptionally fluid situation. Moreover, Patton's picture was current because the Army Information Service eliminated the time lag required for a message to travel from a division operating at the front through the corps headquarters and on to army headquarters. With the Third Army headquarters never remaining at any single location for more than five days during the open-field running that characterized August 1944, the Army Information Service played a vital role in maintaining contact between senior and subordinate units.[50]

The front stabilized on 15 September when Third Army linked up with Seventh Army, closing in from the south. As units established wire networks, the importance of the work carried out by the scouts diminished. Rather than gathering information for Patton, they passed information to the subordinate unit commanders about the "broad picture" and how their corps or division supported the overall mission. The 6th CavReconSqdn rotated with the 28th CavReconSqdn on 21 September. The new squadron operated thirteen detachments in support of ten divisions and three corps. This allowed Third Army to maintain contact with VIII Corps operating on the Brittany Peninsula, four-hundred air miles from the Third Army HQ at Châlons-sur-Marne.

As the Third Army prepared to break out again in November, Patton directed the 6th CavGrp commander to reorganize the "household cavalry" so that it could be run by a single cavalry reconnaissance squadron, which allowed him to build a task force around the 6th CavGrp headquarters. The 6th CavGrp with an attached battalion of Army Rangers and a company of tank destroyers and engineers joined XX Corps and attacked dismounted toward L'Hôpital and the forest of Karlsbrunn on 2 December 1944.[51] Even Patton, strapped for manpower, could no longer avoid committing his last reserve of highly mobile troops to the infantry fight, and the entire Army Information Service discontinued its service to Third Army in December 1944.[52] Between August and October, the 6th CavGrp suffered fifty-eight casualties, roughly split between combat and vehicular accidents.[53]

Patton's Third Army fulfilled the interwar expectation of mechanized ground reconnaissance units in a variety of ways. Although they had been used on exposed flanks extensively, they also led rapid advances. Even if Patton's unorthodox use of the 6th CavGrp reflected the worst interwar abuses of the mechanized reconnaissance units in horse cavalry units, it substantially contributed to his ability to command and control his forces, especially during the halcyon days of August and early September. To the north, LTG Courtney Hodges left the employment of cavalry groups up to his corps commanders.

While Third Army raced off in every direction after Operation Cobra, First Army had to repel the German counterattack at Mortain on 7 August, encircle the German Seventh Army in the Falaise pocket, liberate Paris, and then race for the German border. The 4th, the 102nd, and the 113th CavGrps continued to support the VII, V, and XIX Corps respectively. During the fighting in early August, the cavalry groups' service was little different from the operations they conducted prior to the breakout. Like the cavalry groups in Third Army, the First Army cavalry groups came into their own once the Falaise pocket closed on 21 August.

The Twelfth Army Group commander, GEN Omar N. Bradley, believed that "for all its past glories, Paris represented nothing more than an inkspot on our maps to be by-passed as we headed toward the Rhine," but he could not avoid the "city of light" even if its four million inhabitants represented a major logistics burden.[54] GEN Philippe Leclerc started petitioning Patton, on 15 August, only fifteen days after being committed to combat on the Continent. He wanted his French 2nd AD to have the honor of entering Paris first.[55] Having "liberated and celebrated" across France since he arrived on 1 August 1944, Leclerc was determined not to miss the biggest party of all.[56] Taking matters into his own hands, and without orders, Leclerc dispatched an advanced party of seventeen tanks, ten armored cars, and two platoons of infantry on trucks toward Paris on 21 August, but the French were about to cross paths with an American cavalry group and would have to share some of the glory of liberating Paris.[57]

The 102nd CavGrp arrived in June 1944 in time to share in the hedgerow fighting while supporting V Corps. With the Falaise pocket closed, the group abandoned its previous role during the breakout—of maintaining contact with adjacent units on the corps' flanks—and assumed the mission of leading the 4th ID to Paris. Troop B of the 102nd CavReconSqdn joined Leclerc's French 2nd AD, perhaps as the corps commander's way of keeping track of the French. The 38th CavReconSqdn, encountering limited resistance, secured all the bridges on the Seine and reached Notre Dame Cathedral on the morning of 25 August.[58] Troop B, 102nd CavReconSqdn, raced to Paris with the French 2nd AD "at 50 miles an hour" with "[French] soldiers and ladies drinking in the vehicles."[59] Leclerc ordered the 1st Syrian Spahis a chance to lead the way. With roots reaching back to horses and camels in Syria when World War II began, the 1st Spahis served the same function as a cavalry reconnaissance squadron and used much of the same equipment.[60] Only days later, the Spahis escorted DeGaulle's triumphant return entourage.[61] The lightning dash to Paris, led by the mechanized ground reconnaissance units drawn from two nations but organized along lines of American design, was instrumental in retaking the city.

For the men of the 102nd CavGrp, the stay in Paris was all too brief, as they drew the task of moving forward toward the Meuse River.[62] After V Corps paused while VII Corps changed its axis of advance, the 102nd CavGrp reconnoitered the advance of the 4th ID until it ran up against the Siegfried Line near the Belgian villages of Manderfeld, Holzeim, and Krewinkle on 14 September, just south of where the 4th CavGrp stopped as it too hit the West Wall. All along the way there had been sharp fights with withdrawing Germans, but the group traveled from Paris to the German border in roughly two weeks.[63]

The 4th CavGrp played a vital role in maintaining contact between First Army's VII Corps and Patton's Third Army during the rapid advance in August and early September. Rather than leaving behind a division to fill the growing gap after First Army reoriented VII Corps' axis of advance toward Mons, Belgium, MG "Lightning Joe" Collins used his 4th CavGrp, reinforced with a battalion of light tanks, motorized artillery, tank destroyers, infantry, and three companies of engineers, to fill the growing void. Reaching the Meuse on 3 September, the 4th CavGrp used the river as a natural obstacle and screened the right flank of VII Corps from Mézières to Rocroi.[64] The 4th CavGrp filled this crucial gap until V Corps could be brought back into the line south of VII Corps, thus linking First and Third Armies.

As V Corps took up position between VII Corps and Third Army, the lack of German defensive measures allowed the cavalrymen to cover great distances, up to the point of crossing the Meuse between Dinant and Givet. Once in Belgium, however, they encountered stiffening resistance that forced them to the corps' flanks until their advance culminated in the shadow of the Siegfried Line on 14 September 1944. There they tested the disposition of the Germans and determined the contours of the defense in the vicinity of the Elsenborn Ridge and the small villages of Rocherath, Krinkelt, and Bullingen.[65] Maintaining contact with V Corps to the south, the 4th CavGrp secured the southern flank of VII Corps as it battled its way into Aachen from 16 September until 2 October, thus allowing GEN Collins to concentrate his combat power. V Corps assumed the extensive sector held by the 4th CavGrp as it shifted north.[66]

The 113th CavGrp operated forward of XIX Corps during its northward dash across Europe beginning 13 August 1944 near Mortain. Ordering the group to "fan out ahead of the advance in a fast bold run, keeping well ahead of the skirmish line," Corlett had the 113th precede the advance of the 30th ID, while the 82nd ARB moved forward of the 2nd AD. Starting on 19 August, the 113th covered 106 miles as XIX Corps attempted to cut off the German forces escaping from the Falaise pocket. Aside from reconnoitering the advance of the 30th ID, the group captured small objectives with dismounted attacks supported by the light tanks and assault guns in each of the squadrons.[67] COL William S. Biddle's cavalrymen pursued the Germans across the Seine at St. Germain on 29 August and from 1–2 September led the 30th and 79th IDs as the 2nd AD moved on their left flank. During the first two days of September, the group gobbled up 150 miles and crossed the German border on the afternoon of 2 September.[68] Then, like all other American forces operating in Twelfth Army Group, the gas ran out.

Out of fuel, Corlett visited Biddle's headquarters on 4 September and ordered him to turn his cavalry group due east and clear a twenty-five-mile-wide swath of Belgium all the way to the Prince Albert Canal, approximately 125 miles distant.[69] On 5 September, the 113th CavGrp executed what was later described as a "perfect cavalry mission," as mechanized reconnaissance men moved ahead, days in advance of the corps they supported. The group was finally able to achieve the operational depth envisioned by writers of interwar mechanized reconnaissance doctrine, but only because it lacked fuel to move along the five routes the cavalrymen were clearing. Lack of fuel was not the sole problem the group faced; the rapid advance across Europe had worn the tracks off of Biddle's light tanks, so he advanced them on 5 September with nothing more than wheeled vehicles.[70]

The cavalry group experienced little resistance as it liberated Belgium, ending its first day near the Waterloo battlefield. Belgian "forces of the interior" dealt with German prisoners, and the cavalry group bypassed pockets of resistance as it plunged farther east toward the German border. All Corlett could do while his cavalry group liberated Belgium was listen to the reports coming over the

SCR 399, long-range radio.[71] By the evening of 7 September, the 113th CavGrp reached Hasselt in the north and St. Trond in the south of its assigned zone, only a few miles short of the Prince Albert Canal. With enough fuel to resume movement, GEN Corlett ordered the group to move its northern squadron, the 125th CavReconSqdn, south as the 82nd ARB, with full fuel tanks, raced ahead of the rapidly closing 2nd AD, which was moving up on the cavalry group's left flank. By the end of the day, the entire corps closed on the Prince Albert Canal.[72] Against crumbling German resistance, the 113th CavGrp with its wheeled vehicles raced ahead of the heavier forces, immobilized for lack of fuel as if on a maneuver. Days later, Corlett drew on the same mobility to find a way across the water barrier to his front.

To the south of XIX Corps, VII Corps was able to secure an intact bridge over the Meuse at Liège. With two companies of attached infantry riding on trucks, a tank destroyer battalion, and two companies of engineers, COL Biddle took his cavalry group across the corps boundary, drove thirty-five miles to Liège where he crossed the Meuse River, and then proceeded north with squadrons abreast. With infantry platoons riding assault guns and assault guns attached to the leading reconnaissance squadrons, the cavalry group turned the Germans out of their positions. This allowed the 30th ID to construct a bridge at Visé. With a bridgehead over the Meuse in the XIX Corps sector, the 113th CavGrp advanced on the left flank of the 30th ID as it advanced into Holland.[73] When the British pulled out of the line to the left of XIX Corps in preparation for their ill-fated drive on Arnheim, a fifty-mile gap developed between Corlett's corps and the British. The British supplied a Belgian brigade that Corlett augmented with the 113th CavGrp and an infantry battalion, thus beginning the type of defensive operations characterizing the remainder of the 113th CavGrp's stay in Holland.[74] Augmented as it was, the 113th CavGrp continued to turn the German forces out of their positions with bold movement, more than with fighting prowess. At the opposite end of the Allied line, in the south of France, a similarly augmented cavalry reconnaissance squadron carried out an even bolder maneuver.

The controversial decision to invade southern France went forward against Prime Minister Winston Churchill's wishes, on 15 August 1944. Operation Anvil, or Dragoon as Churchill preferred, featured MG Lucian K. Truscott, a veteran cavalryman, improviser, and amphibious-landing expert, as commander of the VI Corps. Unlike at Anzio, what ensued in southern France was a "wild cat" and not a beached whale. Within days of landing, Truscott unleashed a reinforced mechanized cavalry squadron on a plunge into the German rear. Not focused on reconnaissance, Task Force Butler sought nothing less than to cut off German Army Group G.[75]

Truscott could not plan on the use of his floating armored reserve for a deep thrust inland. The reserve, a combat command from a French armored division, waited offshore, and LTG Patch, Seventh Army commander, insisted upon its return to French control by 19 August for the French drive on

Toulon. Therefore, Truscott had to create his own fast-moving, hard-hitting unit to envelope the Germans.[76] Truscott built Task Force Butler, named for his assistant corps commander, around the 117th CavReconSqdn, a corps asset. The 117th immediately contributed not only its mobility and combat power, but also the staff and command and control apparatus upon which to attach the other assets destined for service with Task Force Butler. And this was accomplished without cutting into VI Corps staff and command and control assets (radios).[77] When activated on 17 August, the remainder of the task force included: an armored field artillery battalion, independent tank battalion less two companies, an infantry battalion, a tank destroyer company, an engineer company, a medical company, and a quartermaster truck company to move the infantrymen.[78]

Task Force Butler advanced north on 18 August toward its first intermediate objective, Sisteron. Having had limited combat experience in Italy, the 117th moved tentatively but picked up speed with Butler's encouragement. By the end of the first day, the cavalrymen captured German LXII Corps' commander, LTG Ferdinand Neuling, and his staff and advanced as far as Digne. Light aircraft assisted the rapid advance by finding bypasses for destroyed bridges and maintaining contact with VI Corps headquarters— now well beyond radio contact—until one of the same light airplanes flew in a long-range radio for Task Force Butler's use. The Maquis, local resistance fighters, established a number of roadblocks oriented on avenues of advance from the Route Napoléon to prevent any penetrations of Task Force Butler's line of communication. Trucks carrying nothing but gasoline ensured that the advance deep into the German rear continued on 19 August.[79]

For the next two days, Task Force Butler, with the 117th CavReconSqdn leading the forces, pushed north, capturing Gap on 20 August, and then raced more than half way to Grenoble. The 36th ID, following in the wake of Task Force Butler, oriented on Sisteron. On 21 August, Truscott ordered Butler to change directions, "go west, young man, go west," toward the heights that dominated the German escape through Montélimar.[80] Although the main body of Task Force Butler was nearly one hundred miles from its new objective, Troop B arrived at a position that gave a full view of the escaping German forces by the afternoon of 21 August.[81] The remainder of the 117th CavReconSqdn and Task Force Butler closed rapidly on Montélimar, where during the next two days, they fought alongside the Maquis and waited for the 36th ID. When the division arrived and disbanded Task Force Butler, the 117th took up a position along the Roubion River, the scene of heavy fighting when the Germans attempted to turn the American flank.[82]

Task Force Butler, even with the support of the 36th ID, lacked the power to close the German route of retreat. Inspired by the fighting in Italy, Truscott managed, however, to inflict serious damage and maintain pressure on retreating German forces, which enabled the Seventh Army to rapidly move up the Rhône Valley and tie in with Patton's Third Army on 11 September 1944.[83]

Men from Troop B were able to join hands with troopers from the 1st Syrian Spahis, the same unit that led the French 2nd AD into Paris, on 18 September as the 117th CavReconSqdn continued to assist the advance of VI Corps.[84] As part of Task Force Butler, the corps reconnaissance squadron had moved 235 miles in four days and fought heroically against superior German forces. Although it had received a number of attachments and did not focus on reconnaissance, the 117th CavReconSqdn's contribution to Task Force Butler offered one of the most exciting examples of what interwar mechanization advocates had hoped to realize. The 117th CavReconSqdn, like the other squadrons operating to the north, was doing far more than just reconnaissance.

As the front stabilized in Lorraine and along the Siegfried Line, a general pattern began to emerge in regard to how corps and division commanders used their mechanized ground reconnaissance assets. Cavalry squadrons became interchangeable with infantry regiments and groups and, at times, with divisions. Screening had long been a cavalry function, but it presumed proximity of larger forces within supporting distance, like the 2nd CavGrp depending on the 4th AD to come to its assistance at Lunéville. Now mechanized cavalry units took up their own defensive sectors and in some cases were committed to offensive operations as dismounted infantry.

Just as the augmentation of cavalry groups, sometimes called task forces, was not uncommon during the offensive phase of Allied operations during the first six months in Europe, the corps commanders applied the same concept to defensive operations. As the VII Corps front grew to thirty-five miles, MG Collins assigned the 4th CavGrp twenty miles of his corps' responsibility. Collins provided COL Joseph Tully, "a great cavalryman and fine fighter," additional artillery, tanks, and a battalion of infantry to round out what had grown into a "small corps."[85] This use of the cavalry group in an economy of force role allowed Collins to narrow his active front to fifteen miles. Relief of responsibility for portions of the 4th CavGrp's sector in September prevented the cavalry group from becoming overstretched in static positions, but in many respects, the worst was yet to come.[86]

The 3rd CavGrp saw offensive action, too. MG Walker formed Task Force Polk on 3 November 1944 from a battalion of heavy field artillery, a battalion of regular field artillery, two tank destroyer battalions, and an engineer battalion, with the 3rd CavGrp as the centerpiece of the robust organization.[87] Task Force Polk's first mission was to secure the town of Berg and the commanding hills around it, which threatened XX Corps' planned crossing of the Moselle in its efforts to reduce the defensive complex of Metz. Lightly held, COL Polk elected to use a single platoon of dismounted cavalrymen to seize the hill. MAJ George D. Swanson, executive officer of the 43rd CavReconSqdn, led the dismounted attack that briefly gained control of the hill. A German counterattack swept the small American contingent off the hill the next day. A combined-arms attack drawing on many of the attachments now found in Task Force Polk retook Berg on 5 November. Task Force Polk then patrolled a twenty-mile sector along the

Moselle, while the remainder of the corps prepared to cross the river.[88]

On 13 November, XX Corps ordered Task Force Polk to follow the advance of the 10th AD already across the Moselle. Task Force Polk crossed the Moselle and took up a position on the left flank of the 10th AD. Beginning on 16 November, the task force attacked north, protecting the flank of XX Corps as it ascended the Sarre-Moselle triangle. Shifting its additional assets between squadrons, the group advanced by bounds. Once the cavalry squadron, using all its assets plus the task force assets, had seized an objective, it halted as the group's other squadron employed all of the additional support to seize its objective. The leap-frog advance ended on 19 November when Task Force Polk ran into the Siegfried Line.[89] Like the 4th and the 102nd CavGrps, which first encountered the West Wall in mid-September, the lightly equipped cavalry group was incapable of further forward progress. Its sister group in Third Army, the 2nd CavGrp, fared no better.

From north to south along the extended American front, cavalry groups served every American corps then operating in Europe. A handful had experienced the struggles in the Norman hedgerows, all had experienced the exhilaration of the breakout and race across Europe, and now all experienced, to some degree, the frustration of being limited to what was primarily an infantry role on the periphery. These experiences generated a number of observations about the techniques, doctrine, and equipment used to move across Europe during the first six months.

From afar, the former chief of cavalry, John K. Herr, kept abreast of the situation in Europe through his son-in-law, BG Willard "Hunk" Holbrook, serving with the 11th AD, but still waiting to be committed to action. Having visited with other former cavalrymen now serving as armored division commanders, Holbrook expressed enthusiasm that these men were using their divisions "much like our old cavalry" with the principal exception being the advantage of their "tremendous firepower." Probably much to Herr's liking, Holbrook spoke of the "present 'cavalry'" being completely dedicated to reconnaissance as "not very satisfactory."[90]

Holbrook, yet to see combat, was not the only man dissatisfied. Two young mechanized cavalry officers, who had seen combat, took exception to an observation expressed by BG (Retired) Hamilton S. Hawkins, in the July–August edition of the *Cavalry Journal,* which suggested that the operations in Europe lacked the participation of "cavalry." In the September–October edition, Hawkins freely admitted that these men had been fighting in Europe and then proceeded to list a number of other actions in the history of the branch in which troopers fought without their horses. He went so far as to blame the prolongation of the American campaign in the Philippines as a result of horse cavalry fighting without their mounts. In 1944, Hawkins was equally "convinced that large forces of cavalry, using horses, could, in combination with mechanized forces, shorten the war there and save thousands of lives."[91] He believed the units that had just raced across France and now confronted the

German West Wall would be better served with the support of "strong horse cavalry units." After all, the Russians were still using horses.[92]

From their conceptualization, visionaries and practical men intended mechanized reconnaissance units to fill the gap between the leading edge of ground forces and the planes that had ranged ahead of the action since World War I. Combat forced improvements in the realm of air-ground coordination, and by July the IX Tactical Air Command directly supported troops on the ground. Operations Order No. 90, Advance Headquarters, IX TAC, 20 July 1944, provided for "Armored Column Cover," consisting of a "four ship flight" flying in support of the moving columns on the ground. These flights passed vital reconnaissance information and attacked "any target which [was] identified as enemy," while focusing their efforts on "the terrain immediately in front of the advancing columns."[93] This set the stage for the rapid advance across France and Belgium, the war of movement long envisioned, that demanded the creation of the mechanized reconnaissance forces to fill the space between the leading edge of the main force columns and the enemy. Commenting on the rapid advance of "Tiger Jack" Wood, an observer noted that "the cub planes [were] worth their weight in gold," moving at the front of the armored spearheads streaking across France.[94] "P. Wood's" only complaint was that as the division commander he needed a faster plane.[95] Planes had also played a vital role in the rapid advance of Task Force Butler in the south of France.

Cavalry group scouts learned to tune the radios in their M8 armored cars to the frequency of the artillery observers flying above. These same observers, at times, directed the advance of the mechanized reconnaissance men operating below. Divisional reconnaissance troops could now expect almost immediate close air support, if it was available, by directly contacting the air-liaison party at the division headquarters. The process was somewhat more complicated for the cavalry groups that might have to relay their request through the divisional cavalry troop or squadron of the closest division to which they were assigned.[96] In the opinion of an armored division combat commander, the best support resulted when the Army Air Force attached pilots to marching columns, because "they were able to talk the language of the pilots in the air and talk them onto the targets."[97]

Despite the general improvement, sharp contacts with retreating German forces beyond Paris continued to result in losses. Lead vehicles rounded corners and were "nailed by an 88." GEN I.D. White later remarked:

> I believe now with our scout helicopters that we probably could have avoided direct meeting engagements with those elements and shelled them with artillery fire and eliminated them without loosing our lead personnel and vehicles. We did not use our light artillery observation planes as much as I think we should have for scouting. One reason was because the Germans had pretty effective low level antiaircraft defense and it wasn't particularly healthy to fly low enough where you would have to fly to observe and locate these weapons.[98]

While there were marked gains in air-ground cooperation, these gains were accomplished by commanders and staffs, not by changes to the organization and equipment found in the mechanized ground reconnaissance agencies. From Army Ground Forces personnel all the way down to the common trooper, there were calls for action to improve or modify existing equipment in the field to better the performance and survivability of the men in combat.

Limitations of the M8 Armored Car came to the fore again during the first six months of fighting in France and Belgium. The car still could not absorb the blow of a Teller mine, but with modifications, such as an additional steel plate welded to the floor, the crew did have a better chance of survival.[99] Like jeeps, armored cars not only received armor modifications, they also received modified weapons mounts and additional storage racks for ammunition and personal items. Still, armed with only a 37mm cannon, the M8 was of little use against the tanks and heavy pillboxes being encountered at the end of September's sprint across Europe.[100] For all its shortcomings, however, the M8 provided a relatively safe platform for the radios essential for requesting support and passing information. The armored car's road speed and greater fuel economy allowed the cavalry groups to move rapidly and farther as the heavier-armored divisions ground to a halt in early September for lack of gasoline. Limited to .50-caliber machine guns or 37mm cannons, scouts relied on other equipment in the squadron to deal with more robust threats. This was especially true for the men who were supposed to be riding in the jeeps and armored cars, because they often found themselves attacking on foot like their brothers in the infantry.

The assault guns found in Troop E of each cavalry reconnaissance squadron remained popular with commanders because of their ability to shoot indirect fire from defiladed positions. Their continued presence in the organization guaranteed cavalry commanders a limited indirect-fire capability when field artillery was not attached to their groups by the division or corps they were supporting. The 117th CavReconSqdn used 105mm M7 assault guns rather than the smaller 75mm M8 assault guns found in the other mechanized cavalry units, and they thought the advantage obvious. Having worked with attached field artillery battalions, the advantage was becoming obvious to other commanders as well. Not only did commanders begin to express a desire for a larger-caliber assault gun, they also expressed concern that they might lose their assault guns once they fielded light tanks with comparable cannons. For this reason, they emphasized the important indirect-fire capability the assault guns lent their units.[101] Given the large amount of dismounted action they had performed during the first six months of war in the European theater, both offensively and later in static defensive positions, the indirect-fire capability afforded by the assault guns proved critical to the cavalry reconnaissance squadrons.

Light tanks continued to gain attention also. During the Normandy phase of the campaign, the commander of the 121st CavReconSqdn commented on the survivability of the light tank, remarking, "Mines are plain hell and don't let anyone tell you a light tank can take a Teller mine. *They cannot.*" Just as units in North Africa and Sicily, the units now in Europe tried to modify their light tanks and other vehicles with additional armor.[102] But mines were not the only factor limiting the mobility of Allied forces in June and July 1944. The greatest inhibitor was the *bocage*. Sergeant Curtis G. Culin, 102nd CavReconSqdn, developed the first device to be affixed to the front of a tank that allowed some mobility through the hedgerows. Although the device was developed in the early part of July, it remained secret until five-hundred additional "Rhinoceros" tanks could be created for use during First Army's breakout, Operation Cobra. It was Sergeant Culin's "American ingenuity" that allowed all armored forces operating in Normandy to "surmount a difficulty" the planners had not anticipated.[103] This was also in keeping with the traditions of the men who had filled mechanized reconnaissance units from the very beginning right down to the time of building their first vehicles up from the chassis in the late 1920s.

The French had little use for the light tanks the United States issued them for use by the 1st Spahis. Aside from being "very noisy, lightly armed and armored," the M5s were notorious for catching fire "at the least impact." The son of a commander of one of the combat commands in the French 2nd AD burned alive in his light tank while fighting in the Forêt d'Ecouves in mid-August before the liberation of Paris. The 1st Spahis lost the majority of their light tanks in combat during August and elected to replace them with medium tanks.[104] At the same time the 1st Spahis were losing their light tanks, the 4th and 6th ADs had adopted the practice of placing their own medium tanks on point, which one observer noted, had "paid them dividends."[105]

American reconnaissance units did not field medium tanks, but started replacing their M5A1 light tanks with the M24 light tanks during the fall of 1944. As far as the 12th Army Group's armored section was concerned, the M24 light tank with its larger 75mm cannon could not be substituted fast enough for the under-gunned M5A1 light tank. War Department officials promised to replace losses with the newer tank, because the priority was first to equip those units still in the process of getting from the United States to the European theater.[106] The M24 proved the "premier reconnaissance tank in all armies," according to those who had probably spent the most time putting together the empirical data for the U.S. Army's effort in Europe, namely, the armored section of the 12th Army Group staff. The M24's mechanical reliability even offered hope that its chassis might serve as the starting point for a new generation of armored infantry carriers.[107]

For all the shortcomings of the light tanks and the inability to field better tanks faster, calls from the field demanded yet another role for light tanks, command and control. Only weeks into combat, COL Reed, commander of the 2nd CavGrp requested a change to the table of organization and equipment for his group headquarters. Specifically, he wanted light tanks included in the headquarters, so that he might be able to accompany "his assault guns and tanks into enemy lines." He believed the "presence of these tanks would greatly increase the speed and efficiency of operation of the group commander and his staff."[108] Much had changed in the cavalry since the interwar years, when commanders willingly abandoned their vehicles for horses. Now not only did the commanders fully appreciate the command and control capabilities afforded them by radio-carrying vehicular platforms, they also sought maximum protection as they led their groups near the front.

In the realm of command and control, there had been a number of modifications and improvisations less dramatic than Patton's "household cavalry." Many of the maps that had facilitated the race across Europe were common roadmaps purchased in England before the invasion. As the advance continued, Americans raided gas stations along the route of advance to fill the bellies of their iron ponies.[109] Motorcycles continued to enjoy some utility for aiding in the command and control of mechanized columns and far-flung corps and cavalry groups during the breakout. One of the pioneers in the field of mechanized ground reconnaissance, BG I. D. White, in command of a combat command in the 2nd AD, often operated from the buddy seat of a motorcycle. This allowed him to get around columns on the narrow Norman roads. He also believed, mistakenly, that he could avoid setting off land mines by riding on the motorcycle rather than in a jeep; plus, it was easier to dismount and get into the ditch when rounds started to fall.[110] The 6th CavGrp, operating Patton's household cavalry, also made extensive use of motorcycles, until cold weather set in and they transitioned to jeeps.[111] Perhaps the decision to use jeeps rather than motorcycles was also influenced by the number of non-combat casualties suffered by the 6th CavGrp in vehicular accidents while carrying out their duties.

Most reports from the field expressed satisfaction with the radios then in use. Rainy weather in France had limited the range at times, and there was a reminder that the radio operators themselves should know more about their equipment, so that they might effect minor repairs.[112] French Spahis, the mechanized reconnaissance agency of the French 2nd AD, were thrilled with the inclusion of radios at every echelon in the reconnaissance organization. Radios represented "an important change from the desert days," when they depended on "different color pennants for signaling!"[113] Fighting along the Siegfried Line forced mechanized reconnaissance units to call for the inclusion of switchboards and additional communications wire. The requirement of an organization designed for mobility to maintain wire communications to higher headquarters, subordinates, and adjacent units was more than the cavalry group headquarters organization could handle.[114] The importance of radios to mechanized reconnaissance, long recognized in the United States, was one area in which mechanized ground reconnaissance units continued to do well, even in combat.

All the dismounted fighting called for the addition of weapons systems not associated with the interwar cavalry in any respect. One group commander, like many others holding large sectors of the Siegfried Line, saw his unit regularly committed to dismounted patrolling. He proposed the addition of sniper rifles as a means to "keep the Jerrys in their holes during the day time."[115] Having already acquired a number of Browning Automatic Rifles (BAR), the same commander suggested this infantry weapon was needed in the mechanized cavalry.[116] Not only was there beginning to be a call for infantry weapons, there was also a call for the addition of entire rifle troops and platoons to the existing organization.[117] Some units had gotten extremely creative in their search for additional personnel.

When it crossed the beaches at Normandy to begin its drive on Berlin, the 1st Reconnaissance Troop brought along an extra rifle squad above the troop's authorized strength. It used even more creative means to maintain the strength of its organization as it advanced across Europe by adding Dutch, French, and Belgian men to the troop. These men were mostly used to man the machine guns and, occasionally, to drive, leaving the radio operation, gunning, and vehicle commanding to the Americans.[118] The additional rifle strength, both the unauthorized Americans and the foreign tag-alongs, was probably useful to the troop throughout the remainder of the European campaign. The local nationals certainly helped by manning the vehicles and providing an increased ability to converse with the locals and hence gather more information, the primary purpose of reconnaissance. When the troop was not conducting reconnaissance or providing security to the division's flank, it often itself employed as an infantry unit.[119] On these occasions the additional rifle strength would have been particularly important to make up for the mounted unit's limited dismounted capability. The experience of the 1st Cavalry Troop was not unique. Many other mechanized cavalry units avoided contact under unfavorable circumstances, especially during the fast-moving days in August and early September, by heeding the warnings of people who relished being liberated from Nazi oppression. In the realm of reconnaissance, cooperative citizens were an incredible force multiplier, but not one that could be counted on as the fight moved to Germany.

The commander of the 121st CavReconSqdn sounded like a rifle battalion commander when he informed an Army Ground Forces observer in August 1944 that there needed to be more emphasis placed on the use of mortars, more training on infantry tactics, more fighting in cities, more reconnaissance by fire, more use of white phosphorus for clearing houses and buildings, and more standard operating procedures for immediate action on contact.[120] Another report, filed in December, also spoke of the static as opposed to mounted performance of cavalry missions in Europe, when it reminded those back in the States to "learn early to dig foxholes and dig them *deep* as it is too late after the artillery begins to fall." Perhaps a jibe at the infantrymen that the troopers now found themselves serving alongside in the foxholes, the report offered that it was "not necessary to eat out of tomato cans, wear muddy clothes and fail to shave to be a good fighter . . . cavalry tactics are sound."[121]

Mechanized cavalrymen had reclaimed their full-fledged combat identity, but seemed to want to make sure they were accomplishing the missions thrust on them with the same style that had distinguished them in the past. Horses or no horses, they were still warriors with more class than foot-sloggers.

Combat command leaders in the armored divisions had mixed feelings about the utility of the cavalry reconnaissance squadrons assigned to their divisions. Since squadrons rarely operated as a unit, it is not surprising that commanders were critical in regard to the combat characteristics of the reconnaissance troops they received. BG Truman E. Boudinot, CC B, 3rd AD, remarked that the current mechanized cavalry reconnaissance squadrons had "no combat power," and because one had to "attack to get information," leading the attack with light vehicles was "suicide."[122] A commander with the 6th AD saw the only utility of such squadrons in finding alternate routes, because to place them on the main axis of advance was sure to cause a delay for the advancing friendly force.[123] Commanders in the 5th AD echoed these views, seeing some utility for the race across France against light resistance, but concluding that the "present reconnaissance squadron is not a combat unit."[124] With a full complement of mechanized infantrymen, light and medium tanks, and the full weight of the division's artillery assets to back them, it was easy for the commanders in the combat divisions to discount the capabilities of the relatively lightly equipped cavalry squadrons.

Often held in low esteem, reconnaissance squadrons in the armored divisions may have reflected another problem, poor training. As late as August 1944, the commanding general of the Armored Center at Fort Knox, Kentucky, concluded that "almost without exception, inspections by this office reveal that the training of the reconnaissance squadrons of armored divisions is not up to the standard of the other units of the division."[125] In essence, the Armored Center concluded that reconnaissance squadrons were poorly trained because they were not getting enough of the division commander's attention and were not under the constant supervision of the commanders they could expect to work for in combat.[126] The army's rapid expansion also played a role. In the 41st CavReconSqdn, organic to the 11th AD, the squadron executive officer who oversaw much of the unit's training before it was shipped to Europe, had only graduated from West Point eighteen months prior to pinning on his major's oak leaves.[127] Though the rapid rise in rank and the commensurate authority and responsibility that came with it were exceptional in this case, they nonetheless placed an individual with extremely limited experience in a critical position as a unit trainer.

The "cavalry group" organization, in contrast to the regimental organization, started to come under fire during the first six months of combat. McNair's pooling concept, when applied favorably, lent to cavalry groups attachments that allowed them to better accomplish their missions, but when groups lost their squadrons to independent missions, group commanders viewed the concept with disdain. COL S.N. Dolph, serving with the 102nd CavGrp, saw the return to a regimental headquarters as the most appropriate solution. The regimental headquarters could manage the organic tank destroyers, engineers, and liaison planes that assets groups had come to depend on but were not guaranteed to them under the pooling concept. Dolph also wished to consolidate logistics and support resources, found in each squadron, under regimental control.[128] Such consolidation would force corps commanders to employ regiments as organic units, not as independent squadrons.

During the first six months in Europe, every type of mechanized ground reconnaissance unit saw wartime service in nearly every capacity imaginable. Cavalry group commanders recognized the ability of their units to do far more than just reconnaissance and realized that the reconnaissance they performed often required fighting. To this end, the old cavalrymen who commanded the groups started reclaiming their former branch identity, even if doctrine prepared by their own branch suggested their inability to carry out fighting missions. COL Dolph observed, "We have performed all the cavalry missions listed in the field service regulations except withdrawals and delaying action. I believe mechanized cavalry is perfectly capable of performing these missions; we should not limit ourselves to reconnaissance."[129]

COL Joseph Tully, 4th CavGrp commander, echoed these sentiments:

> Experience in the campaigns of Western Europe has proven the doctrine of "sneaking and peeking" by reconnaissance units to be unsound, as we have had to fight to obtain information in practically every case. Our training back in the states and in England was guided by the belief that we would have to fight for information. Extensive training in "combat" reconnaissance exercises has paid dividends.[130]

Even if COL Tully anticipated the need to "fight for information," the doctrine the army went to war with did not.

The first six months of war on the European continent were laden with more successes than disappointments, accompanied by plenty of irony with regard to the employment of the mechanized ground reconnaissance units so long in development. The corps cavalry regiment concept worked well, and in this John K. Herr could take pride, even if the cavalry groups lacked real regimental identities and horses. For the first time during the war, if only for a few days in September, the corps cavalry groups gained the operational depth long envisioned during the interwar years. True, it was a lack of gasoline that allowed them to fill the gap between air reconnaissance and pursuing divisions, but they met the challenge. Though not fully codified, and worked out largely on a unit-by-unit basis, cooperation between ground reconnaissance units and eyes in the sky was improving. Equipment shortcomings remained as troopers continued to improvise in the field, while the army acted by fielding improved light tanks. The organization of the cavalry groups with all arms worked well even with a shortage of riflemen and no regiment with which to identify. Patton had, in the spirit of the worst interwar abuses of mechanized ground reconnaissance units, used an entire cavalry group for little more than its radios. His malice aforethought was justified when one considers the contribution the "household cavalry" made to commanding and controlling four corps moving in four different directions, but all into the rear of the enemy.

The greatest irony of all was the ease with which the cavalry groups took to combat. With the pioneers of mechanized ground reconnaissance now leading the Mechanized Force, those that led the Cavalry Branch rump represented those who remained loyal to their branch and horses to the bitter end. When Marshall took their horses in 1942, they had no alternative but to learn to deal with all-mechanized regiments that gave way to groups. The horses may have been gone, but the prejudices held by doctrine persisted and continued to identify horse cavalry as the fighting arm, whereas mechanized cavalry was almost solely limited to reconnaissance. Steeped as they were in the traditional sense of cavalry, the mechanized commanders had no reservations about dismounting to fight and abandoning their specialized role. They were eager to carry on all the cavalry missions and were unwilling to abide by the doctrinal constraints their own beloved horse cavalry instincts had placed on the men who rode iron ponies between the wars. Although the cavalry groups could not fully take back the missions now performed by the armored divisions of the Armored Force, they could easily claim the ability to do far more than passive reconnaissance.

Fighting was not always a matter of choice; it reflected the hard realities of the "ninety-division gamble." Infantry divisions launched December attacks at 75-percent strength for lack of replacements. This lack of manpower was due to an upsurge in casualties, largely a function of the bloody fighting in the Hürtegen Forest and some 12,000 non-battle casualties resulting from trenchfoot.[131] Although economy of force measures enacted in September allowed corps commanders to maintain pressure on the withdrawing Germans, the practice gave way to the dangerous habit of filling extended gaps with lightly armed cavalry groups and squadrons. Safe behind the "West Wall" and closer to logistics support, German forces were far from beaten. There had been "shades of Jeb Stuart" in August and September, but the stagnant front of October, November, and December created a recipe for disaster.[132]

★ **PART TWO** ★

The second purpose, which will be equally important, is to use the armor to destroy the large enemy counter-attacks which we shall indubitably receive. For this purpose they should attack parallel to the original front and at the shape of the bulge which the enemy will make when he counter-attacks.[133]
—LTG George S. Patton, Jr., to GEN Lesley McNair, 4 September 1943

"Can't Lt, too busy shooting Germans."[134]—*Staff Sergeant Woodrow "Pappy" Reeves, 14th Cavalry Group, in response to Lieutenant West's order to seek safety inside his light tank at the Battle of Poteau, 18 December 1944*

★   During the first fifteen days of December 1944, very little aerial reconnaissance occurred in the First Army sector mainly because of limited hours of daylight and poor weather.[135] What the planes could not see beyond the Siegfried Line was equally inaccessible to the troops, who had confronted extensive field fortifications and endured poor weather for months. Nonetheless, American forces in the center of the First Army sector continued to push east toward the Roer River dams in the V Corps sector, as VII Corps supported their attack from the north while VIII Corps remained idle. North of First Army, Ninth Army rested on the Roer River and awaited better weather and attachment to the British Twenty-First Army Group before resuming the offensive.[136] South of First Army, Patton's Third Army prepared to launch a general offensive that promised a three-day "air blitz, followed immediately by a ground assault."[137] Yet even as this offensive was shaping up, Patton found his army 12,000 men short and had begun the process of "cannibaliz[ing] headquarters and anti-tank gun sections to provide infantry riflemen."[138] On 16 December, Patton commented in his diary that he had taken "another 5% out of the Corps and Army troops to make infantrymen."[139] The lack of infantry replacements continued to stretch American forces, and this was particularly evident in First Army's VIII Corps sector.

The Americans were not alone in planning for offensive operations, nor did they completely have the initiative. In a last desperate gamble, Hitler defied what Allied commanders would have thought a more prudent course of action. Rather than assembling a mobile force to strike at the inevitable penetration of the Siegfried Line, Hitler launched his own three-army offensive in an attempt to avoid unconditional surrender. The brunt of this attack fell on First Army's VIII Corps and along the boundary between VIII Corps and V Corps.[140] Paralleling the line on the map that defined the corps boundary was a distinct terrain feature, the Loesheim Gap. In 1914, German cavalry pushed through this gap in the otherwise dense Ardennes region to rapidly gain the Meuse River, and Field Marshal Erwin Rommel had done the same in 1940.[141] A single mechanized cavalry reconnaissance squadron of the 14th CavGrp occupied the Loesheim Gap in December 1944 with COL Mark A. Devine, Jr., in command of the group. A week before disaster befell him and his unit, Devine wrote to an old acquaintance from the First World War, John K. Herr. His letter provided a clear indicator of the conditions in the 14th CavGrp prior to the Battle of the Bulge. Relating that his "squadrons [had] been filling gaps in the line, covering flanks, etc.," Devine was particularly proud that in the two months his unit had been in the line, the Germans had captured just "one prisoner," and only because he was "wounded and couldn't be gotten out at that time."[142] After two months of continuous service, Devine finally gained control of both of the squadrons normally assigned to his cavalry group and was preparing to rotate the 18th CavReconSqdn out of the line, after he refitted the 32nd CavReconSqdn, which was about to rejoin the group. His rotation plan left a single squadron to patrol the high-speed avenue of approach that ran through his sector. Rather than being dismayed by the extensive sector he was responsible for, Devine was impressed with his unit's ability to patrol the sector, especially at night, and with the unit's facility in covering "frontages which would have been considered excessive for units many times their strength." This caused Devine to remark, "in plain language the cavalry trooper doesn't have an equal."[143] The untested cavalry group commander remained upbeat and confident that even though his unit was doing nothing but "work on the line," it would be prepared for "end plays" and "open field running" when the right time arrived. Devine expressed satisfaction with his command and expressed no desire for horses. As if to elicit a bit of envy, Devine closed his letter to Herr with the news that he was on his way to where he knew Herr and "every soldier would like to be going at this moment." Devine was off to patrol the Siegfried Line, not on horseback, but aboard an iron pony.[144] Little did he know what awaited his well-trained mechanized cavalry squadrons operating on what was self-admittedly an overextended front.

The 18th CavReconSqdn took up its positions in the Loesheim Gap on 19 October 1944, only one day before the 106th ID deployed from the United States. The 106th ID later replaced the 2nd ID and assumed the extensive sector that stretched eighteen air miles, and deployed two of its infantry regiments atop a significant salient in the German lines commonly known as the Schnee Eifel. The "Golden Lions" formed an elevated island of American resistance looking down into the Loesheim Gap. The 18th CavReconSqdn served as the vital link with the 99th ID, across the corps boundary north of the Loesheim Gap. Troop B, separated from the rest of the squadron, monitored the avenue of approach that ran along the southern end of the Schnee Eifel.[145] Thus, only a single cavalry reconnaissance squadron in conjunction with the 106th ID's own mechanized cavalry reconnaissance troop guarded the approaches that, if exploited properly, could lead to a swift double envelopment of two infantry regiments atop the Schnee Eifel.

While the 18th CavReconSqdn was attached to 2nd ID, the cavalrymen elected to defend the Loesheim Gap from eight strongpoints centered on village crossroads. They maintained their assault guns in battery southwest of the squadron headquarters in Manderfeld, where the light tank company also remained ready to support the outposts. The cavalrymen had inherited their outposts from an infantry unit and dubbed the positions, often dominated by high ground, "sugar bowls," since they would be sugar to attack.[146] With the assistance of an attached tank destroyer company equipped with towed three-inch guns and a battalion of supporting field artillery, the 550 men dismounted the machine guns from their vehicles and took up the tasks of conducting limited patrols and ambushes between and forward of their positions. The 18th CavReconSqdn also developed two hundred pre-planned artillery targets in conjunction with 2nd ID. If the Germans attacked the exposed cavalrymen, the veteran 2nd ID intended to support them with an immediate counterattack from the Schnee Eifel. When the 106th ID took over the sector on 11 December, it made no similar plans with the 14th CavGrp, which officially assumed responsibility for the Loesheim Gap the same day.[147] COL Devine, having just regained control of his second cavalry squadron, the 32nd, did not change the disposition of the 18th CavReconSqdn.

From the German perspective, the fresh 106th ID fell within the boundaries assigned to the Fifth Panzer Army. Field Marshal Hasso von Manteuffel assigned the sector to the LXVI Corps, which allotted the task of making the initial penetration of the Allied position in the Loesheim Gap to the 18th Volks Grenadier Division. The boundary between Manteuffel's Fifth Panzer Army and Sepp Dietrich's Sixth Panzer Army bisected the 18th CavReconSqdn's sector. Dietrich planned to lead his attack with elements from the 3rd Parachute Division, backed by forty assault guns in the Fifth Panzer Army portion of the Loesheim Gap and with Kampfgruppe Peiper following the 3rd Parachute Division in the Sixth Panzer Army's portion of the gap. The 18th CavReconSqdn, less Troop B out of sector to the south, was about to face significant opposition. The Germans had created the single greatest overmatch of forces along their three-army front in the Loesheim Gap.[148]

Within the 18th CavReconSqdn's sector there was little to indicate what was about to happen. There had been an eerie silence on the German side of the lines for two nights and an ambush force of cavalrymen from the Afst outpost encountered thirty Germans, more than ever before, in the early morning hours of 16 December 1944. Then, at 0530 that morning, the Germans illuminated the sky with spotlights, followed by the ripple of flashes as tube after tube of artillery and rocket fire arched skyward.[149] Limited visibility prevented the widely separated outposts from immediately engaging the advancing Germans, but increasing daylight allowed the small outposts to kill Germans by the hundreds with direct and indirect fire as they often marched forward in rank and file. But undulating terrain between the individual outposts provided the Germans myriad routes to bypass and envelop the American positions.[150] The islands of resistance held temporarily, but ammunition ran low and attempts by the light tank company to escort the resupply of ammunition were mixed as German assault guns, prowling behind what had been the front line, interdicted movement to some outposts.[151] The situation grew dire.

By 1100 hours that morning, elements of the 32nd CavReconSqdn started to arrive in the Loesheim Gap. They had been conducting extensive maintenance fifteen miles behind the front west of St. Vith in the town of Vielsalm. The 14th CavGrp warned them at 0600 and then ordered them by 0930 to move to the front. The 32nd CavReconSqdn deployed its troops from the key bridge at Andler in the south, which sat astride the fastest route to St. Vith, north to Manderfeld. The fresh squadron's assault guns joined the fight with indirect fire, while other elements' efforts to retake Krewinkle and Loesheim immediately failed.[152] Running out of ammunition, unable to get the 106th ID to counterattack down the Schnee Eifel into the flank of the deepening German penetration, and unable to gain the support of four VIII Corps artillery battalions, COL Devine ordered the beleaguered outposts, in a sea of Germans, to withdraw northwest to the Manderfeld Ridge.[153] Some units managed to fight their way out intact, but others surrendered or infiltrated to the rear in small groups.

Unable to hold, the 14th CavGrp reformed along the Manderfeld Ridge, reaching from Andler in the south to Hosheim in the north. Helpless to receive guidance from the 106th ID commander during the night of 16 December, COL Devine made the decision to withdraw what remained of his group farther west, but initially north of St. Vith.[154] The same morning the 14th CavGrp lost contact with 99th ID to the north when *Kampfgruppe Peiper* rolled through Honsfeld and forced Troop A, 32nd CavReconSqdn, to retreat west in total disarray.[155] Troop B, 32nd CavReconSqdn, lost Andler when German armor overwhelmed them.[156] Devine ordered the remaining members of his group to a new line north of St. Vith, near Wallerode and Borne.[157] During the next two days the 14th CavGrp operated on the north side of what was becoming a shrinking perimeter around St. Vith. A nighttime ambush contributed to COL Devine's complete breakdown and forced him to relinquish his command to the group operations officer, who was soon relieved by the 32nd CavReconSqdn commander. Given the cavalry group's condition, VIII Corps attached it to 7th AD, which ordered the group to consolidate what remained into a single reconnaissance squadron. Ultimately, the reorganization mattered little, because the reconfigured squadron was divided among the three task forces that maintained the St. Vith perimeter until 23 December.[158]

The 14th CavGrp suffered horrendous casualties, losing 20 percent of its officers and 33 percent of its enlisted men, as well as 53 percent of its vehicles.[159] On 30 December 1944, it reverted to First Army control and moved to Tongres, Belgium, "for rehabilitation." The First Army refilled the empty saddles of the 14th CavGrp, but not to full strength because of the overall shortage of personnel. Equipment shortages forced each squadron to be reconstructed with two reconnaissance troops rather than the three normally assigned. First Army attached the emaciated cavalry group to the XVIII Airborne Corps effective 25 January 1945. More equipment arrived in early February and the ranks were fully filled, allowing the 14th CavGrp to rejoin the fight when it crossed at Remagen four days after the leading elements of 9th AD captured an intact bridge over the Rhine River.[160]

Having been dealt a poor hand, the 14th CavGrp performed admirably within its limited capabilities. The 32nd CavReconSqdn had moved swiftly to the front on rehearsed routes. Troopers sitting in the sea of Germans rained down shell after shell of artillery and mortar fire on rehearsed targets. When overcome, the group withdrew along routes planned before the crisis, leaving in their wake the shattered remnants of the 294th and 295th infantry regiments. The 14th CavGrp fought hard, but it could not withstand the might of two panzer armies converging on it in the narrow Loesheim Gap. Even so, the group contributed to the disruption of the German plan that depended on the rapid seizure of the critical road junction at St. Vith.

Soft-spoken and often overlooked, LTG William Hood Simpson's Ninth Army, north of First Army, committed five divisions to First Army's fight during the first week of the December crisis. These divisions helped contain the gaping hole in the American lines created by the German onslaught, while Simpson also took control of First Army's VII Corps' defensive frontage as it counterattacked to the south.[161] Unlike Simpson, who responded willingly and rapidly with 7th AD and additional divisions, Patton initially protested and scoffed at Bradley's request for his 10th AD.[162] By quickly reversing his initial assessment and implementing a plan he and his staff had started developing in anticipation of such an eventuality, Patton's Third Army counterattack, coupled with the heroic stand of the 101st Airborne Division, grabbed the headlines. In the process, despite the fact that the Battle of the Bulge was far more than Patton and paratroopers, the terms "Blood and Guts" and "Screaming Eagles" defined what most people recall of this epic struggle. Just as mechanized cavalrymen absorbed the initial blows delivered by the Germans, so they also led the advance of the friendly forces that erased the German salient. In the process of leading counterattacks, mechanized ground reconnaissance units did more than their fair share of fighting, too.

From the north, Simpson's 7th AD, and later 2nd AD, moved rapidly to the sound of the guns. In each instance, the performance of the cavalry reconnaissance and armored reconnaissance battalion assigned to each division served as brilliant examples of what could be expected from mechanized ground reconnaissance units under extreme conditions. With few maps, little understanding of the developing situation, and no mission other than to get out in front and show the way, the 87th CavReconSqdn led the bulk of 7th AD to St. Vith. From their position in Holland, the 87th CavReconSqdn marched across the First Army's chaotic rear area on the westernmost of the two routes allotted the 7th AD.[163] After leading the march that started around 0500 hours, 17 December 1944, Troop B, 87th CavReconSqdn, was the first 7th AD unit to arrive at St. Vith, around 1300 the same day. With the division's combat commands trailing, Troop B deployed directly to the east of St. Vith, dismounted, and formed a defensive line with its six officers and 136 enlisted men. When the other two reconnaissance troops arrived, they deployed north of St. Vith and attempted to reestablish contact with the 14th CavGrp.[164]

The 87th CavReconSqdn fought tenaciously, having the additional support of its parent division's combat commands, which brought medium tanks, field artillery, engineers, and mechanized infantrymen into the line alongside all the other soldiers who manned the perimeter, including the remnants of the 14th CavGrp. In one instance, an M8 Armored Car from

Troop B managed to destroy a German Tiger tank, the terror of the western European battlefield. Equipped with only a 37mm cannon, the armored car closed to within twenty-five yards of the Tiger and unleashed three rounds from the 37mm cannon into the thin armor on the tank's backside. Although the tank erupted in flames, such heroics did not come without a cost.[165] On 21 December, a German penetration cut off Troop B, which had deployed directly to the front at nearly full strength. By the time BG Bruce Clarke ordered the troop to withdraw, it had lost every officer, and only forty-seven of its original 136 enlisted men remained.[166] Even though the 7th AD and the remnants of many other units that formed the St. Vith perimeter ultimately retreated west across the Salm River on 23 December, they had denied the Germans the use of the vital road junction at St. Vith, which American forces had expected to control by the end of 17 December.[167] The mechanized cavalrymen of the 87th CavReconSqdn played a significant doctrinal role in leading the relief force and then went well beyond their doctrinal expectations in helping to maintain the St. Vith perimeter.

The use of mechanized reconnaissance units was only slightly different on the south side of the bulge. After talking with GEN Omar Bradley on 18 December, Patton started moving III Corps, commanded by MG John Milliken, to the threatened sector.[168] Before going to meet Eisenhower and Bradley at Verdun, Patton ordered his staff to develop a number of options for the employment of Third Army. He even went so far as to develop code words with his chief of staff, BG Hobart Gay, to activate each plan over the telephone without compromising secrecy.[169] At the meeting, Patton promised to be ready to attack north with the Third Army by 22 December 1944.[170] Patton commented in his diary at the degree of surprise among the others, when he declared he could counterattack that fast with three divisions, namely, the 4th AD, and the 26th and 80th IDs.[171] Also included in the planned counterattack, but not mentioned in Patton's diary entry, was what had been his "household cavalry," the 6th CavGrp.

As 4th AD counterattacked into the exposed German flank, relieving the "Battling Bastards" encircled at Bastogne, a gap developed between the 4th Armored and the 26th ID. To prevent the German Seventh Army from exploiting the opening, Patton planned to use cavalry.[172] The 6th CavReconSqdn maintained contact between the 4th AD and the 26th ID, while its parent HQ, the 6th CavGrp, advanced on the left flank of the 4th AD with attached engineers, tank destroyers, and artillery, to screen the exposed flank of the 4th AD should the Germans counterattack from the west.[173] Although follow-on forces relieved the 6th CavGrp by 31 December 1944, the cavalrymen were happy to have abandoned the dismounted infantry attacks they conducted in the Saarbrucken sector before being diverted north.[174] One of the units that replaced the 6th CavGrp on the left flank of III Corps was the 11th AD.

Arriving at the port of Cherbourg on 23 December, the 41st CavReconSqdn of the 11th AD did not pause as it crossed the docks and headed north to the ongoing fight in Belgium. By Christmas Eve the squadron reached the Meuse River and waited for its division to join it.[175] Days later, the 11th AD joined Third Army's attack on the southern flank of the German salient. New to combat, the mechanized cavalrymen soon experienced dismounted combat with the support of their assault guns and light tanks, but during this initial combat Troop A experienced particularly heavy losses on 15 January 1945.[176] Relieved from the line on 16 January, the squadron moved to the Belgian town of Bertogne to await its next mission, which was not long in coming.

Ever the cavalryman, and regardless of his thoughts about mechanized reconnaissance units, John K. Herr's son-in-law, BG Willard "Hunk" Holbrook, delivered to the 41st CavReconSqdn what he termed "an excellent reconnaissance squadron mission."[177] With the squadron commander away with Troop C, the squadron's executive officer, MAJ Michael Greene, learned that he was to lead his unit ten miles behind German lines to link Patton's Third Army with the First Army of Courtney Hodges near the town of Houffalize. An hour later, in growing darkness and unable to reach the squadron commander, Greene led his small task force, composed of a single reconnaissance troop, the assault gun troop, and the light tank company, toward his new objective.[178]

Undeterred when he was thrown clear of his half-track as it struck a mine, Greene led the night-long march over ice-covered roads and through thick forests. Much of the advance took place at the pace of a walking man, since darkness—combined with the distorted view of the landscape covered with eighteen inches of snow—forced the lead platoon leader to rely heavily on compass bearings to keep the mounted columns on course. Two miles short of its objective, the squadron encountered another natural obstacle, an ice-covered hill. After two hours, the squadron's tracked vehicles managed to tow the remaining wheeled vehicles to the top of the rise.[179] Having overcome the many natural obstacles, all that remained were the Germans.

MAJ Greene and LT Eugene Ellenson, who had led the dismounted advance, approached the city-limit sign of Houffalize, congratulating themselves on accomplishing the mission. Noticing movement on an overlooking hilltop, the two officers advanced to investigate and were greeted by the barrel of a German machine gun. Only the fast action of a sergeant, who shot off a burst of suppressive fire from his watching armored car, saved the officers. Brought back to the reality of war, Greene spent the rest of the morning deploying his combined-arms force. Scouts dismounted on the high ground overlooking the town, and assault guns took up position on the road and began to shell the town's German defenders, while the light tank company made a brief foray into the center to discern the strength of the garrison.[180] By 1000 hours a cavalry patrol from the 41st CavReconSqdn linked up with a patrol from 41st Infantry Regiment from First Army's 2nd AD, thus reconnecting First and Third Armies.[181] The "bulge" was pinched out.

Under the worst conditions imaginable, the mechanized cavalry reconnaissance squadron accomplished its mission. Hampered as it was in terms of mobility, when it did arrive the squadron brought with it sufficient combat power to maintain the connection to First Army. No interwar horse cavalry unit could have deployed the squadron's collective firepower of twenty-five 37mm cannons found on the light tanks and armored cars and 75mm cannons on the assault guns. A month later, this was lost on the man who ordered the "excellent reconnaissance squadron mission." Mechanized cavalry reconnaissance units' contributions to turning back the Germans during the Battle of the Bulge were not limited to fighting delays against heavy odds, leading the advances of armored divisions, or establishing and maintaining contact between units. The cavalry groups in particular contributed in other meaningful ways.

When Patton turned his attention to the growing German penetration, shifting his axis of attack ninety degrees, he used three cavalry groups, the 2nd, 3rd, and 6th, to cover his army's rear as he attacked north.[182] The 6th CavGrp, accompanying III Corps, quickly found itself in the Allied rear, but the other two cavalry groups remained at risk in the same way the 14th CavGrp had been, had it not been for some reinforcements. Believing that "the only place the Germans could hurt us is the front held by COL [James] Polk," Patton ordered GEN Walker, the XX Corps commander, to conduct a reconnaissance of this sector. Walker was to identify suitable positions from which to defend and block the roads, should the Germans try their luck at the Third Army flank.[183] Walker's neighbor to the south had been GEN Middleton's XII Corps, but Patton pulled XII Corps out of the line and moved it north of XX Corps to take over for the crumbling VIII Corps. Middleton took the 2nd CavGrp with him when he assumed his new sector and assigned them the responsibility of guarding his exposed right flank as he concentrated on holding the south shoulder of the bulge. Using the Moselle River as a natural obstacle and reinforced with a combat engineer battalion, the 2nd CavGrp, Task Force Reed, concentrated not on reconnaissance, but on establishing a defense in depth.[184]

The Battle of the Bulge revealed both positive and negative characteristics of the mechanized cavalry units that fought there. The 14th CavGrp had little chance against the tidal wave of Germans that crashed over it on 16 December. In light of the overwhelming forces that flowed by the islands of resistance it had created, the 14th CavGrp could not expect the same kind of relief that saved the 2nd CavGrp when it was attacked at Lunéville in September 1944. Unlike other places along the front, the 14th CavGrp could not take advantage of any natural obstacles in the Loesheim Gap. Rather, it sat astride one of the traditional invasion routes through the Ardennes. The concept of pooling also played a factor, as did the inabilities of the infantry commander to whom COL Devine's 14th CavGrp was attached days before the attack. Pooling allowed the dispersal of Devine's squadrons on separate missions. He had only regained command of the 32nd CavReconSqdn days before the attack. To his credit, even though they were conducting maintenance, he had the men reconnoiter the routes used on the morning of 16 December to reinforce the rest of the cavalry group as it fought to slow the German advance. MG Alan Jones, who commanded the 106th ID, was so new to combat he had been on a troop transport coming across the Atlantic when the men of the 18th CavReconSqdn took up their positions in the Loesheim Gap. Only his infantry regiments, under very little pressure on 16 December, and the corps artillery assets at his disposal were capable of saving the 14th CavGrp, but he was not up to the task of orchestrating their employment. The commander's inexperience, combined with his failed appreciation for the limited capabilities of a single squadron of mechanized cavalry, led to a disaster that cost Jones the bulk of his division.

The mechanized cavalry reconnaissance squadrons leading the modern-day cavalry to the rescue, in the form of armored divisions, again proved their worth. They negotiated routes under the most extreme conditions and fought with tenacity when they arrived at the front. Especially in the case of the 7th AD at St. Vith, the speed with which it moved into the sector from a far distance had a devastating impact on the timetable the Germans had to maintain if they were to be successful. Following cavalry reconnaissance squadrons and armored reconnaissance battalions, more substantial combat power of the armored divisions moved to the crisis. Unlike the 14th CavGrp, armored division reconnaissance units knew help was always close at hand.

Having contained Hitler's last gamble, Allied forces still had to defeat the German war machine on its home turf. On 7 February, Patton visited the 2nd CavGrp. Like other cavalry groups not directly involved in the counterattack to reduce the German penetration, the 2nd CavGrp contributed to the overall effort in an economy of force role. Patton found it still fulfilling this role and "was very much pleased with his [COL Reed's] method of occupying the line." Perhaps wanting 2nd CavGrp to avoid the same fate that had befallen the 14th CavGrp, Patton noted in the same entry the continued availability of the 4th AD to rush to the support of the cavalrymen should they come under serious attack.[185] Eleven days later, the 2nd CavGrp undertook the one significant operation that later caught the attention of the General Board for the period between the end of the Battle of the Bulge and the final pursuit across Germany.

Emblematic of the contributions being made by other mechanized cavalry units along the front as the Allies gained the ground between the West Wall and the Rhine River, the 2nd CavGrp, heavily reinforced and known as Task Force Reed, did not lead the advance of the XX Corps. Rather, two troops from the 2nd CavReconSqdn gave up vehicles for boats to conduct a supporting attack across the Moselle River on 19 February 1945. The remainder of Task Force Reed guarded the river as assault forces crossed at Ehnen, Luxembourg, and moved dismounted to their objective at Wincheringen. There, drawing on indirect fire support from beyond the Moselle and direct fire support from the towed three-inch guns they brought with them, the cavalrymen secured the town and waited for the leading elements of the 10th AD to relieve them.[186] Task Force Reed was not the only cavalry group making dismounted attacks in the Third Army sector.

For much of February, the 3rd CavGrp, with an additional artillery battalion, tank destroyer battalion, and engineer battalion, continued to patrol along the Saar River as the rest of XX Corps struggled to clear the Saar-Moselle Triangle. Stretched thin along the river, it too had resorted to a series of strongpoints, with each squadron's light tank companies held in mobile reserve. To maintain contact between the outposts, the cavalrymen patrolled among the German minefields. On 8 February 1945, the 3rd CavGrp finally received the long-awaited equipment they needed to perform their job better and more safely: saddles and bridles. Concluding it was better for a horse to step on a landmine than a man, the troopers had begun conducting mounted patrols between their positions. Fully equipped troopers riding bareback welcomed the arrival of bridles and saddles.[187] Although horses had not provided a stealthy means of getting behind the German lines on the other side of the river, they had provided the means to avoid a mine strike. But the return to horses was short lived.

MG Walton Walker, XX Corps commander in Patton's Third Army, created the first modern mechanized cavalry brigade, but with a decidedly different purpose from what was envisioned for the 7th CavBde(M) during the interwar years. In early March, while 3rd CavGrp troopers enjoyed patrolling on horseback, XX Corps was still trying to advance beyond the Moselle River and return to swifter advances aboard more contemporary mounts. In order to pull infantry divisions out of the line for the new attack, Walker used the 3rd CavGrp to maintain contact with XII Corps to the north and secure the line of departure for the upcoming offensive. Third Army then assigned the newly arrived 16th CavGrp to XX Corps, which Walker in turn attached to 3rd CavGrp, thus creating the 316th Provisional Mechanized Cavalry Brigade.[188] One could only speculate if McNair might ever have envisioned this as one possible result of the pooling concept.

With two complete cavalry groups, COL "Jimmy" Polk deployed all four squadrons on a line along a seven-mile sector. Polk's brigade also included an attached battalion of field artillery, two supporting battalions of field artillery (one in direct and the other in general support), an engineer battalion, a company of self-propelled tank destroyers, and an air-support party. Beginning on 13 March 1945, the 316th Provisional Cavalry Brigade attacked across its seven-mile front for the next eight days. Using all these assets, the brigade managed to gain ten miles, which would have been remarkable for World War I but fell far short of interwar expectations. Most of the fighting occurred dismounted with accompanying light tanks, just as one would expect to find in any infantry or armored division's attack.[189] The provisional brigade was "cavalry" in the traditional sense of the fighting Cavalry Branch that had been largely replaced by the doctrinally forced constraints of the largely reconnaissance branch that emerged in 1942. Regardless of doctrine, XX Corps formed a cavalry brigade from the two cavalry groups, not to achieve greater reconnaissance capability, but solely to enhance the fighting capacity of the unit making an attack in support of the main effort.

Two officers to the north of COL Polk's modern mechanized cavalry brigade had a great deal of experience with horse cavalry and the unresolved identity crisis being played out in Europe. These officers took the time to write to the last chief of cavalry about the current situation. Coming out of the intense fighting of the Battle of the Bulge, BG Willard A. Holbrook, still serving as the commander of a combat command in the 11th AD, wrote to his father-in-law, John K. Herr, about the conditions confronting American forces. Citing the "patch work" of heavy forests that characterized the terrain between Belgium and the Rhine River, Holbrook was convinced that horse cavalry could have made an important contribution to the ongoing campaign.[190] Herr's sharpest pen during the interwar horse-mechanized debates, COL Wesley W. Yale, commanded another combat command in the 11th AD. He echoed Holbrook's sentiment, citing the frustration of seeing it take seventy-two hours for his men to close up on the flanks of any unit once it penetrated the Siegfried Line. Yale had a solution to the current predicament confronting the Allies in the West.

> With no joking whatever, the old Garry Owen with JKH at the head could go through this Siegfried affair in damned near an hour with negligible losses. What a wonderful bet we missed in those days. A tank-horse outfit would be invincible. And what we wouldn't give for a regiment here and now![191]

Currently leading his own combat command, Yale could not deny the role played by his tanks, but in alluding to the tank-horse outfits called for in the pages of the *Cavalry Journal,* he knew how to pull at the old man's heartstrings. February's frustrations were about to give way to March's exhilaration as the 11th AD forged ahead beyond the Siegfried Line. Letters from March and April, even with efforts to assuage Herr's feelings, revealed a different tone.

After the 11th AD fought through the Siegfried Line, it advanced rapidly to the Rhine. Holbrook wrote his father-in-law:

> I will never forget the thrill, that exaltation of feeling, after twenty six years of service to be able to exploit the breakthrough. When I hit Mayen I felt as if I could throw away my map which failed to cover the final objective anyway. I was fighting over familiar territory. Many scenes were the same twenty years afterwards. Only the mounts differed.[192]

Indeed, the mounts were different and had allowed Holbrook's combat command to cover thirty miles in twelve hours. Even after writing about the exhilarating feeling of the pursuit, albeit on "different mounts," Holbrook must have felt compelled to offer the old man some solace. He remarked that he and "Wes" both believed they could have "used horse cavalry in nearly all phases of the action we have seen." There was no doubt in Holbrook's mind that the horses could have maintained the pace of the "break through to the Rhine." Having said that, even the son-in-law of the "Czar" had to admit that the horse cavalry would have been deficient in one very important aspect, firepower. The "German guns" with their high velocity had destroyed American tanks with ease until heavier armor-plating was introduced. As some indication of the seriousness with which Holbrook now took the matter of horse cavalry's relevance on the battlefield, he appended the following sentence to the paragraph discussing German firepower: "Cavalry would have had to get some really powerful weapons to be successful."[193]

COL Yale's combat command managed 120 miles in four days as it advanced to the Rhine. Unlike Holbrook's letter with references to horses, Yale's included the touches of humor that distinguished his personal letters to his old boss, but he also offered something very telling, even if it was intended to be glib. After detailing just how the "looting" continued to be excellent and how he planned to "clean out" one of "the most famous wine cellars in Europe," the former horse cavalryman noted that if they would only "give us Pershing tanks," the war would "be over in a week."[194] Calling for heavy tanks, with 90mm cannons and increased frontal armor, in essence echoed Holbrook's comment that the horses might have been able to keep up, but they would have needed far more firepower than even the mechanized cavalry units were capable of bringing to the fight. Although the two, who only a month before had written in support of the horse, now called for more firepower, there was one man who refused to give up the fight.

From retirement, Hamilton Hawkins, in the March–April 1945 edition of the *Cavalry Journal,* expounded on a common theme, horse cavalry. He was willing to acknowledge the accomplishments of the mechanized cavalry units serving in Europe and the 1st CavDiv serving in the Pacific theater without its horses, but he refused to give up the notion that the United States continued to apply a "self-imposed disadvantage that may have handicapped operations in some theaters."[195] He cited his belief that the inclusion of horses in the existing mechanized reconnaissance formations would have made these formations "even more useful at less cost in men if they had been supported by or combined with squadrons of horse cavalry."[196] He went on to suggest that the employment of a horse cavalry corps equipped with bazookas and mortars could have conducted a more effective delay during the Battle of the Bulge and that lack of horse cavalry prevented the Fifth Army from exploiting its success in Italy. He lamented that the "American public" remained "oblivious" to cavalry operations "despite the wonderful success of Russian cavalry."[197] Although the American public was not about to come nose to nose with Russian horse cavalry, many mechanized cavalrymen were drawing closer to that meeting. Yet in doing so, they were growing less inclined to support the return of horses to their formations.

The long, slow fight to the Rhine culminated on 21 March 1945, when British and American forces from north to south were all poised to cross the river. Having captured a usable bridge at Remagen on 13 March 1945, Eisenhower had limited the efforts of LTG Hodges to expand the bridgehead to twenty-five miles in width and ten miles in depth.[198] The Rhine crossings that followed, combined with the breakout from the Remagen bridgehead, led to the encirclement of the Ruhr pocket, a race to the Elbe River in the north, and a breakout that carried Patton's Third Army and Patch's Seventh Army into Czechoslovakia and Austria in the south. What amounted to little more than a month of fighting revived memories of the dash across France and Belgium in August and September 1944. The last month of combat also imprinted on the minds of the men who would write the final report on mechanized ground reconnaissance units, by now synonymous with mechanized cavalry, the strengths and weaknesses of those units' organization, doctrine, and equipment.

From this period and as part of the reduction of the Ruhr pocket, the General Board chose to highlight one particular offensive operation of the 87th CavReconSqdn. Like so many of the historical vignettes later chosen to illustrate the multiple uses of mechanized reconnaissance organizations, this one featured dismounted combat. Ordered to seize a defended town, the 87th CavReconSqdn, by now veterans of months of campaigning, used all its resources to accomplish the mission plus the additional firepower afforded to it with an attached field artillery battalion. Since it was the divisional reconnaissance squadron of an armored division, its squadron was able to dismount three reconnaissance troops, which it put on line while the fourth troop continued to support a combat command on a different mission. Light tanks drawn from Company F supported each of the dismounted cavalry troops, as the assault guns provided indirect fire support rather than accompanying the advance.[199] Similar to the kind of tank-infantry cooperation that had developed in infantry and armored divisions, the mechanized reconnaissance men continued to prove equally capable of employing the same technique with assets wholly organic to their squadron.

As April progressed, the American drive toward the east gained momentum, and GEN Holbrook and his combat command in 11th AD became part of the wave of American forces breaking over Germany. Nearing the end of the pursuit, Holbrook wrote Herr another letter about conditions in Europe. He remarked that under the leadership of MG Holmes Dager, the 11th AD was "pushing along at a great rate," paced by radio-directed orders straight to the combat commands he and COL Yale led.[200]

Another factor contributed to the rapid advance of Holbrook's combat command. Reflecting a far different attitude than he had in his February letter, Holbrook embraced the men who since March, when the 11th AD had begun to pick up speed, had been attached to his combat command, often in at least troop-sized strength.[201] He wrote, "My cavalry has been performing superbly. I usually send it out about two hours ahead of the combat command."[202] Holbrook was now gaining the full benefit of the time-space advantage of having rapidly moving, radio-equipped scouts warn him of trouble ahead, so that he could take appropriate measures before he crashed into the problem with the bulk of his combat command. Usually assigned in troop-sized elements, the reconnaissance troops brought more assets than the armored cars and jeeps found normally in their organization. Like the armored reconnaissance battalions that started the war and lived on in the 2nd and 3rd ADs, the cavalry troops leading combat commands generally brought along at least two assault guns from Troop E of the cavalry reconnaissance squadron. Additionally, they benefited from the tank-killing capabilities of an attached tank destroyer platoon and the obstacle-reducing capabilities of an attached engineer platoon.[203] Holbrook concluded that this robust collection of assets conducted reconnaissance "like horse cavalry except that their firepower is much greater and they are slightly more road bound."[204] In a reversal of his February opinion, Holbrook had come to believe his fully mechanized cavalry reconnaissance force was "more suitable," now that "daily bounds" were forty to sixty miles.[205]

Although he did not mention it in the letters to his father-in-law, there was another possible explanation for Holbrook's shifting position on the continued utility of the horse in modern war. As part of the most mechanized and motorized military force the world had ever seen, Holbrook had come into contact with the real German army. Despite popular misconceptions associated with blitzkrieg, the German army was largely horse-drawn, the result of a conscious decision based on the availability of natural resources.[206] Thus, Holbrook must have been exposed to the impact of modern weapons on large horse-based formations. By April, he would have known, in particular, about the devastating air strikes—called by COL Yale's CC B's Tactical Air Liaison officer—on retreating Germans trapped on the west side of the Nahe River after all the bridges were destroyed. The 11th AD's after action report records that "scores of German horses, horse-drawn transport, motor vehicles and enemy soldiers" died in the narrow streets of Kirn on 18 March 1945.[207] Although Holbrook later found it hard to separate captured Austrian cavalrymen from their mounts "until they [the horses] had been groomed," far more than sentiment for the beloved horse informed his view on modern mechanized ground reconnaissance conducted without horses.[208]

The large bounds that Holbrook wrote about were common across the American front in April. In the north, the 11th CavGrp with an attached tank destroyer battalion rode on the left flank of the advancing XIII Corps. In twenty-one days, the group ranged the 378 miles from where they crossed the Rhine River to the American limit of advance on the Elbe River. Aside from reconnoitering the advance of the 84th ID through "virgin enemy territory," the group conducted a number of other important missions during what amounted to its last month of combat.[209] The rapid advance entailed bypassing a number of pockets of enemy resistance, out of which German soldiers began to prey on the XIII Corps' lines of communications. Corps commander LTG A.C. Gillem directed his cavalry group to take responsibility for escorting supply convoys in particular sectors that had been subject to frequent attacks.[210] The group's activities were not limited to reconnaissance and security.

One of the group's most important tasks was maintaining contact with the British Second Army, which was also advancing rapidly across northern Germany. As the gap between American and British forces lessened to as little as fifty miles, the cavalry group maintained contact through radio liaison teams and personal visits by the group's squadron commanders and their staffs. Upon arriving at the Elbe River, the 11th CavGrp drew the responsibility of establishing liaison with another ally, the Soviets, including the III Cavalry Corps.[211] Although it finished the advance by linking the eastern and western fronts, the dash from the Rhine to the Elbe had not been without cost, even against what was termed light resistance. Sharp encounters, often ambushes, cost the cavalry group five light tanks, six armored cars, and twenty-three jeeps.[212] Riding on point in a jeep remained a hazardous place to be for mechanized scouts from the beginning to the end of World War II.

With the race to the end in full swing, George S. Patton, Jr., took the time to write to and send along a captured German pistol to an old cavalryman, MG Kenyon A. Joyce. Patton told Joyce that leading his "pack . . . racing for the kill" was the 11th AD and COL "Jimmy" Polk's 3rd CavGrp, which had just gobbled up all the terrain between Nuremberg and "the bend of the Danube River."[213] The same action captured the attention of the General Board after the war as an example of a cavalry group conducting offensive operations, in this case a pursuit.

Characteristic of the organization of so many cavalry groups during the last six months of war, the 3rd CavGrp depended on a number of attachments to enable it to pursue the Germans south to the Danube River at Regensberg. Polk's cavalry group included two companies of self-propelled tank destroyers, a company of combat engineers, an entire Ranger battalion, and an additional artillery battalion in direct support, though not attached. In two days the 3rd CavGrp moved nearly forty miles, confronting multiple manmade and natural obstacles and limited enemy resistance, while securing a number of key bridges for the infantry division following it. The pressure the group applied kept the Germans from establishing an effective defense north of the Danube River.[214]

Advancing south into Bavaria, the 106th CavReconSqdn of the 106th CavGrp experienced its last real combat as it led XV Corps rapidly down the autobahn to Munich in an attempt to regain contact with retreating German forces. As night fell, with Munich in sight, Troop C and two light tanks from Company F drew heavy fire from self-propelled guns, tanks, and small-arms fire from infantry just nine kilometers from their objective. The engagement cost the troop four dead, four armored cars, and four jeeps. In bitter fighting with diehard Nazi troops, the 45th ID, for which the 106th CavReconSqdn had reconnoitered, took Munich by nightfall 30 April.[215] The 106th CavGrp's last sharp encounter was not unique, nor was the advance it made during the final days of the war, when compared to other cavalry groups. Nonetheless, from 2 May to 4 May, the cavalry group advanced from Munich, captured a Hungarian division of 8,000 soldiers, crossed the Inn and Salzach Rivers, and was present for the formal surrender of Salzburg, Austria. To crown this accomplishment, the 106th CavGrp liberated Leopold, King of Belgium, from his Nazi captors in a villa outside of Salzburg.[216] In covering seventy miles in two days, the fast-moving cavalry group lent a modern meaning to the notion of a cavalry pursuit.

During the waning days of World War II, a most symbolic event took place in the struggle between horse and machine. Just as the 106th CavGrp had liberated a nobleman, the 2nd CavGrp liberated a herd of noble beasts, the breeding stock of the Piber Austrian Lippizanner horses, which had long supplied the Spanische Reitschule in Vienna. In what amounted to a partly comic, partly downright dangerous, and partly remarkable event, the men on iron ponies saved the animals they had replaced.

An interrogation conducted by the 2nd CavGrp on 25 April 1945 led to an intense negotiation with a German unit seeking refuge in a hunting lodge on the Czechoslovakian border. After an exchange of "harmless gun fire" on the morning of 26 April 1945, one of the negotiated terms, the German unit promptly surrendered. Among the prisoners was a general officer who shared the cavalry group's commander's passion for horses. The German general showed COL Reed pictures of the Lippizanners and Arabs located at Hostau, Czechoslovakia, which coincidentally housed hundreds of Allied prisoners of war. Agreeing that it would be bad for the Soviets to capture the horses, the two men conjured a plan to save both the steeds and the prisoners.[217]

After asking Third Army for permission to save the horses and receiving a terse response from Patton—"Get them. Make it fast! You will have a new mission"—the operation commenced. The first danger lay in the Nazi SS division lodged between the cavalry group and the horses. In an age when upper-class America shared the burdens of war by serving in the front ranks, none less than a senator's son, CPT Thomas M. Stewart, rode through enemy lines with the captured veterinarian from the horse farm.[218] Once through the lines, he gained the consent of the Hostau garrison to be liberated by the Americans, and then returned to the 2nd CavGrp aboard a German motorcycle with sidecar. On the morning of 28 April, a small task force—built around Troop A, 42nd CavReconSqdn, some elements of Troop C from the same squadron, an assault gun platoon, and a light tank platoon—fought its way through to Hostau. There it liberated nearly four hundred American, British, Polish, and French prisoners, and the prized horses. The Axis garrison was as varied as the breeding herd, consisting of Germans, Czechs, and White Russian Cossacks who had taken up arms against the Soviets. What ensued quickly lent a comic opera feel to the entire operation.

Just as Patton had promised, the 2nd CavGrp renewed its drive, which carried it into Czechoslovakia near Pilzen. Troop A joined this advance leaving CPT Stewart only a platoon of light tanks and an assault gun platoon to secure the Hostau breeding farm. With little alternative, the intrepid young captain formed a multi-national defense force to fend off the attack launched by the SS soldiers on 30 April. While American tanks and assault guns delivered a devastating base of fire, regular German soldiers stood shoulder to shoulder with the Poles who had been their prisoners just days before to collectively counter the fanatical Nazi attack.[219] Nazi threats were short-lived, but soon gave way to nascent Cold War threats.

Fearful that the Soviets and Czech communists, both of whom had begun expressing an interest in the horses, might occupy the Hostau breeding farm, Patton ordered the evacuation of the animals to Schwarzenburg, Bavaria. Third Army issued orders that gave horses, not vehicles, priority road use as German, Cossack, and even a few American, cowboys served as outriders to drive the herd west in small groups interspersed with jeep-mounted cavalrymen. Reminiscent of the horse-mechanized days, trucks moved young colts and mares heavy in foal along with German and White Russian women and children. Safe in the West, 215 of the horses returned to Austrian control while the majority of the herd moved to German breeding farms in Hesse. The liberators shipped some of the best horses back to the United States for use as future cavalry mounts.[220] Although he had commanded a fully mechanized cavalry group and been in combat since August 1944, the horse-loving cavalryman had to intervene to save the proven technology that he was still not sure had been fully eclipsed. COL Reed never rode one of his war trophies while on active duty in a tactical unit. He did, however, continue to ride the daughter of one of the horses he rescued at Hostau, an Arabian, on a daily basis years after the rescue.[221]

The last six months of war in Europe beginning with the Battle of the Bulge in December 1944 and ending with victory in May 1945 continued to inspire comments by Army Ground Forces observers. Men in the field carrying out the missions assigned to mechanized ground reconnaissance units continued to have their own observations about what was good and bad about the organization, equipment, and doctrine prescribed for their units. Many of these observations found their way into reports prepared after the war and into the General Board findings that attempted to consolidate and capture all that had been learned.

Air-ground coordination continued to improve during the last six months of fighting. During the Battle of the Bulge, air observers played a critical role in the defense of Monschau, directing artillery fire that drove the Germans south and thus helped hold the northern shoulder of the deepening bulge.[222] Although the General Board's report celebrated this defensive stand, it ignored the integration of liaison aircraft in helping to establish and maintain the defense. In GEN Truscott's VI Corps, the home of innovative uses for all types of cavalry and light observation aircraft, the trend continued.

The 117th CavReconSqdn's command post was often less than ten miles from the airfield of the habitually attached observation aircraft. The liaison between the pilots and Troop E, which provided the squadron's indirect fire support with its assault guns, remained outstanding. This assured Troop E's ability to place accurate and timely fire on German positions spotted from the air. Although liaison with the ground troops was not as good, the pilots flying in support of the squadron continued to make every effort to warn ground troops of impending dangers lurking around the next bend in the road, such as well-sighted anti-tank guns. When radios failed, swooping passes over ground scouts conveyed the message.[223] The unique relationship found in the 102nd CavGrp and in the 117th CavReconSqdn may have been rare in comparison to the other cavalry groups, but the presence of liaison aircraft, even if not working directly for mechanized cavalry units had become ubiquitous in the skies over Europe. Lucky were the cavalrymen who had the planes working for them in the capacity imagined since the end of World War I, but still not fully realized in the realm of mechanized ground reconnaissance.

Having barely eliminated the "Bulge," the 12th Army Group commander, GEN Omar N. Bradley, found the time to write a personal note to GEN Patton about the misuse of the light reconnaissance planes that had become so vital to the American war effort. Its brevity and levity merits full citation:

> Dear George,
> Game wardens in Luxembourg have contended that our troops are shooting deer there illegally from the air. The attached letters would indicate that the reconnaissance planes have been employed. Will you look into this matter and reply to the enclosed communication from the Saint-Hubert Club of Luxembourg.
> Sincerely,
>
> O. N. BRADLEY
> Lieutenant General, USA[224]

While upset game wardens managed to capture the attention of the 12th Army Group commander still a few months away from victory in Europe, perhaps the ultimate misappropriation of a light reconnaissance plane captured the attention of a veteran cavalry group commander just after VE Day.

In early May 1945, COL Biddle stood beside a road watching his 113th CavGrp move to the west to rejoin XIX Corps north of Frankfurt am Main. As he watched his troopers pass, Biddle grew more and more disconcerted with the amount of enemy equipment, especially vehicles, that his men had taken as their own. While he thought about the situation something caught his attention in the sky above him. There flying over his column was a German Fieseler Strorch airplane, the German equivalent of the light observation and liaison plane. With the time since the end of hostilities barely measured in days, sight of the German plane startled Biddle. His surprise only increased when he later learned his troopers had taken upon themselves to rectify one important organizational deficiency of the mechanized cavalry group. The man who piloted the plane was a sergeant from the 113th CavGrp.[225]

In the 6th CavGrp, COL Edward M. Fickett, spelled out to an Army Ground Forces observer in considerable detail exactly why the ad hoc solutions that had worked in some units were not any more acceptable than the extreme measures taken in the 113th CavGrp after VE Day. During the last six months of war, the cavalry groups almost always had an attached field artillery battalion in direct support. With the artillery battalion came the light planes used to adjust rounds onto target. No one denied the utility of this capability, but as COL Fickett explained to the Army Ground Forces observer, he needed planes solely dedicated to reconnaissance and liaison. When planes had been diverted from their artillery-spotting functions to conduct these tasks, fire support had suffered. The observer took this issue to the Third Army staff, who agreed that they too recognized the problem and saw the only solution as the addition of organic light aircraft to each mechanized cavalry unit. Patton's staff further concluded this might create the need for a "pilot and observer training course for cavalry officers."[226] Airplanes were not the only missing components on the minds of cavalry commanders in 1945.

If the reports had seen the mechanized cavalry commanders calling for infantry weapons during the first six months of the European campaign, COL Polk with the 3rd CavGrp appealed to an Army Ground Forces observer for infantrymen. He did this before XX Corps brigaded his cavalry group with another to form the provisional mechanized cavalry brigade that conducted what was largely a dismounted attack. Identifying the lack of infantrymen as one of two "critical deficiencies," the other was a lack of communications wire, phones, and the men to maintain them, Polk proposed the addition of a rifle squad to each platoon. To assure equal or greater mobility to the cavalry platoons mounted on jeeps and armored cars, Polk asked that the infantry squad ride on a half-track.[227] Just as the cavalry groups could not depend on the use of light aircraft to support their reconnaissance needs, the infantry squads could not depend on the attachment of the most precious fighting commodity, riflemen. In each case, the cavalry commanders sought to secure the resources they deemed necessary by ensuring that these resources were organic to their units. This was contrary to the notion of pooling that they had come to depend on for additional fire support, engineer support, and tank destroyer support, which they also asked to be permanently included in the cavalry group.[228]

Equipment also drew attention. The basic technique for reconnaissance continued to rely on heavier weapon systems, such as armored cars or tanks, establishing an "over watching position" while jeeps moved forward in search of the enemy or an enemy response. If no fire was "drawn" the entire force bounded forward "bit by bit under cover of the supporting weapons," until the entire process could be repeated.[229] Because of the hazards that had long existed on point, troopers continued to seek the additional protection still needed by jeeps. During the same month the 3rd CavGrp received its saddles and bridles, its efforts to find suitable armor plating for its jeep-mounted scouts failed.[230] The last combat fatalities in the 113th CavGrp occurred when two troopers riding in a jeep in search of Russians east of the Elbe River struck a mine.[231] With tactical mobility only surpassed by the dismounted man and horse, the jeep remained a mixed blessing for mechanized reconnaissance units throughout World War II.

Fielding of the M24 light tanks, a vast improvement over the M5 light tank taken to war in 1942, continued during the last six months of campaigning in Europe and continued to be well received by the troopers. Although the M8 Armored Car still received fair marks for its ability to deal with German armored cars, it was generally held in low regard. The trend to replace the M8s with light tanks during the first six months of fighting continued during the last six months.[232] Having come to value the abilities of his mechanized cavalry support, GEN Holbrook believed the M8s were "worthless" and noted that the 41st CavReconSqdn was replacing them with light tanks for greater "cross country mobility."[233] The army failed to field an improved armored car, the M38, before the fighting ended. The M38 had been destined to carry a 75mm cannon, far more fire power than the 37mm-equipped M8, and comparable to the newly fielded M24 light tank. With victory secured in Europe and the Pacific by the end of 1945, the army discontinued its plans to field the M38 in 1946.[234]

If the greatest acceptance of the current organization and performance was found with the divisional cavalry troops, which in some cases were proving themselves very adept at fighting as well, there was continued discontent with cavalry group doctrine and organization. Group commanders continued to call for organizational changes such as the inclusion of liaison planes, infantrymen, engineers, tank destroyers, lighter radios for dismounted operations, rifles instead of carbines, bayonets, and field packs instead of musette bags. From the field, group commanders wanted Army Ground Forces to understand that they were "performing all cavalry type missions" and that "the teachings of cavalry manuals which limited the mechanized cavalry employment to that of reconnaissance" was wrong.[235] In the view of the commanders, the proposed organizational changes were "essential if the groups were to be of maximum value."[236] The mechanized cavalrymen were not looking for new missions; the missions had already found them, and these missions were familiar in the traditional sense of cavalry, in which men on horses provided what had been known as the power element, and dismounted fighting was the norm. Stripped of the riflemen who had ridden the horses, yet now finding themselves being called on to do the same missions they had once performed, they asked for two things: the resources to do the missions better, and most important, acknowledgment that the doctrine guiding their employment was flawed. Forced to trade horses for machines, the old horse cavalrymen wanted the record set straight about their performance and wanted to be able to perform better in the future when called on to conduct missions that entailed far more than reconnaissance.

While Bradley had worried about the safety of deer in Luxembourg in February 1945, Patton was turning his attention to the future of the army in the postwar era. He envisioned four branches based on his war time experience: Infantry, Armored Cavalry, Artillery, and Anti-Aircraft. "Armored Cavalry" was Patton's answer to the question posed by 12th Army Group about the establishment, by law, of a separate arm "composed primarily of tank units."[237] The notion of armored cavalry represented the verbal reconciliation of the divorce that had torn Cavalry Branch apart in 1940 when the Armored Force emerged as its own entity, but not as a branch. While the Armored Force had been unable to gain branch status, Cavalry Branch clung to life on the deeds of troopers fighting as infantrymen in the Pacific and in Europe, from the backs of iron steeds when not fighting on the ground as traditional cavalrymen. It was the very nature of the fighting in Europe that celebrated the need for the reconciliation to take place. The next major step in this process occurred during the remainder of 1945. Officially the United States Army, and unofficially the last chief of cavalry, reflected on the culmination of a process and product whose genesis was rooted in the attempt to capture what had been learned during the last world war.

6

The Last Cavalry War

I, General Herr, explained to General Marshall the inadequacy of vehicular units for reconnaissance in most terrains, and personally tried to dissuade him from his intention to abolish the horse and substitute completely the vehicular units. He said the weight of evidence was against me. I retorted that not the weight but the volume was opposed and that he must know very well that the persons opposed knew nothing about the subject. He said that I was placing myself in the position of knowing more than anyone else about the merits of the case. I replied that this was right; that I did know more because it was my business to. He said my cause lost some standing because of my vehemence. He did not like my great earnestness. It ended up by my telling him that I thought he was making a great mistake. He has made a lot of bigger ones since.[1]

—Major General John K. Herr, Retired

General Mission, Tactical Doctrine and Technique, and Future Role of Mechanized Cavalry

(1) Major General Gay (Chief of Staff, The General Board) asked whether consideration had been given to eliminating the word "mechanized" in the designation "mechanized cavalry."

(2) It was unanimously agreed by the visiting officers:

(a) That "mechanized cavalry" should be designated as "cavalry."

(b) That horse cavalry should be designated as "cavalry (horse)."[2]

—*Extract from Minutes of Conference on Mechanized*

Cavalry, 27 November 1945

★   The war in Europe was over and in yet another "war to end all wars," horse cavalry remained largely absent from the action. And once again, in its finest tradition, the United States Army, including the residual Cavalry Branch, sought to capture the lessons learned at the cost of blood and treasure. Many leading cavalrymen gathered at Bad Nauheim, Germany, to consider the findings of cavalry group and army after action reports. In the process they continued an effort begun when they stormed ashore on D-Day—to recover their full identity as cavalrymen with or without horses. Ironically, some of their findings echoed the prewar opinions of another man undertaking his own private campaign to restore his credibility and what he believed was the proper role and organization of Cavalry Branch. John K. Herr waged his campaign with letters and attempts to gain the support of Congress. Never having given up on the horse, and with the war concluded, he started to marshal the evidence he needed to restore cavalry to its proper place. The last cavalry war was a contest of ideas. Ultimately the vision of the mechanized men for the future of Cavalry Branch trumped Herr's dream. A year after the victory in Europe, the reconciliation of the Armored Force and Cavalry Branch took place as the former absorbed the latter. Yet even as the iron pony, in a variety of breeds, emerged from the interwar years and World War II as the best means of conducting reconnaissance for fully mechanized and motorized armies—and the horse continued to see limited service—the hope held by some for its wholesale return one day never diminished.

A study of Herr's personal correspondence reveals that he allowed former colleagues and subordinates, many of whom had become leading commanders, the latitude to fight and win their war; he left them alone, whereas he did maintain personal correspondence with the likes of his son-in-law and Wesley Yale. As the action in Europe culminated, and before all attention shifted to the watery expanses of the Pacific, Herr sought answers to important questions that fueled his desire to undo the damage done during the dying days of his watch. In the first of a series of letters, Herr asked George Patton his "$64.00 questions." Unabashedly, Herr opened his inquiry with the statement that he and his longtime mentor, Hamilton H. Hawkins, had done their best to "nurture the belief in horse cavalry" in the pages of the *Cavalry Journal* throughout the war. Now they were placing their faith in "renowned warriors" to lend a voice to their "unshaken faith in cavalry, in spite of the stupidity of those who have sought to discredit it." The matter was of such grave importance to Herr that he singled Patton out as the "#1" cavalryman and one who had "achieved great distinction" during the war. Were Patton to "testify," the believers would "live." Were he to withhold his experience, Herr was convinced "we will die."[3]

> The Questions:
> 1. How do you stand on the necessity for cavalry (horse) in our Armies? If in favor amplify by giving your reasons.
> 2. What effect in your opinion would the presence of a cavalry corps have exercised on the operations in North Africa? Could Tunis and Bizerte have been captured quickly?
> 3. What effect would cavalry and pack trains have had on your Sicilian operations? What would you have added to accomplish the best results?
> 4. What effect would the addition of a cavalry corps or more have had on our operations in France and in your move toward the Rhine?
> 5. With a cavalry corps or more could you have gone right on to Berlin when you had to halt for gas, etc.?
> 6. Do you consider vehicular reconnaissance adequate?
> Note: in my opinion our reconnaissance groups are quite inadequate without horse elements. Porteed if need be!
> I might go so far as to say that the Armored Divisions should also have porteed cavalry.

In closing, Herr told Patton that he looked forward to hearing the "salty response" to his questions "[r]ight from the horse's mouth." But in a note of seriousness, Herr informed Patton that he "must know how you are voting." Perhaps taking into consideration some of Patton's off the battlefield problems during the war, Herr added in a postscript that all responses were to be strictly "confidential and not to be quoted during the war," unless authorized by Patton. A second postscript spoke on an even more personal note offering that he, Herr, was glad to see that Patton's son-in-law, "Jonnie Waters . . . a fine man and grand cavalryman," was doing well. Waters had recently been freed from imprisonment in Germany, having been captured at Kasserine Pass in 1943.[4] Unwilling to count solely on Patton, Herr covered his bets with additional letters.

The same day, Herr composed a letter to the war's greatest cavalry innovator, LTG Lucian K. Truscott. Again, he made his purpose perfectly clear and his point of attack the same.

> I am convinced that the suppression of our cavalry was a stupid and hideous mistake, and I have been endeavoring to keep alive the flame of faith in our horse cavalry. . . . It seems to me that these vehicular units are completely

inadequate. They are largely bound to the roads, and in fact our armies have been for the most part road bound. I will go so far as to say that in my opinion, not even the Armored Divisions are capable of executing reconnaissance for thenselves [*sic*], and are greatly in need of porteed horse cavalry. The same of course goes for the reconnaissance groups.[5]

Herr's continued faith in the use of porteed horse cavalry was, and continued to be, fueled by letters from his son-in-law, Willard Holbrook, which also expressed the ideas of Wesley W. Yale.[6] Herr's continued attack on mechanized units continued to be mobility, which he sought to restore with the return of the horse.

Herr flattered Truscott in the one manner in which he would find common cause with the veteran commanders of World War II drawn from the ranks of the prewar army.

I have, of course, been highly gratified to note your great success, not only because I ear-marked you as perhaps the best bet from the cavalry, but also because the success of polo stars in war has confirmed my theory that polo is our very best school for the development of leaders in war.[7]

Having launched his first round of letters, Herr waited for sympathetic replies. He did not find whole-hearted support for the restoration of the horse in the postwar army, but he did find a common ground with regard to the importance of equestrian sports.

Patton's initial response was a mixed blessing for Herr's campaign and a boon for historians, because the letters between Herr and Patton are not included in Patton's published papers. Patton, the man most synonymous with armored warfare during and after World War II, favored retaining a horse cavalry division but remained convinced that the army must otherwise develop improved mechanized regiments. The most important improvement Patton thought necessary was the addition of a rifle troop to every squadron.[8] He supported this overall stance by responding to Herr's original list of "$64.00 questions."

In response to Question 1 about the necessity of horse cavalry in the postwar army, Patton was consistent with his interwar opinions. He acknowledged the need for the retention of at least one horse cavalry division: "should we be so unfortunate as to have to fight another war and have to do this in our own country, we would have an excellent mounted unit immediately available." Patton believed that shipping horses to a distant theater was probably no longer an option, but with the retention of a single horse division the nation would at least possess "men and officers trained" in the use of horses, since there was now in the army "practically no horse knowledge in the enlisted men or officers of our Armies."[9] Ever the nineteenth-century romantic, Patton, having led the earliest American tank attacks on foot during World

War I, and having driven corps and armies across North Africa, Sicily, and Western Europe, could still not fully appreciate the changes occurring in society around him. Even COL John Considine recognized as early as the first horse-and-mechanized regiment experiments in 1940 that the man on the street in America was now more comfortable with engines than oats.

Patton offered Herr his greatest hope when he remained true to early comments about the utility of horses in North Africa and Sicily. Patton stated unequivocally that: "Had we possessed a horse cavalry division in Tunisia, I believe our situation would have benefited. I do not believe that there was room or water available for a cavalry corps." One can only speculate that Patton would have been happy to have the additional rifle strength in Tunisia given the extensive dismounted fighting required to force Axis forces into capitulation. It does not appear that Patton saw the use of a horse cavalry division as an alternative to the mechanized reconnaissance forces that did see action in Tunisia. Patton was equally unequivocal when remarking that in Sicily, it was still his "considered opinion . . . no German would have escaped from the island" had horse cavalry been available.[10] What was not clear, beyond his comments on the utility of pack animals, was exactly what if any impact horse cavalry would have had on the reconnaissance effort.

If Patton's response to Herr's second and third questions had provided hope that, in Patton, Herr would find a strong advocate for the return of a strong horse cavalry force, Patton's response to question four cast doubt. In France, Patton was quick to point out, "a horse cavalry corps would not have been of any use . . . as the speed was too great and it could not possibly have kept up."[11] Patton was even more critical in responding to question number five, pointing out that it would have been unlikely the horses of a cavalry corps could have even arrived in theater by the time his Third Army's "gas failed on August 27th." More than a function of getting to the fight, Patton concluded that they would have lacked the "sufficient striking power and rapidity of movement to throw mechanized troops off balance."[12] Simply, Patton would not have substituted horses for his own 2nd, 3rd, 6th, and 106th Cavalry Groups during the race across France. Patton's conviction in the utility of his mechanized reconnaissance units was clear in his response to question six, which addressed the utility of vehicular reconnaissance.

Patton disappointed Herr, but his answer was not based in the reality of the war in Europe. Patton offered that "against motorized and mechanized armies vehicular reconnaissance is adequate," but went on to remark that faced with an opponent who fights on foot or "depended on animal transportation," horse reconnaissance would once again "be necessary."[13] This comment was fraught with conflicting ideas. From its meteoric rise in the eyes of the world's collective military observers and opponents, to its dying days in 1945, the German army had depended far more on regular leg infantry and horse-drawn transportation than had its opponents. Patton did lend limited support to the reconstitution of some porté units, so that "officers making close reconnaissance prior to an armored attack" might have a means of getting around.[14] If Patton had largely failed to provide Herr the support he sought in his effort to restructure Cavalry Branch around large and small horse formations at every echelon, Patton offered a gleam of hope.

The very best reason Patton could come up with for the retention of some horse cavalry was that it might foster a restoration of "polo for the whole army," which along with football, he thought—concurring with Herr—to be the "two best sporting preparations for battle."[15] Patton had shared this same line of reasoning with another officer during the summer of 1945 writing to the deputy chief of staff of the army, General Tom T. Handy that,

> This sounds, I admit, as if I were too Arm conscious, but I am not. The Divisions commanded by artillerymen and cavalrymen have in the majority of cases been superior to the Divisions commanded by infantrymen. This is not only true in the armored divisions but in the infantry divisions. Of course, for God's sake, do not show this to the Old Man! [George C. Marshall, an infantryman][16]

Patton believed mounted sports like polo, which he ranked on a par if not superior to football, were "a training school for commanders, because they have to think and sweat at the same time."[17] Even if he was inclined to keep a few horses around, Patton recognized "it would be very difficult," but he mentioned it to his friend the deputy chief of staff, back in Washington, "for whatever it is worth."[18] Patton advocated the retention of the horse not chiefly as a means of reconnaissance or as the foundation of a combat arm, but as a way for the leadership laboratory of polo and other equestrian sports to produce capable combat leaders for the entire army well into the future.

In closing his initial response to Herr, Patton directly addressed Herr's appeal that the great cavalry commanders during the war lend their valued opinions to his efforts to see horse cavalry restored. Pulling no punches, Patton informed Herr that had he, Herr, "accepted the command of the armored corps when I [Patton] tried to make you do it, the cavalry would now be in a much better position than it is." Perhaps in an effort to take some of the sting out of this comment, Patton, just as Herr had, commented on the qualities of the other man's son-in-law, remarking that BG Willard "Hunk" Holbrook had done "a very good job and I enjoyed having him with me."[19] Patton did not have to wait long for a response to his answers.

Herr swiftly wrote to Patton that, "It is encouraging to note that you at least leave the horses [sic] head in the door of the tent."[20] Never one to surrender the initiative when confronted with answers that did not fully support his position, Herr referenced his original questions and commented on Patton's responses. He accused Patton of having bought into the "moss-grown fable" that shipping horses and fodder to a distant theater was impractical.[21] He challenged Patton's assertion that there had not been enough water in Tunisia by citing the presence of considerable cavalry at the Battle of Zama, but was pleased with Patton's comment that horse cavalry would have been useful in Sicily. Herr noted that the blame should reside with "the lack of foresight possessed by McNair and Marshall" and brushed off Patton's response to question four about the utility of horse cavalry in the campaign across France by simply discounting the long hard fighting as little more than a "strategic march; little opposition."[22]

Herr expressed serious displeasure with Patton's assertion that horse cavalry lacked the capacity to confront mechanized forces and invoked the Soviet experience. He reminded Patton that "the Russians didn't know that horse cavalry lacks sufficient striking power and rapidity of movement to throw mechanized forces off balance, when the 1st Cavalry Guard Corps under General [Pavel] Belov, south of Moscow, destroyed General [Heinz] Guderian's famous army consisting of the 17th AD, the 29th Motorized Division and the 167th Infantry Division."[23] He also disagreed with Patton's assertion that mechanized reconnaissance had been adequate. Citing his own "research," Herr remained convinced that the mechanized cavalry troops in the divisions and groups lacked the "powers to bull forward in a road war nor the cross country mobility to find out what is blocking them." He stated his belief that "they have lost more vehicles than will ever be admitted."[24] In this respect, Herr's "research" must have ignored the letters from his son-in-law stating that the old horse cavalry would need a lot more firepower to compete on the modern battlefield.

Regardless of research, in his last surviving letter to Patton, Herr could not let pass the suggestion that it was he, John K. Herr, who should have taken an Armored Force assignment for the betterment of Cavalry Branch. Much of the closeness of the prewar cavalry community had been destroyed in the divorce that rocked the branch in 1940. Herr blamed Patton for the state of cavalry and for his own refusal to take the proffered command of the 1st CavDiv instead of a brigade in the newly formed 2nd AD in 1940. Herr further believed that even had he taken the Armored Force post, "Marshall would never have given me the power that was essential" to command. He was "sure that [he] could never have tolerated him [Marshall] and his evasive methods."[25] One can only speculate about what might have happened had Herr taken Patton's suggestion, but history clearly recorded the impact of Patton's decision to take a lesser command in 1940. The letter also offered a glimpse of Herr in a larger context than that of just a man obsessed with horses, and it provided some idea of what might have been had Patton lived longer. For their sharp disagreement on an issue so obviously important to Herr, the former chief looked forward to hearing the former subordinate's "profane and salty comments on this [the cavalry issue writ large] provided I [Herr] could swear back at a full General."[26]

Patton's last letter to Herr followed quickly and was filled with some of the color that made "Old Blood and Guts" an American icon. Ever the student of military history, Patton immediately took Herr to task for challenging his line of reasoning in regard to water in North Africa. He reminded Herr that the "battle of Zama was fought before the Arabs got to Tunisia," and that at that time Tunisia remained "heavily wooded and well watered." Then in typical Patton prose, he reminded Herr that it had been the Arabs who "cut down all the trees" and that this meant that there was "no water."[27] What really grabbed Patton's attention was Herr's continued faith in the Soviets for no other reason than their use of the beloved horse. Patton, as only Patton could, replied:

> I believe that the Russians are great liars as what they did with horse cavalry, particularly since I have seen units that they call cavalry. It has been my experience in this and the last war, and also the experience of horse cavalrymen in the last war, that reconnaissance cannot be secured by looking. You have to fight for it, and we have found that groups of two peeps and one armored car can get ample reconnaissance and suffer extremely small losses.[28]

Getting the last word, Patton informed Herr that "I don't agree with you one God damn bit, and look forward with pleasurable anticipation to thrashing it out in the Army and Navy Club or some other secluded spot."[29] One can only speculate just how colorful that meeting would have been had the war's most famous, and now mechanized, cavalryman gone head to head with Jonnie K. Herr.

Efforts to enlist the support of Truscott earned Herr a copy of Fifth Army's report to Army Ground Forces in regard to the future of cavalry, since it is filed among his papers with Truscott's comments. Truscott, having fought across the mountains of Sicily and Italy, was upset to find that the Mediterranean Theater Observer Board United States Army (MTOUSA) stated that "no need exists for horse cavalry except for pack animals." Truscott opined that the words had been "inserted by one who has not had the intimate experience with the Sicilian and Italian campaigns that has fallen to my lot." He detailed how during the Sicilian campaign, infantrymen had been converted to cavalrymen mounted on captured horses. Though recruited from a body of men with some horse experience, their lack of formal training led to high losses of the horses, especially among the pack animals.[30]

Most of Truscott's remarks spoke to his belief in the need for traditional horse cavalry and its traditional missions. He said nothing about the utility of the mechanized reconnaissance units that saw extensive service in Sicily, Italy, and southern France. The closest he came to focusing directly on the reconnaissance issue involved the use of provisional horse cavalry troops and mounted battery that served with 3rd ID. He credited them with not only covering the flanks of the division's advance from Battipaglia to Mignano, Italy, but also with helping to "outflank enemy delaying positions behind obstacles in the mountain defiles."[31]

Truscott made additional claims that indirectly cast a serious shadow on the decision to convert all horse cavalry units into mechanized cavalry and armored units. In Sicily, in full agreement with Patton, Truscott held that had the "First Cavalry Division been available in the Seventh Army when we began to advance from the Licata bridgehead, we could have prevented the escape of the German forces from Messina," but had not been able to do so for a want of mobility. Instead, the 1st Cavalry, less its horses, was campaigning in the Pacific theater as line infantry. Making the claim about the use of horse cavalry in Sicily seem almost minor, Truscott, basing his comments on his experience in Italy, saved his greatest damnation for those who had eliminated the traditional horse cavalry:

> I am of the opinion that, had a regiment of well-trained American cavalry been available to me when the Third Division began its advance north from Battipaglia, or even when we crossed the Volturno, the battle of Cassino would never have taken place, Anzio would have been unnecessary, and the Italian campaign might have terminated many months before it actually did.[32]

Given the carnage associated with the fight before Monte Cassino and Churchill's "whale" instead of "wildcat" at Anzio, Truscott's words merit consideration.

Like Patton, Truscott advocated the retention of horse cavalry for use in "our own hemisphere" and "under conditions where armor and other transportation cannot be employed effectively."[33] His return to the common themes of pre-mechanized cavalry—that the horse merely provides the means of transporting the man and the firepower to the proper place on the battlefield—demonstrates his interest in the restoration of the traditional combat missions of the cavalry. He further asserted these ideas in calling for technological advances in "radio communications" and "recoiless weapons" that might offer new avenues for the now defunct horse cavalry, advances unavailable at the beginning of the war.[34]

Herr's timely response included a comment similar to one he had made to Patton, admitting he had "been waiting for some nice famous cavalryman with plenty of glory of service under his belt to come back and take over this job, which I have been holding only until I could achieve a substitution of such a one. Looks like you might be it."[35] If Truscott's views had been worth waiting for, as Herr collected the data he would need to resurrect the old horse cavalry, the views he received from another famous horse cavalryman were not, when they arrived more than a year after the end of hostilities in Europe.

MG Ernest N. Harmon weighed in during the summer of 1946 and demonstrated the poignant transformation of an interwar cavalryman into a complete mechanization advocate. Having spent more time in combat as the commander of an armored division than any other general, Harmon saw no room for horses in the postwar army, with the exception of the limited number being used in the constabulary force he commanded in Germany in 1946. Even the horses included in his constabulary organization were of limited utility, because average patrols were covering from "fifty miles . . . and some as high as 180 miles a day."[36] At least Harmon had the decency, at the beginning of his letter, to warn Herr that he did not expect his views on horse cavalry to be appreciated.

Harmon did not believe horses were worth the shipping space, given their limited utility. He stated, "The horse simply cannot stand up against the carnage of modern warfare. . . . Mechanized troops, on the other hand, traveled down road under very heavy artillery fire without stopping and with very few casualties."[37] Harmon, who saw no service in Sicily, admitted that horses might have been of some use there, but rhetorically asked "how horses could have gone through Italy if the tanks couldn't," given that all the routes were well covered by artillery, machine gun, and tank fire. Harmon offered that the 1st AD had forged ahead only by "smash[ing] our way through these [passes] with steel and gunfire."[38] Then Harmon played a trump card few others could claim, when he reminded Herr that, "in World War I, I fought with the horse cavalry and had the definite feeling when the war was over the horse cavalry was obsolete from the standpoint of playing a major role in another war."[39] Harmon was certain that the horse had no future in modern warfare, but he also shared doubts about the continued utility of the very tanks he had commanded so successfully.

Touching on a debate, which rages into the present, Harmon saw little hope for the future of the tank in the never-ending evolutionary cycle of armored protection versus tank-killing rounds.[40] His views were tainted by the poor performance of American tanks during the war, and he credited the superior number, not quality, of American tanks for allowing Allied units to maneuver on the enemy for killing shots. In language sacrilegious to proponents of the need for ground combat power, Harmon credited much of the Allied success to air power, commenting that Herr would really have to "come over here and see it to appreciate" its "destructive capability."[41] Harmon neither provided Herr with support for retaining the horse, nor did he directly attack the role of mechanized ground reconnaissance assets as he called into question the future utility of tanks. Given that Harmon wrote his letter to Herr with the full knowledge that the last chief of cavalry would be less than pleased, he was probably little surprised by the response he received.

Herr told Harmon that Patton and Truscott agreed with him [Herr], inferring that these other cavalrymen saw the matter more clearly than Harmon did.[42] He then gave Harmon the standard treatment, returning to common themes such as the inadequacy of the mechanized cavalry groups because of their limited mobility and how horses could have easily extended the capabilities of road-bound forces. Herr pointed out to Harmon a point all too often missed by Americans. In returning to the topic of the eastern front, Herr admonished Harmon not to

> assume because you won a war against a fragment of the German army, largely
> as you say because of overwhelming air domination, and in a theater where

exists the best net-work of roads in the world, and where the Germans had no cavalry to delay and cut to pieces the motor elements of the Patton pursuit columns, that these conditions will apply ever again.[43]

Only the future would reveal if Herr was correct.

Juxtaposed to the end of Herr's letter, which read, "With warm personal regards . . . Faithfully Yours," his last sentence was filled with the undying vehemence of the horse versus mechanization debate that had consumed him for nearly a decade. To Harmon he offered, "I regret not that you say what you think but that what you think may be used by the War Dept. to sustain it's [sic] mistaken purpose to assassinate the branch which gave you birth."[44] Herr failed to realize that the men left to carry on the traditions of the divided branch he bequeathed to them were more in line with Harmon's views than with those of Patton and Truscott. Herr's unending criticism of the mechanized cavalry reconnaissance effort blinded him to the fact that these men too wanted a return of the traditional Cavalry Branch that had created their professional identity. Like Harmon, however, and regardless of the soft spot in their hearts for horses, the realities of war left modern cavalrymen asking not for horses to regain their identity, but for more power with which to fight.

Herr was unable to put the fruit of his efforts to work until 1947, when he was called to testify as a witness before the United States Senate. At hand was the matter of returning Hungarian horses captured at the end of World War II, and Herr pulled out all the stops during his testimony. He damned the War Department for retiring him in 1942 and dismantling the horse cavalry. He cited intelligence data prepared by the General Staff G2 in 1946 that saw horse cavalry as essential if the United States were to campaign in Yugoslavia, Poland, and the Soviet Union, and showed that even the Chinese (Nationalists) were learning the hard way about the value of horse cavalry. He singled out "mechanized reconnaissance units," organized against his will, for special criticism. In a prepared statement, he used the material from Patton and Truscott to buttress his case. Directly related to the matter he had been called to testify about, Herr recommended retaining the Hungarian horses, since "horses of this type not only upgrade the entire horse breeding industry in this country, but it would be particularly valuable in case the Cavalry is revived, for upgrading Cavalry remounts."[45] It was not the last time Herr tried to reestablish the horse cavalry, but it was the last time he appeared before Congress.

While Herr had conducted his personal campaign by gathering opinions from senior commanders who still saw limited roles for the horse in the postwar army, the men who had worn crossed sabers during the European campaign waged the only fight that ultimately mattered. The army provided these men with the official forum to write the history that confirmed them as the winners in the long contest between horse and machine.

Patton, a self-described student of history since the age of sixteen, intended to capture the essence of the epic contest in which he had just participated. His desire was to illuminate the experiences of those who had done the fighting— men in units like the 2nd, 3rd, and 4th Cavalry Groups—since it had been his experience that there were many books on war, but few on "fighting." Men who knew anything about fighting were "either killed or [were] inarticulate."[46] Although Patton did not live to see the results of his last command, the Fifteenth Army collected what had been learned during World War II and converted it into General Board and Theater reports. Although Patton's new command had no maneuver units and only created mountains of paper that he feared might never be read, the mission of his command served his own interest in military history and finding out about "fighting."[47] In preparing the report on mechanized cavalry reconnaissance units, the Fifteenth Army was well served by a small collection of articulate cavalry group commanders who had seen plenty of fighting and lived to record their thoughts.

COL William S. Biddle, commander of the 113th Cavalry Group from Normandy to the Elbe, oversaw the efforts of the mechanized cavalrymen. In late September 1945, orders detached him from his position as the assistant division commander of the 102nd ID for ninety days of service with the Fifteenth Army. Arriving in Bad Nauheim, Germany, Biddle found a staggering array of talented officers who had also been charged with drawing lessons from the recent war in order to "construct a guide for future wars."[48] Specifically, COL Biddle worked for the Armored Section of Fifteenth Army, which was filled with former cavalrymen. Together they prepared a report that addressed the missions, tactics and techniques, organization, and equipment of mechanized cavalry units.[49] The final product, published under the auspices of the General Board, came out as Report No. 49 of the *Mechanized Cavalry Units,* or simply "the Biddle Report."

In producing such a detailed analysis, a unique group of cavalrymen assisted Biddle, and they sorted through a series of reports that already reflected large amounts of staff work focused on the same subjects Biddle was to address with his report. Given the mass exodus of units and personnel from Europe as the army withdrew and downsized as fast as it could after having defeated the Axis powers, Biddle was fortunate to secure the help he did. COL Charles Reed, commander of the 2nd Cavalry Group and no stranger to combat, served as Biddle's principal aide. Joining the group very late in the process was LTC Harry W. Candler. Candler had been the "leading from the front" commander of the 91st CavReconSqdn when it made its first attack in North Africa dismounted. After leading this squadron to the conclusion of a successful campaign in North Africa, Candler served on the faculty at the Cavalry School before returning to combat with the 11th Cavalry Group as it fought to the Elbe with Simpson's Ninth Army. LTC George Benjamin served with the 85th CavReconSqdn, and First Lieutenant Donald Burdon came to the team from the 6th Cavalry Group, Patton's "household cavalry."[50] With the exception of Candler, who had actually commanded a tactical unit in North Africa, the board members' experience reflected a bias based on their service in Western Europe, which led them to ignore what had been learned in the Mediterranean when they prepared their final report. The army studies and the after action reports largely used to prepare the final report, in conjunction with the questionnaires the board mailed to the commanders of groups, squadrons, and divisional troops, reflected this same bias. Additionally, there were fewer respondents from the cavalry reconnaissance squadrons assigned to armored divisions and the cavalry troops assigned to infantry divisions.[51] Although Biddle and his group addressed smaller mechanized cavalry units, their primary effort dwelt on the cavalry groups, a reflection of their own combat experience.

Biddle's most important finding shaped all other aspects of the report. He and his group concluded that mechanized reconnaissance units had done very little reconnaissance, the job they were doctrinally tasked with and in theory organized to perform. This finding mirrored the after action report of the First Army that noted, "Campaigns in Western Europe proved the doctrine of 'sneaking and peeking' by reconnaissance units to be unsound." German units had consistently forced the mechanized cavalrymen to "fight to obtain information."[52] In the process of fighting to obtain information, men in the cavalry groups did most of their fighting on foot, in fact nearly twice as often as they performed their missions mounted.[53] "Handicapped" by a "lack of organic strength and firepower," almost all missions the cavalry had performed required some form of reinforcement.[54] Biddle's board rejected the doctrine America had taken to war and reaffirmed as recently as June 1944.

Rather than trying to affix blame for their primary conclusion, COL Biddle and his board focused on righting the wrong with better doctrine and organization. They did not even consider the last chief of cavalry's argument that "many high commanders" had arrived at the wrong conclusions about what the reconnaissance force should look like, since war games had rewarded a race to contact, and each side was confident that nothing operated to their front.[55] Instead, the board argued for a view of mechanized ground reconnaissance units—cavalry—that was more encompassing and would include performing "most of the traditional combat missions of the cavalry," found in *FM 100-5, Operations*. The missions enumerated in the manual were the same operations the mechanized cavalry groups had performed during the European campaign: offensive combat, exploitation and pursuit, seizing and holding key terrain until the arrival of the main force, reconnaissance, security, delaying, and liaison.[56] Rather than maintaining separate doctrinal manuals for mechanized cavalry units, the board recommended a return to *FM 2-15, Employment of Cavalry*, revised to reflect the mechanized nature of the new Cavalry Branch and with rewritten "provisions [particular] to horse cavalry."[57] The board concluded "the mission of mechanized cavalry should be combat."[58] Cavalry was reborn, or at least came one major step closer to reconciliation with the Armored Force, which had ridden away with the choice cavalry missions in 1940.

Concluding that they had performed traditional cavalry missions to some extent, albeit usually with reinforcements, Biddle and his board examined the shortcomings of the current mechanized cavalry organizations—designed for reconnaissance—and made recommendations for how to recast them for their larger combat role. Herr may have correctly gathered that wartime conditions would not allow the headlong dashes that characterized interwar maneuvers, but mechanized cavalrymen did not seek Herr's solution, the horse. Biddle's report was not as blunt as First Army's assertion that although there was great need for additional troopers to provide the dismounted firepower characteristic of the old horse cavalry, there was no need for the horse.[59] Loathe to make any new organization "unwieldy by adding excessive personnel and vehicles," Biddle's most important organizational change called for additional dismounted riflemen.[60]

Although no one debated the need for additional rifle strength in mechanized cavalry units, there was debate about how such strength should be added to the organization, and this immediately led to another organizational issue. The cavalry group concept had never been popular, and commanders often expressed their dissatisfaction with losing their regimental identities—an internalized loyalty inculcated on often remote and austere interwar cavalry posts. First Army's report damned the group organization for cavalry during the European campaign, believing the flexibility anticipated by the group, rather than the regimental organization, was not worth the cost to the individual trooper. No longer able to feel the "unity, esprit de corps, history and morale" associated with membership in a cavalry regiment, modern mechanized cavalrymen were being shortchanged.[61] Those who sought a return to regiments had a friend in the man bearing overall responsibility for the findings of the General Board. Patton grew concerned that a number of cavalry reconnaissance squadrons were "being alerted for redeployment in the Pacific, but that the Group Headquarters" was not accompanying them. Commenting that they had been "habitually used [as] Cavalry Groups, which should be called 'Cavalry Regiments,'" Patton believed the army was committing a "great tactical error to contemplate the employment of single Cavalry squadrons."[62] Patton found the current number of riflemen "inadequate" and strongly believed cavalry groups had "more elan than have infantry units."[63] This élan must have been the key element that allowed cavalry groups, at least in Patton's opinion, "to get forward without getting hurt."[64] With no one taking up General McNair's argument for grouping, Biddle and his board recommended the replacement of groups with regiments composed of three mechanized squadrons.[65]

The decision to include three squadrons in each regiment was directly influenced by the more contentious issue of how to add the rifle strength.[66] Herr had argued for three squadrons in the corps reconnaissance regiment,

TABLE 2—Types of Missions Conducted by Mechanized Ground
Reconnaissance Units in the European Theater of Operations

% Time Spent Performing Specific Mission	Defensive Combat	Special Operations	Security	Offensive Combat	Reconnaissance
Cavalry Group	33%	29%	25%	10%	3%
Cavalry Squadron (Armored division)	11%	48%	24%	4%	13%
Cavalry Troop (Infantry division)	4%	39%	50%	1%	6%

Source—U.S. Army, Forces in the European Theater of Operations, The General Board, *Mechanized Cavalry Units*, Study No. 49, 7–8.
Offensive Combat—Includes attacks, pursuits, and exploitation.
Defensive Combat—Includes defense, delaying, and holding key terrain until the arrival of follow-on forces.
Security—Conducted for other units. Includes blocking, moving and stationary screens, protecting flanks, maintaining contact between larger units, filling gaps.
Special Operations—Includes acting as mobile reserve, providing rear-area security, operating as Army Information Service.

but his vision included two squadrons of men on horses and one on iron ponies.[67] Admittedly, Marshall and McNair's decision to convert the horse-mechanized regiments into all-mechanized regiments contributed to the lack of riflemen that Patton later viewed as one of the salient deficiencies in the organization of the existing squadrons.[68] He related this general theme to Herr: "The present cavalry regiment *mez.* [inserted by hand] is efficient but it is too weak in dismounted rifle power. Each squadron should have an additional troop composed of dismounted cavalry armed with rifles and light machineguns."[69] Biddle's board, with Patton's assistance, formalized the call for more riflemen.

Biddle's primary task was to select one of the recommendations put forward by the field armies or some combination thereof. First Army proposed adding a squad of riflemen to every platoon while Seventh Army called for two squads. Third Army suggested adding an entire troop of riflemen to each squadron. Biddle decided to add a troop to each squadron and a squad to each platoon, but this resulted in a cavalry regiment with more than three thousand men, thus violating the other guiding force of reorganization that endeavored to keep cavalry units small enough to avoid becoming unwieldy.[70] At this point the question of what kind of reconnaissance unit armored divisions should have in the future came to the attention of military authorities, who provided the solution.

A conference of armored division commanders convened in November 1945 to discus the organization of their units in the years ahead. The commanders proposed to formalize the trend of replacing armored cars with light tanks in their reconnaissance squadrons. The armored division commanders called for a reconnaissance squadron built around four reconnaissance troops and an assault-gun troop. Each reconnaissance troop was to have nothing but light tanks and jeeps. Biddle, who was present at this conference, voiced his concerns about the proposed organization's lack of dismounts and what he termed a "lack of sustained mobility."[71] Although the committee overruled Biddle's objections, Patton vetoed its proposal and sided with Biddle.[72] Biddle's intervention reflected his combat experience and what he was trying to do in a bigger sense. Although he recognized the shortcomings of the armored car, he valued its ability to carry his group across Belgium when the light tanks had been unable to keep up for lack of fuel and worn tracks. The need for riflemen to dismount in furtherance of the reconnaissance mission was consistent with his campaign to restore the full cavalry identity to the now all-mechanized cavalry force. The armored division commanders sought a more robust reconnaissance organization that would not get in the way when it encountered resistance. These sentiments reflected what Charles Scott had predicted between the wars and what he reconfirmed after he returned from North Africa where he had observed the British.

Patton's intervention led to a reorganized three-squadron regiment; individual squadrons would each have two instead of three reconnaissance troops with an infantry squad added to each platoon. Squadrons would retain their assault-gun troop and light tank company, and would add an entire troop of riflemen. The subtraction of a reconnaissance troop from each squadron, from the wartime three troops per squadron, ensured the regiment would not become unwieldy. Squadron commanders would look to regimental commanders to support them and to rotate their troops in and out of the line.[73] Unlike wartime group commanders, regimental commanders would be more assured of having the forces they needed to accomplish their missions because their squadrons would be organic and include the riflemen needed to perform more than passive reconnaissance. COL Mark Devine's 14th Cavalry Group certainly would have had more options under this organization, but one can speculate where Army Ground Forces would have found the riflemen to help out the cavalry, when it could not keep pace with the demand from the infantry divisions. With war over and without the constraint of the "90 Division" gamble, which had governed the number of ground combat divisions the United States would field during World War II, Biddle and his board were free to speculate and propose to their hearts' content. Reconnaissance squadrons proposed for service with armored divisions emerged with the same organization, except they now had three reconnaissance troops rather than four and gained a rifle troop

TABLE 3—Comparison of Fire Power in a Cavalry Reconnaissance Squadron and an Infantry Battalion

Weapon	Cavalry Reconnaissance Squadron	Infantry Battalion
75mm gun (light tank)	17	0
75mm howitzer (assault gun)	6	0
57mm gun	0	3
37mm gun (M8 scout car)	37	0
81mm mortar	0	6
60mm mortar	27	9
.50-caliber machine gun	25	6
.30-caliber machine gun	122	22
.30-caliber Browning Auto Rifle	0	45
.30-caliber rifle	126	624
.30-caliber carbine	394	249

The General Board concluded that although a cavalry squadron could generate 200% more firepower than an infantry battalion could with 75% of the strength in manpower, it was dependent on "fairly short fluid fighting . . . done from vehicles or at short distances from them." Extensive use of cavalry reconnaissance squadrons for dismounted combat forced them to abandon the firepower advantages their vehicles afforded them and led to calls for weapons like the Browning Automatic Rifles (BAR) and a higher percentage of rifles to carbines. Data derived from U.S. Army, Forces in the European Theater of Operations, The General Board, *Mechanized Cavalry Units*, Study No. 49, Appendix 8.

to offset their loss.[74] Ironically, the riflemen insisted on by the cavalrymen had been part of the armored reconnaissance battalion before World War II. The idea of riflemen riding in the van on vehicles had been developed at Fort Knox, yet now it was the men with a greater affinity for Fort Riley who insisted on bringing back this old idea.

The issue of mechanized ground reconnaissance units for infantry divisions had never been as contentious. Throughout the war, divisional cavalry troops earned high praise, so there were fewer issues to deal with on 20 November 1945 when the Fifteenth Army Operations Division, G3, convened a conference of former infantry division commanders.

Again, without manpower constraints, the commanders concluded that a single troop of mechanized cavalry had not been adequate to support a division's needs. As evidence that troops had generally performed well, the majority of the commanders requested the augmentation of the existing troop by providing it with similar resources found in the wartime cavalry reconnaissance squadrons, such as light tanks and assault guns. Biddle, who attended this conference, recommended a two-troop cavalry squadron instead, later approved by Patton.[75] Again, Herr could say, "I told you so." He believed the single reconnaissance troops, which had been added to the relatively new "triangular" infantry divisions, were insufficient in size, forcing "combat team commanders" to form their own "reconnaissance troops . . . consisting of Jeeps, Bantams, and weapons carriers filled with infantrymen and supporting weapons."[76] Of course his proposed solution had been the horse cavalry regiment.[77]

In every instance, Herr wanted more cavalry, especially horse cavalry, to ride in support of every echelon of army forces, all the way from proposing that each field army have its own division of horse cavalry and a mechanized reconnaissance regiment down to each infantry division having its own horse-mechanized reconnaissance regiment.[78] If Herr and his supporters had dreamed big before the war in attempting to secure the future of Cavalry Branch, another group of men after the war had grand aspirations to redefine Cavalry Branch. Had Herr known what they were attempting, he would have been proud.

On 27 November 1945, Biddle assembled a remarkable number of former cavalry group commanders. They included: now BG Joseph Tully, who had commanded the 4th Cavalry Group from D-Day to VE Day; COL Edward Fickett, who had commanded the 6th Cavalry Group in its role as the Army Information Service and in combat; COL John C. McDonald of the 4th Cavalry Group; COL Larry Smith, who took over the 14th Cavalry Group after its demise in the Ardennes; and COL Garnett Wilson of the 115th Cavalry Group.[79] Collectively, these mechanized cavalry commanders provided an enormous amount of input for COL Biddle's report, and many of their key ideas were preserved in the annexes of the final draft. Having concluded that their units had always been used to "the limit of their strength," the former commanders sought solutions that exceeded the recommendation for a three-squadron regiment.[80]

Referencing the creation of the provisional mechanized cavalry brigade in Patton's Third Army, the conference members called for brigading two of the newly organized cavalry regiments with supporting troops, "for assignments to the army and employment when a light and fast but strong striking force was needed."[81] It will be recalled that the provisional cavalry brigade was employed in what was largely a dismounted attack. Having tossed out that idea, however, they went even further. If each army needed a brigade of mechanized cavalry, then why not a division? Less the horses, their proposed cavalry division with two mechanized cavalry regiments and a regiment of "armored infantry or dragoon regiments and supporting troops" was clearly an effort to build the kind of organization that would be almost on an equal footing with the wartime armored division.[82] Their dream was short lived. Patton torpedoed the division concept and allowed that perhaps the brigade concept merited testing. He also insisted on the addition of tank destroyers to each squadron. The cavalry commanders had been divided on this issue, because tank destroyers had been readily available throughout the war and had become all but a permanent attachment in many cavalry groups.[83] In respect to the future, the most important organizational recommendations to emerge in COL Biddle's final report were that cavalry regiments replace groups and that each regiment be composed of three combined-arms squadrons. Such an organizational pattern would ensure each regiment's ability to perform the revised doctrine largely with organic assets.

Absent from COL Biddle's recollection of the late November conference of cavalry commanders was the need for reconnaissance aircraft in each group or regiment. Long one of the most crucial missing pieces of all mechanized reconnaissance organizations, the issue received scant attention in the final report, a single sentence offering nothing more than "[t]hat liaison aircraft should be provided in mechanized cavalry units."[84] Perhaps the reason there was so little discussion of such an important matter was because only three days before the conference the army finally added two liaison airplanes to each cavalry squadron. The army did this when it published the third change to table or organization for cavalry squadrons in June 1945. There was no longer a need for all the ad hoc organizations and field-expedient methods that had attempted to overcome this incredible deficiency, because doctrinal and organizational codification finally captured what had been practiced at Fort Knox since the 1930s.

Like doctrine and organization, the report's section on equipment restated the same concerns expressed throughout World War II in observer and after action reports. For its findings, the General Board drew heavily from the questionnaires returned by mechanized reconnaissance commanders and from the discussions that occurred at conferences of the armored division commanders and the cavalry commanders.[85] In these questionnaires and discussions, attention was focused on vehicles, weapons, and communications, with vehicles generating the most comment.

The Biddle Report supported retention of wheeled armored reconnaissance vehicles with the caveat that whatever the army finally selected should be an improvement on the M8 Armored Car.[86] This reflected the long-held belief in the interwar mechanized community that technology must continue to improve to satisfy the needs of the cavalry trooper, whereas the horse advocates also pointed to technological shortcomings in vehicles for retaining horses. Now the roles were almost reversed as mechanized cavalry leaders looked to technology to provide them a better vehicle, while armored division commanders recommended replacing the cars with proven technology: fully tracked vehicles. Cavalry group commanders appreciated the range and road mobility of wheeled vehicles although quickly acknowledging the poor off-road characteristics of the current armored car.[87] Armored division commanders were far more apt to replace the armored cars with light tanks or fully tracked personnel carriers. I.D. White, who finished the war commanding 2nd AD, but had been one of the key personalities in the development of mechanized reconnaissance at Fort Knox, Fort Riley, and then at Fort Benning after the creation of the Armored Force, no longer saw the need for "huge cruising distances" or "quietness of operation." He now advocated nothing but fully tracked personnel carriers for accompanying riflemen and light tanks and assault guns to assist their advance.[88]

The role of the jeep received scant mention. There was no debate over its future in mechanized reconnaissance organizations. The only minor issue to emerge was centered on whether the jeep should be armored. From North Africa to the Elbe River, jeep-mounted scouts had confronted mines and small-arms fire and developed a number of techniques to improve their survivability, ranging from sandbags to armored windshields. Not all cavalry commanders agreed that armored protection was even necessary, though the constant effort of the troops in the field to protect themselves speaks for itself. LTC John F. Rhoades, commander of the 4th CavReconSqdn during the war, reasoned that "the first burst usually misses, and without armor plate there is no tendency to huddle behind the jeep—it isn't meant to be a tank, armored car, or large gun platform."[89] In the end, troopers received some solace in the board's recommendation that in the future jeeps have some type of lightly armored windshield with wings to protect the occupants from "frontal or near frontal small-arms fire."[90] The question of providing the men who rode in the van with some form of protection to survive an expected encounter with the enemy was at least resolved on paper if not in practice.[91]

Other observations and recommendations about equipment sought to improve mechanized cavalry units' ability to perform the traditional missions they hoped to regain with the inclusion of additional riflemen. The board report cited the need for a larger-caliber assault gun, 105mm instead of the 75mm gun they had used throughout the war. Calls for fewer carbines and more rifles, including Browning Automatic Rifles, reflected fresh memories of extensive dismounted fighting. Calls for better and lighter radios, and more long-range radios recalled the heavy reliance of the mechanized reconnaissance units on this form of communication to synchronize their efforts and fulfill their information-gathering and reporting mission. Like the attention given to infantry weapons, calls for additional wire and telephones again reflected the amount of time mechanized reconnaissance units had spent in static positions, instead of conducting mounted operations. A tinge of interwar language entered the board findings related to what kind of vehicle the recommended rifle squads should use. All agreed they wanted a full-tracked armored vehicle, but some continued to express reservations about such a vehicle's ability to operate over extensive distances and to achieve a degree of silence in operation.[92] One particular observation about equipment found the board reasoning along the same lines as John K. Herr, and it exposed both parties' interest in maintaining mobility with as much firepower as possible. Herr and the board both pinned high hopes on further developments in recoilless weapons technology, because such technology seemed to offer lightweight tank-killing capabilities.[93]

Members of the General Board on mechanized cavalry were largely products of the interwar army. They had not set out to disregard the flawed doctrine that was supposed to guide their actions, but rather they accepted willingly the missions that came to them and carried them out to the utmost of their abilities. In doing so, they exposed the greatest weaknesses in the mechanized cavalry's organization and doctrine. Equally important, no longer able to fall back on the horse as the best means of making up for exposed shortcomings—even if they had desired to—the mechanized reconnaissance men implemented the needed changes with the assets on hand, and after the war they advocated additional measures needed to maximize the proven potential of such reconnaissance. They had demonstrated that mechanized cavalry was far more capable and versatile than the narrowly defined list of capabilities the prewar doctrine prescribed and were confident that what they had done proved they were cavalrymen in the most traditional sense, mounted and dismounted fighters, not mere collectors of information. They represented all that was left of the old branch, and all that was left were those mounted on "iron ponies." Gone was the horsy and greasy divide, and what they sought was the redress through doctrine and organization that would end the interwar debates. They submitted their draft findings to Patton on 20 December 1945, complete with historical vignettes prepared by LTC Candler.[94] Struck by a life-threatening embolism that very same day—a complication of the paralyzing injury he sustained on 9 December—Patton never read the report, because he died the next day.[95] The final draft was approved by the acting board president in early January 1945.[96] Biddle's work was done.

While Herr undertook his personal campaign to restore the proper place of the horse, and Biddle's Board attempted to capture the specific lessons learned about mechanized reconnaissance units in Europe, there remained the unresolved issue of the future of Cavalry Branch. An article in the September–October edition of the *Cavalry Journal* revealed the tension of the unresolved Cavalry Branch identity crisis. With his article titled "Let's Face the Facts," COL Roy W. Cole, Jr., observed that Cavalry Branch was nothing but "infantry, armor, and 'spare parts.'"[97] The problem, COL Cole observed, was that until the army resolved the future of Cavalry Branch, veteran officers stood to suffer. Those who had fought as infantrymen in the Pacific with the 1st CavDiv deserved the opportunity to use their hard-won expertise to advance their careers inside Infantry Branch, if there was to be no independent Cavalry Branch worthy of rebuilding with their talents. The same was true of the former cavalrymen who had fought in armor units. Cole said nothing of the "spare parts," perhaps unable to imagine a Cavalry Branch built solely around the types of mechanized units that had seen so much combat in the European theater. He could envision no future for the "spare parts" except absorption by the Armored Force or the restoration of a mounted arm "second to none in personnel, arms, equipment and indoctrination."[98]

The same issue of the *Cavalry Journal* featured the last article prepared by horse advocate, Hamilton Hawkins. Convinced to the end that the use of horses would have "saved both time and losses," he abruptly conceded that future discussions on the subject were "academic" and perhaps limited to "military students and historians."[99] What had caused Hawkins to step down from the pulpit from which he had so strongly advocated the continued military role of the horse, particularly in the field of reconnaissance? Hawkins conceded that with the advent of the atomic bomb, "it may be useless to speculate on these subjects," since "no man" could tell what the future held, and that age-old question of how to stop war "might already have been found."[100] Having fought valiantly for his trusted companion, Hawkins was unwilling to continue the fight now that the battlefield had transitioned from being mechanized to nuclear. He slipped completely away from the debate in 1950, dying at the age of seventy-eight.[101]

Even if Hawkins was no longer willing to fight for the continued use of the horse in the pages of the *Cavalry Journal,* his inspired subordinate John K. Herr was willing to defend the old Cavalry Branch. Hearing rumors of a potential merger of what remained of Cavalry Branch and the Armored Force, Herr took up his pen and addressed the editor of the *Cavalry Journal* in the May–June 1946 edition. In an effort to "release the historical truth," Herr published a number of the memoranda exchanged between himself and George C. Marshall and the Army G3 between May 1939 and June 1940.[102] It demonstrated Herr's continued belief that the Armored Force had been unfairly taken from Cavalry Branch. It also demonstrated his frustration and perhaps anger with what he knew was about to transpire. He was not alone in his last-ditch effort.

In what would be the final issue, the *Cavalry Journal* continued to slake the thirst of its old members' desire to read about horses. Just as the significant cavalry commanders during World War II had grown to see the continued use of horse units for leadership training and not specifically combat or reconnaissance, the last edition of the *Cavalry Journal* introduced its readers to a similar line of reasoning. "The Horse's Place in Our Future," put some of the burden of the emerging Cold War firmly on the back of the horse. Citing the new world in which every citizen "without regard to age, sex, or station" might become a soldier or serve on the home front, the article opined that everyone must be prepared to serve.[103]

> In such a world; the physical fitness and aptitude of our citizens become of paramount importance. To build and maintain a strong, vigorous, active manhood and womanhood is both a necessity and an obligation. Because it

involves our national security, the obligation is properly that of the agency charged with our national defense.[104]

Believing that World War II had exposed softness in the average American citizen as a result of living a "modern life with its abundance of luxuries," salvation could only be found in getting people "into the country," where they could develop "stamina, self reliance, and mental alertness."[105] The horse and "wholesome sports," such as jumping, hunting, all forms of racing, polo, rodeos, and other physical activities, provided the best way to ensure a ready population. Of course, such a venture would require the maintenance of the Remount System, which would have the added benefit of helping supply the lesser-developed Allies with the horses they needed. Maintenance of the Remount System would also ensure a ready supply of horses for use in the Western Hemisphere.[106] Perhaps the author's concern for the health of the manhood and womanhood of the nation was little more than a stalking horse for the restoration of the traditional horse cavalry?

There was little possibility of this happening. I.D. White returned to Fort Riley as the Cavalry School commandant during the summer of 1945. No one was happier to have one of his own calling the shots at the home of cavalry than MG Charles L. Scott, who continued to command the Armored Center at Fort Knox.[107] White, no longer an outsider from Fort Knox, as he had been in the late 1930s when he served on the faculty of the Cavalry School as the instructor of mechanized reconnaissance, was in charge. His assistant commandant was Vennard Wilson, who had commanded the 106th Cavalry Group during the fight to liberate Europe.[108] During White's tenure, Fort Riley experienced sweeping changes.

In the fall of 1946, GEN Jacob L. Devers, now the commander of Army Ground Forces, disestablished the Cavalry School and recast it as the Army General School Center and the Center for the Aggressor Command.[109] White left little in his papers to indicate any displeasure with this redesignation of his command. After being a star polo player at Norwich and serving in horse cavalry units, he had come under the influence of the evolving ideology at Fort Knox when he was the aide to the first post commander there, and he expressed some sentiment for the changing of the times. In an address to the National Horse Association dinner in New York on 3 November 1946, White acknowledged the role of equestrian sports in "the development of the quick-thinking, aggressive, and versatile officer."[110] That was the extent of any connection he drew between horses and the fighting that took place during a war. The balance of his remarks only lamented the difficulty the army would have in the future in trying to field competitive equestrian teams: "[W]ith no horse-mounted units in the Army, it is natural that the young officers will turn to other forms of sport."[111] No mounted units also meant it would be much harder for the army to maintain "suitable mounts . . . for training for high grade competition."[112] Winning horse shows would be more difficult in the future, but not winning wars.

White did more in 1946 than carry out the orders of Devers to disband the Cavalry School and speak to the National Horse Association. He provided the leadership that created the first symbolic measure to reconcile the divorce that had torn Cavalry Branch apart in 1940. As the president of the Cavalry Association, White called a meeting of the association's members on 8 July 1946. The purpose of the meeting, which drew 613 members in person or by proxy, was to amend the association's constitution. The War Department had already decided to form a new branch that encompassed all armor units, mechanized cavalry units, and horse cavalry units, by linking them with their common doctrines of "aggressive mobility combined with great fire power."[113] The War Department had concluded, and so had the executive committee of the Cavalry Association, that "We may thus regard animal elements, mechanized cavalry and armored units as the three components of the Armored Cavalry."[114] This was the beginning of the process Patton had predicted in February 1945, when he called for the combination of the Armored Force with the "Cavalry Groups" as the basis of a new branch.[115] Hoping to pull all of its former members closer together within a common professional association, the executive committee decided it would be better to put the name change to its membership, since the War Department had already made its decision to do the same. All that remained was for Congress to ratify the decision, but by then, if the association did not act, the substantial number of its members now associated with the Armored Force and mechanized cavalry units might elect to form its own association. When the votes were tallied, 440 members voted in favor of the name change that 160 members sought to avoid.[116]

Unsurprisingly, John K. Herr "deplore[d] any such abandonment of our cavalry," especially since there had been "no legislation making effective the stupid effort."[117] MG Verne D. Mudge responded to Herr's letter and criticism of what was about to take place. Mudge promised that before the vote was taken, Herr's letter and the views it expressed would be distributed and read to the "Executive Council." Mudge further thanked Herr for his continued "interest and leadership" and promised to write again after Herr's letter had been considered. Based on the outcome of the voting, Herr's letter had little impact, but Mudge's original reply, before the vote had taken place, holds a clear indicator of Herr's position on the outcome. Scrawled at the bottom of Mudge's letter to Herr appears the following note: "Mudge never wrote to me again as he indicates he would. He wrote a miserable apology published in the Cavalry Journal trying to justify the action taken in changing its name to armored association journal."[118] Herr never forgave Mudge and the others like him who saw and seized the future. Mudge drew Herr's ire again in the future.

In the first edition of the *Armored Cavalry Journal,* the Armored Cavalry Association promised to continue to satisfy its former readership's interest in "the horse and all interests pertaining thereto either from a military standpoint or otherwise."[119] As proof of the commitment to the proud traditions of the cavalry, "OLD BILL" (a Remington-inspired drawing of a Great Plains cavalryman) continued to "occupy his usual place on the cover," but as "a good Cavalryman could never stand still," the recast journal was moving ahead with the changing times.[120] The subject matter for the first edition bore out both promises, but clearly demonstrated the new direction of the Armored Cavalry. The first seven articles were either excerpts taken directly from the historical vignettes included in COL Biddle's official findings or otherwise related to mechanized cavalry or armor operations. One had to search to the bottom of the first edition's table of contents to find a two-page article dedicated to Japanese horse cavalry.[121] "Armored cavalry" started to replace "cavalry" in August 1949, when official correspondence described the branch as "an arm of mobility, armor-protected firepower, and shock action," capable of all types of combat in combination with other arms.[122] This represented the next step in fulfilling the complete reconciliation of the branch that the Cavalry Association had predicted in 1946 when it called for the name change.

The last step occurred in 1950 when the *Armored Cavalry Journal* became *Armor Magazine*. This name change reflected the legal change that had finally taken place when the 1950 Army Organization Act designated *Armor* as the new branch that "shall be a continuation of the Cavalry."[123] The new journal captured the complete reconciliation of Cavalry Branch and the Armored Force allegorically with "Old Bill" superimposed on one page of the announcement and an M26 Patton tank superimposed over the text on the facing page.[124] There was no more identity crisis, there was no more compromise, there was no more debate. The armored cavalry regiments that had begun to reappear in 1948, as the constabulary regiments in Europe took up their combat roles again, now had a home in a new branch.[125] The conversion of constabulary regiments back into modern armored cavalry regiments also meant the final disappearance of the horse from conventional military operations.[126]

When John K. Herr died in 1955, his obituary reminded readers that "[n]o one ever loved his chosen branch more, nor fought for it harder."[127] Herr continued to correspond with like-minded individuals who blamed Marshall, McNair, and the "Benning Boys" for the ultimate destruction of the once proud Cavalry Branch built around the noble horse.[128] The Korean War gave Herr his last opportunity to bring back the horse cavalry. Having failed during the late 1940s, Herr finally found a congressman willing to put forward a House resolution to reestablish the horse cavalry. Herr wrote Congressman Carl Vinson, chairman of the House Armed Services Committee, that he was "delighted" to be given the opportunity for a "hearing in an open court," and that he looked forward to confronting "any or all" of the issues so "cleverly concealed" by "the infantry generals, including Marshall and Eisenhower."[129] Herr must have been immeasurably disappointed with Vinson's reply, which stated, "there can be no prompt action on the part of the Committee on this matter."[130] Yet, only a few days later, Vinson sent Herr another note with a copy of what would become House Resolution 5156, a measure to "reactivate the Mounted Cavalry as a basic branch of the United States Army, and for other purposes."[131] It was the second attempt of the bill's sponsor, Congressman Daniel J. Flood of Pennsylvania, to introduce the measure.[132] When neither bill succeeded, Congressman Flood, sounding much like Herr, still believed that because "the Pentagon is blithely unaware that 75 per cent of the world's land is mountainous," the nation was wasting billions of defense dollars while leaving itself "unprotected in the vital realm of cavalry." Flood concluded that "for lack of the horse, the war may be lost."[133] Unwilling to let this fate befall the country and again unable to move Congress, Herr took up his own pen one last time.

Two years before his death, Herr coauthored a book, *The Story of the U.S. Cavalry, 1775–1942*. Clearly he saw the end of his branch coinciding with his own dismissal and the decision not to employ horses during World War II. With the Korean War still underway as he prepared the manuscript, he convinced GEN Jonathan "Skinny" M. Wainwright to write the book's forward. Wainwright had surrendered American forces in the Philippines at the beginning of World War II and emerged from his captor's cage the recipient of the Medal of Honor. He had also been the last American commander to lead horse cavalry in war, the Philippine Scouts of the 26th Cavalry Regiment. Wainwright wrote:

> While the peculiar characteristics of mounted cavalry, in its ability to operate over any terrain, in any weather, and at night, have to some extent been taken over by armored troops and the air force for purposes of reconnaissance, the cavalry must be considered superior to the air force in bad weather, at night and in heavily wooded country. Off the road and in mountainous country, armored troops usually cannot operate, and the horseman can go anywhere. . . . Long have I advocated the retention of at least one full-strength mounted cavalry division. . . . Such a cavalry division can operate in any weather, in any country, in any climate, and in any terrain.[134]

During the same year his book was released with Wainwright's endorsement, Herr made his last attempt to be heard before the Congress. A man who had helped regain the same islands surrendered by Wainwright, Herr concluded, stymied this attempt.

Among his papers, John K. Herr, the last chief of cavalry, left a statement he prepared on 25 August 1953, upon learning that once again he had missed his chance to appear before the Senate Armed Services Committee. In the one-page document he recounted his failed efforts to be heard in the 1940s, except for his 1947 testimony, which he briefly recounted. In the statement, Herr concluded his latest inability to gain the ear of the Senate was "*mal odorous*," since a certain individual had led him to believe there was plenty of time for Herr to be heard. It just so happened that this individual was retired MG Verne D. Mudge, the same man Herr crossed sabers with over the renaming of the Cavalry Association in 1946. Mudge worked for the Senate Armed Services Committee in 1953. Herr could not understand why this former cavalryman had not insisted that Herr "be heard" or at least warned that schedule for testimony had been changed, so that Herr "could in person hasten there."[135] Although he concluded the entire matter was a "profound mystery," one who has spent any time with Herr's papers can readily recognize Herr's view of this as mounting evidence of the vast conspiracy that had led to the end of his beloved horse cavalry. When Herr died, so died even the faintest hope that horses might one day supplant machines for ground reconnaissance, or so it seemed.

William S. Biddle's contribution to the most sweeping document about mechanized ground reconnaissance was incredibly important and deserves special consideration in the context of the remainder of Biddle's life. History has recorded well the efforts of the early pioneers in the mechanization process and how these efforts led to the creation of modern armored divisions. Although Patton often gains far more attention than justified, given his fence-sitting posture through most of the interwar years, many are familiar with the names Adna R. Chaffee and Daniel Van Voorhis. For them and for men like Charles L. Scott, Bruce Palmer, I. D. White, Willis Crittenberger, and Robert Grow, mechanized reconnaissance was merely an important enabler to a more important end, an all-mechanized force capable of performing the repertoire of traditional cavalry missions on the modern battlefield, regardless of the mount used. Still fewer recognize the name John K. Herr, the last chief of cavalry, unless perhaps they are students of the interwar years. Herr represents the extreme position, some might argue a reactionary one, for those who sought the retention of the horse at all costs. As history reveals, he was joined by other notable names in advocacy for maintaining a few mounted units and the equestrian sports he so loved. What then of Biddle?

Biddle, and to a lesser extent Charles Reed, may be seen as the last of the real horse cavalrymen and the first of the modern mechanized men. Biddle did not see combat with the armored force or with the 1st CavDiv in its dismounted role in the Pacific, but he clung to his branch identity even when the last of the horses went by the wayside in 1942. He had spent his entire life aboard steeds of flesh, beginning on his mother's ranch in Oregon as a boy, then as a commissioned cavalry officer from West Point in 1923, and later with interwar horse cavalry units on the Great Plains. He rode across Europe on an iron pony, yet refused to accept the narrow role his beloved horse cavalry branch had prescribed for the unit he commanded. He never looked back. His leadership of the board of officers tasked with critically examining the role of mechanized cavalry afforded him the opportunity to inject the smallest wedges of sentiment to keep alive the hope of the horse in some capacity, as his peers Willard Holbrook and Wesley W. Yale were attempting to do in the 11th AD. Herr had not seen firsthand what Biddle had witnessed and would never be able to understand how a man like Biddle could turn his back on the beloved horse. Herr would have been wrong to assume that Biddle rejected the horse. Biddle never gave up his passion for horses. Just as Charles Reed continued to ride one of the offspring of the horses he saved at the end of World War II, Biddle maintained an active association with equestrian sports. He served as the chief riding instructor at the Rock Creek Stables near Alexandria, Virginia, until 1973. He was a member of the board of directors for the Washington International Horse Show and was a leading American Horse Show Association judge until incapacitated by illness in 1976.[136] Despite his undying passion for horses, Biddle put his shoulder not only to the task of winning World War II with an all-mechanized cavalry reconnaissance force, but equally to the task of improving the force for the future.

John K. Herr was wrong when he concluded in his book that 1942 marked the end of the United States Cavalry, but he was correct on a variety of other issues that emerged after World War II. Herr's organizational vision of cavalry units before the war closely mirrored the findings of the General Board after the war. What Herr never realized was that his desire to keep horses in modern American cavalry forces undermined the doctrine, organization, and wartime capability of the mechanized ground reconnaissance units whose growth he could not curb. Mechanized ground reconnaissance units attempted to fulfill the role established for them at the close of the last world war. Interwar doctrine anticipated their ability to range between the leading edge of the main body and the outer limits of the umbrella of visibility created by air power. The expectation that there would be room to maneuver in this space between the main bodies of opposing forces had minimized the perceived importance for these specialized units to be able to fight. This expectation, in combination with the efforts of those like Herr to minimize the role of mechanized ground reconnaissance units, resulted in flawed doctrine and organization. Yet, even with the flawed doctrine and organization, men like Biddle resurrected the United States Cavalry, which Herr argued had expired in 1942, by proving its ability to do more than just reconnaissance, because obtaining vital information often required fighting. Wartime efforts of the mechanized cavalrymen and the conclusions they drew from their experience formed the basis of American cavalry doctrine and organization for the remainder of the twentieth century. These men earned the right to be called simply "cavalrymen," mechanized was now implied.

Conclusion

I am advising a man on how to best employ light
infantry and horse cavalry in the attack against
Taliban T-55s (tanks), mortars, artillery, personnel
carriers and machine guns—a tactic which I think
became outdated with the invention of the Gatling
gun. . . . We have witnessed the horse cavalry
bounding overwatch from spur to spur to attack
Taliban strong points—the last several kilometers
under mortar, artillery and sniper fire.[1]
—*Attributed to an American Green Beret operating in*
Afghanistan in 2001

★ From beginning to end, airplanes changed the manner in which armies conducted reconnaissance during World War I. Even so, when the war ended, there remained a conviction that roles for horse-mounted units continued to exist. These roles included reconnaissance and anticipated a continued need for horse cavalry as a full-fledged combat arm. American military innovators recognized that something was needed to fill the gap between the far-reaching capabilities of airplanes—to render strategic reconnaissance deep behind enemy lines—and the leading edge of ground reconnaissance forces. Armored cars, with road speeds and a range greater than horses or dismounted men, seemed like a viable solution. Horse-mounted units continued to be considered indispensable for ground reconnaissance, since only they still provided true all-weather, day-and-night, cross-country traversing reconnaissance, in the minds of many soldiers. The decade leading up to World War II pitted armored cars, an emerging technology, against horses, a proven technology. Only after the United States entered World War

II did the machine completely earn the job of performing reconnaissance for the army. Yet, it took all of World War II to end unequivocally a future role for the horse in anything but the most special circumstances. The tension created in the struggle to preserve a role for horses had a direct bearing on the organization, doctrine, and capabilities of the fully mechanized reconnaissance units that fought in Europe during World War II.

After World War I, the American army, and Cavalry Branch specifically, recognized the potential value that mechanization offered in the field of ground reconnaissance and combat, yet nearly ten years passed before any meaningful work was conducted to that end. Forced into action by the secretary of war, Dwight Davis, after his visit to England in 1927, Cavalry Branch's initial contribution was the formation of an armored car troop. Having served its purpose near Washington in the first short-lived mechanization experiment, the armored car troop disappeared to Texas where it was largely ignored. Poland's defeat in 1939 forced the horse cavalry community to fully embrace mechanization, albeit in a subordinate role designed to maintain the viability of horse cavalry. Until its demise, the Polish cavalry's organization and doctrinal approach was similar to that being advocated in the United States.

The second armored car troop, raised for the next short-lived mechanized experiment in 1930, formed the nucleus of what was soon to become Cavalry Branch's mechanized cavalry brigade. With no other alternative, the American pioneers tested and developed methods to improve the capability of mechanized ground reconnaissance. They had much larger goals for their brigade, and good scouting and timely reporting only played one part in attaining their ultimate objective, an equal status as a combat force on a par with the lone horse cavalry division. It was not until 1940 that the men at Fort Knox escaped the shadow cast by the 1st Cavalry Division. Events in Europe finally forced the nation's gaze away from the Mexican border, where the horse cavalry division had long served, to Europe, the most likely place the new Armored Force might see service. Regardless of where the Armored Force was deployed, the work conducted at Fort Knox during the interwar years had done the most to advance the capabilities of mechanized ground reconnaissance units.

The impact of the Great Depression cannot be ignored in the story of the development of the reconnaissance units destined to serve America's army during World War II. The Great Depression constrained the military's budget and size, thus limiting the opportunity to conduct large-scale testing until war loomed on the horizon. By then the Germans, who started with similar ideas after World War I, had already taken war to a new dimension. But lack of funds had little to do with the constraints men placed on their minds, although funding for armored vehicles and the expense of maintaining them often entered the lexicon of those opposed to further mechanization. Ironically, much of what was spent was intended to assist and extend the military utility of the beloved horse, most notably the horse-and-mechanized regiment experiment.

At the lowest level of interwar mechanized reconnaissance development, troopers in the field continually disproved their supposed inabilities and used their imaginations to improve the equipment on hand. In the early years, this was often equipment they built with their own hands. At the middle level of the debate, men like Charles Scott, Adna Chaffee, Willis Crittenberger, and Daniel Van Voorhis struggled against the establishment to advance their ideas. Even war in Europe did not change the vision of the most senior leader in Cavalry Branch, MG John K. Herr, an undying advocate for retaining the horse for modern military operations. For men like General George C. Marshall and his associate General Leslie J. McNair, war cleared away any questions about what kind of eyes the army must have as it advanced to confront Axis forces. Looking back on the entire process, one cannot help but wonder how much better off the men going to war in 1942 would have been if the energy invested in protecting the role of the horse had been devoted to improving the capabilities of the existing mechanized reconnaissance technology and doctrine. Mechanized reconnaissance units were victorious in the contest to find the best eyes for the army not because the event had been rigged as Herr suggested, but because the concept proved its worth in every experiment short of war. The foundation of this success was built by a small cadre of men who were dispersed by the army's rapid expansion—beginning in 1940—throughout the large force subsequently created by the arsenal of democracy.

The biggest impacts of the constant efforts to preserve a role for the horse were flawed mechanized reconnaissance doctrine and units not properly organized and equipped to fulfill their missions. What had been developed between the wars was built on assumptions about the characteristics of the next war, chiefly the ability to maneuver. This meant that mechanized reconnaissance units should be able to avoid fighting the enemy, because there would always be space to find a way around. The men at Fort Knox questioned this notion as it related to reconnaissance and led the way in developing more robust agencies to survive first contact and if necessary fight for information, but even their efforts were incomplete. The other assumption, held especially within the horse cavalry community, was that horse-mounted units still represented the real combat potential of any cavalry organization. When horses finally disappeared from the last hybrid mechanized ground reconnaissance units in 1942, so did the preponderance of the units' dismounted capability. If interwar assumptions about the ability to maneuver were true, the dismounts would go unnoticed. But the realities of war replaced peacetime assumptions with cold, hard facts; mechanized reconnaissance units had to fight for information and the dismounted troopers were sadly missed.

A new generation of men had to learn for themselves as they attended their lectures beyond the plains of Fort Riley, the Mexican border, and the wooded hillsides of Kentucky. Shortcomings of the mechanized reconnaissance effort during the North African campaign were commensurate with the performance of the untested army. Divisional reconnaissance troops received relatively high marks, whereas the 81st ARB was unable to provide adequate warning about the disaster remembered today as the Battle of Kasserine Pass. Missing from the North African campaign was a corps-level mechanized reconnaissance unit. Although the 91st CavReconSdqn saw extensive service in North Africa, it never did so in a corps cavalry capacity. What the 91st CavReconSqdn did demonstrate was that in the hands of an interwar horse cavalryman, units organized and equipped primarily for mounted reconnaissance were still capable of fighting for information. In many cases, mechanized cavalrymen fought alongside the infantry.

The terrain in Sicily and Italy dictated the nature of the fighting and largely diminished many of the mechanized reconnaissance units' capabilities. Campaigning in Sicily and Italy did provide some hope to the horse advocates, now watching from the sidelines, that the horse was not completely finished. Calls from horse cavalry stars, Patton and Truscott, for large horse cavalry units to serve in their traditional combat roles reinforced this hope. What the sideline advocates failed to realize was that no one was willing to give up any other unit, mechanized or otherwise, in order to gain the use of a horse-mounted unit. Moreover, as Truscott proved in Sicily and Italy, mounted units could be improvised when needed. For the mechanized reconnaissance troopers serving in Sicily and Italy, there were few opportunities to realize the sweeping war of maneuver anticipated during the interwar development of their units' organization. More so than in any other European theater, they continually demonstrated their ability to contribute to the fight, even if in a dismounted capacity. When the front did reopen for brief interludes of maneuver, they were equally prepared to conduct mounted operations, though often on the flank.

There were minor changes to mechanized ground reconnaissance doctrine and the organizations tasked with carrying out reconnaissance missions. Even so, the doctrinal language, as late as June 1944, still heralded the combat capabilities of the horse and the need for mechanized reconnaissance units to avoid combat if possible. Organizational changes, most notably the conversion of mechanized corps cavalry regiments to cavalry groups, were more a reflection of McNair's pooling concept than representative of changes based on lessons learned in combat. The Cavalry School at Fort Riley used the actions in North Africa, selectively, to reinforce existing doctrine. It did the same with Sicily. But in both cases the number of mechanized reconnaissance units committed to combat was limited, and the conditions were unique in contrast to where performance would matter even more, Europe.

War on the Continent was the ultimate test of all that had been developed and accomplished since World War I. With the exception of the heady dashes during the breakout in August and early September 1944 and the final race across an exhausted Germany in April 1945, war shattered illusions of maneuver. Rarely did the units developed specifically for mechanized ground reconnaissance conduct reconnaissance. Instead, they acquitted themselves well while fighting dismounted, a trend witnessed since North Africa. Even if they saw little service performing purely reconnaissance roles, because of their inability to overcome German resistance and their failure to realize interwar assumptions about maneuver, mechanized reconnaissance units made a considerable contribution to the conduct and outcome of the European campaign. The "90 Division Gamble" placed a premium on the limited number of American combat forces raised to win the European crusade. Mechanized cavalry units organized for reconnaissance often performed what would today be defined as economy of force missions—using the least number of people to accomplish the mission to allow massing of forces elsewhere. The term in use then was "gap filling," and the high degree of mobility and firepower, found in each mechanized cavalry squadron, allowed commanders, from division to corps level, a flexible solution to many tactical problems. On occasion, such as the German thrust through the Loesheim Gap in December 1944, this had serious consequences, but in most cases it worked.

Mechanized cavalry units contributed on the European battlefield by doing more than reconnaissance, because their leaders rejected the flawed doctrine intended to limit their role. Ironically, wartime commanders were drawn from the same horse-dominated Cavalry Branch that had prescribed such a limited role for mechanized reconnaissance units while it attempted to preserve a role for the horse. Now without horses, the old cavalrymen often employed their old doctrine—the doctrine of horse cavalry as a combat arm—with their new organizations, equipped with various breeds of iron ponies. Like the cavalrymen who had already departed for the Armored Force, they learned to accomplish all the old combat missions with different means. In the process, they, the old horse cavalrymen, resurrected Cavalry Branch. When the war ended, they demanded that the doctrine change to reflect what they had done and that their units be organized and equipped to better accomplish the missions that had been theirs to perform. They had no interest in keeping the mechanized elements of their branch weak to maintain a role for the horse; rather, they rejected the horse and demanded their old identity as a branch fully capable of combat, since it took lots of combat to gain the information they were supposed to provide the units they served.

The reconciliation of cavalry and armor made sense in the longer view of Cavalry Branch's history. Cavalry, once a true combat arm, would again be based on the changes called for in its organization after the war. Armor, the offspring of cavalry and usurper of the combat missions coveted by its parent, no longer looked so different from its parent, Cavalry Branch, once post-World War II mechanized cavalry units received heavier tanks, dragoons (dismounts), and other combat multipliers. Just as it was before the introduction of mechanization, those who rode in the van were not all that different from those who followed. Again, cavalry fully anticipated the need and requirement to be able to fight for information.

John K. Herr had also been right, even if his only solution, the horse, was a flawed one on the modern battlefield. American units never raced to meet advancing Germans, according to his criticism of the interwar maneuvers that had led to such a poor showing of his horse cavalry in combat and reconnaissance roles. The only dashes occurred when the Germans collapsed, and it seems unlikely that American pursuit could have been maintained with hay-burning units, horses, rather than with the vehicles that ground to a halt for lack of fuel in the first instance, and for political reasons in the second. The tyrannies of logistics present in World War II remains today. Herr was also correct when he predicted that given the "parking spaces of all this horde of vehicles," the day would come when it would become "unhealthy by reason of bombers" to be part of an armored division. Herr acknowledged that this day would take some time, since "no one is adequately prepared." Fortunately for the United States, from the end of the Cold War and on, all other nations with armored divisions have had to worry about the threat from above, but not American soldiers. Were that day to arrive, anticipating parity of air power, Herr believed horse cavalry would once again be in vogue. This was because only horse cavalry, with its ability to disperse off roads and move cross-country would allow it to evade the growing tide of air power.[2] It does not appear any nation is currently applying Herr's solution.

Herr also saw in air power the means to cut free from the most cumbersome aspects of the horse cavalry division as it existed. Air-delivered logistics, combined with the restoration of pack-trains, would allow the division to cut loose from its motorized logistics tail. The promise of close air support and bombers negated the need for many of the ever growing anti-tank systems the horse cavalry division equipped itself with on the eve of World War II, as it attempted to fend off armored threats. Again, Herr was partially correct. Today air power allows the United States to project and sustain military might to every corner of the globe, but not to supply non-existent horse cavalry divisions conducting operations independent of fixed lines of communications and logistics.[3] Air power remains a crucial supporting force to help with the destruction of enemy armored forces, but even Herr should have recognized, as actions in Afghanistan and Iraq have proven in the first decade of the twenty-first century, that air power cannot linger indefinitely over the battlefield or provide support under all weather and light conditions, to say nothing of enemy countermeasures from below. Thus, even with the United States dominating the air, there is still the need for ground units able to destroy the enemy's armored vehicles with weapons systems found within their organization.

Herr's hope was not realized when he wrote to General Innis P. Swift in 1941, "If we can get by this period of ignorance and prejudice and prevent these shortsighted gumps from wiping u[s] out of the picture in their mistaken belief that the iron horse replaces one of flesh and blood, we will surely come into our own."[4] The "iron horse" prevailed, but Herr would have been proud to know that in some capacity his beloved horses continued to serve the army he loved equally as much. An article recently appeared in the

Cavalry Journal (a modern nostalgia-driven magazine not to be confused with the army's oldest professional journal of the same name, later transformed into the *Armored Cavalry Journal* before becoming *Armor Magazine,* and now the *Armor & Cavalry Journal*) that described how the 10th Special Forces Group trained at the M Lazy C Ranch near Lake George, Colorado. There, the Green Berets learned mule-packing, but they also practiced firing their carbines and pistols from the backs of horses. In addition, they worked on the use of open formations, Herr's solution for how horse soldiers could overcome the lethality of the modern battlefield, and picked up the "old Indian trick of obtaining" horses through stealth. The basis for much of this instruction was derived from pre-World War II cavalry manuals. Even with the enhanced capabilities of global positioning systems and night vision equipment, things Herr could not begin to imagine, it seems unlikely any technological breakthrough could have preserved a role for horses except under the most extreme circumstances found in such places as Afghanistan.[5]

Although Green Berets used horses in Afghanistan, there has been no call for a renewed debate that would once again pit the horse against the machine. The emerging debate pits manned air and mechanized ground reconnaissance units against unmanned machines and passive intelligence-gathering platforms. Many of the old arguments of the interwar years are being dusted off and used in creative ways. Language once reserved for horse-mounted scouts is now being applied to manned reconnaissance units.

> Many Army professionals agree that the ground scout is the most efficient, high resolution, all-weather, contiguously operating, on-site intelligent decision-making, intent-determining, and most timely terrain retaining information asset for the commander to answer critical information requirements (CCIR).[6]

For now, every effort seems to focus on harnessing myriad emerging technology and developing the means for human scouts to enjoy the maximum advantage provided by the machines, just as mechanized ground reconnaissance once served the horse.

In other ways the old arguments are even more important as the army continues to plan and execute its transformation simultaneously. One of the goals of the transformation is that the army will emerge with a largely homogeneous force, the Future Force, built on a family of vehicles yet to be developed, the Future Combat System. The Future Force, equipped with a family of air-deployable vehicles, expects to operate in an environment of information superiority. Having perfect knowledge, a laudable goal but not likely, will in turn help the force compensate for the inherent lack of survivability of the Future Combat System, part of the trade-off for air deployability. Critics of this approach question the ability of remote sensors and unmanned platforms to penetrate all types of terrain, with a special concern for urban terrain, just as advocates for retaining the horse pointed out that no vehicle could go everyplace a horse could go.

There is no irony in the fact that one uniformed critic of the current faith in information dominance to avoid fighting under unfavorable circumstances was himself the direct beneficiary of lessons learned during World War II. Colonel H.R. McMaster—a captain in 1991—commanded Troop E, 2nd Squadron, 2nd Armored Cavalry Regiment, during the First Gulf War in 1991.[7] Fulfilling the role it was designed for in 1943 as the 2nd Cavalry Group—a corps reconnaissance regiment—the 2nd Armored Cavalry Regiment served as VII Corps' cavalry regiment when it advanced into Iraq as part of General H. Norman Schwartzkopf's Napoleonic manoeuvre sur les derrières. Two days into the war, McMaster found his troop at the head of his squadron in search of Saddam Hussein's Republican Guard. In a storm of "blowing sand and swirling mist," McMaster found the enemy and in less than thirty minutes proceeded to destroy more than thirty Iraqi armored vehicles.[8] The "Battle of 73 Easting" validated what had been learned during World War II: ground reconnaissance units would find the enemy by bumping into them, sometimes under the most extreme conditions. Having found the enemy, McMaster had the ability to deal with it in a manner far different from the way his regiment had dealt with the Germans at Lunéville in September 1944.

Today's critics also question the ability of the one-size-fits-all approach to deal with the full spectrum of war, a by-product of a homogeneous force. Again, concern centers on whether medium-weight vehicles, such as the Stryker combat system, can handle the common threats on today's battlefield that no amount of information dominance can completely overcome.[9] If the .30-caliber bullet was the interwar benchmark for what a minimum amount of armored plating had to stop to insure the survivability of mechanized reconnaissance men to live long enough to pass on what they had discovered in a surprise engagement with the enemy, then the ubiquitous rocket-propelled grenade (RPG) and Improvised Explosive Device (IED) must be today's standard. The army has already been forced to add special armor packages to Stryker vehicles to meet this minimum threshold.[10] More importantly, an on-going study of the recent combat in Iraq in 2003 has revealed that at the squadron level and below, little has changed since World War II, with respect to finding the enemy. In 2003, as in World War II, the enemies were found when units "literally ran into them."[11] Until ground reconnaissance units can consistently find the enemy short of "running into it," they will continue to need the firepower and protection required to survive first contact.

The experience of World War II bore out the fact, just as recent operations in Afghanistan did, that having a variety of types of ground reconnaissance units was useful, but the decision to field an all-mechanized ground reconnaissance force during World War II was not unwise. Commanders were still able to satisfy special needs by dipping into the past, using horses when needed or employing creative organization of assets on hand. A Future Force built with certain assumptions about the nature of war in the future might be sadly unprepared for the reality of war. Between the world wars there was a common belief that the next war would be one of maneuver. This in part drove the doctrinal expectation that mechanized ground reconnaissance units would be able to find a way around the enemy, to "sneak and peak" without fighting. Fortunately when this proved unfounded, the men in those units could draw on their doctrinal past to see them through the crisis. With current trends already shaping the army, soldiers in the Future Force may not have the luxury of falling back on old doctrine or creative task organization with existing equipment.

In the end, with full knowledge of how the story ended, it is far too easy to ridicule the most rabid advocates of the horse. They honestly believed that retaining important roles for horse cavalry was in the nation's best interest. In doing so, they misdirected the effort to prepare for the war that arrived, not the one they anticipated, leading to unnecessary harm for the men destined to ride machines instead of horses in the quest for information about the enemy. The men at Fort Knox who demanded that technology rise to meet their expectations were the ones to be emulated. Unconstrained by the need to preserve a role for the horse, they did not limit their expectations of what could be accomplished by mechanized ground reconnaissance units. They could also move ahead with their experiment in full confidence that if they got it wrong, there was still a substantial portion of the army that was changing less rapidly and could act as a safety net if need be. They also had the luxury of the Atlantic and Pacific Oceans to delay the arrival of forces from any other nation but Mexico and Canada. The men at Fort Knox did their utmost to see that their reconnaissance units had what they needed and were more willing to admit they might need to fight to accomplish their mission.

Then, as now, war remains a human endeavor. Until the army develops a remote sensor capable of divining intentions and reading minds, there will be a need to close with the enemy to determine his plans. The army cannot afford to become wedded to the means, be it horse, armored car, or UAV. Rather, it must remain focused on the task at hand and apply the best combination of systems, employed by people, with an effective doctrine capable of deriving the maximum potential from both the humans and machines. It will not be easy for today's descendents of the first men to ride on "iron ponies" to keep alive the spirit of the cavalry as new technologies emerge. Just as no one wants to be compared to John K. Herr when questioning the ability of emerging sensor technology to transform tomorrow's battlefield, the doubters are justified in asking their hard questions. Too much blood has already been spilled by former cavalrymen while disproving interwar beliefs. The development of mechanized ground reconnaissance between World War I and the end of World War II provides rich material for consideration even today as the army continues to transform itself.

Notes

Introduction

1. James Pickett Jones, *Yankee Blitzkrieg: Wilson's Raid through Alabama and Georgia* (Lexington: University Press of Kentucky, 2000), 1.

2. Ibid., 7.

3. Ibid., 185.

4. Michael Howard, *The Franco-Prussian War: The German Invasion of France, 1870–1871* (New York: Dorset Press, 1990), 157.

5. Andrew J. Birtle, *U.S. Army Counterinsurgency and Contingency Operations Doctrine, 1860–1941* (Washington, D.C.: Center of Military History, 1998), 57–60.

6. Ibid., 71–72.

7. Ibid., 201–7.

8. U.S. Army, Forces in the European Theater, The General Board, *Mechanized Cavalry Units,* Study no. 49 (Washington, D.C.: Government Printing Office, [May 1946]) (hereafter cited as The General Board, *Mechanized Cavalry Units,* no. 49), 4.

9. U.S. Army, *FM 100-5, Field Service Regulations, Operations* [hereafter *FM 100-5, Operations*] (Washington, D.C.: War Department, 1944), 8. I have purposely used the 1944 edition of *FM 100-5, Operations,* the army's principal statement on doctrine, because it demonstrates the continued belief that horse cavalry was a combat arm.

10. Ibid.

11. Ibid.

12. Ibid., 51.

13. Cavalry Board Report to Adjutant General, General Headquarters American Expeditionary Forces, 24 April 1919, General Headquarters A.E.F. folder, box 13, entry 39, Record Group 177, National Archives II. The board was appointed by Special Order 44.

14. Ibid., p. 27.

15. Ibid., p. 31.

16. Ibid.

17. The General Service School, *Tactics and Techniques of Cavalry* (Fort Leavenworth, KS: The General Service School Press, 1921), 18–19.

18. Ibid., 19.

19. Mildred Hanson Gillie, *Forging the Thunderbolt: A History of the Development of the Armored Force* (Harrisburg, PA: Military Service Publishing Co., 1947), 115. Gillie's definitions are the same as those used in the *Cavalry Journal* throughout the interwar years. Mechanization is defined as the application of mechanics directly to the combat soldier on the battlefield. Motorization is defined as the substitution of motor-propelled vehicles for animal-drawn vehicles or pack animals in the supply echelons of all branches of the army. Using these definitions, all reconnaissance units, regardless of whether they are mounted on wheeled and/or tracked vehicles, armored and/or unarmored vehicles, will be referred to as "mechanized."

20. LTG John L. Ryan (Ret.), interview by Timothy Nenninger, 15 June 1967, Indian Harbor, Florida. The term "combat car" was used to avoid violating the federal law that limited experimentation and developments in tanks to the purview of Infantry Branch. "Combat car" will be used throughout the remainder of the text, but the reader should

envision a tracked vehicle more commonly referred to as the tank, even though there were some differences in the machines used by the cavalry and the infantry.

21. Cavalry Board Report, April 1919, p. 27.

22. Thomas W. Collier, "The Army and the Great Depression," *Parameters* 18 (September 1988): 102.

23. Ibid., 102–3.

24. Ibid., 103, 108; and Robert K. Griffith, Jr., *Men Wanted for the U.S. Army: America's Experience with an All-Volunteer Army between the World Wars* (Westport, CT: Greenwood Press, 1982), 164.

25. Collier, "The Army and the Great Depression," 106.

26. Ibid., 104.

27. LTG Robert W. Grow (Ret.), interview by Timothy Nenninger, 10 June 1967, Falls Church, Virginia. Grow's remarks were offered in the context of his recounting the visit paid to Fort Knox by the German General Staff in 1933. Grow was confident that it was the Germans who had learned from the Americans. He commanded the 6th Armored Division in France and Germany during World War II.

28. U.S. Army, *Table of Organization No. 2–25, Cavalry Squadron, Mechanized* (Washington, D.C.: Government Printing Office, 1 April 1942).

29. Mary Lee Stubbs and Stanley Russell Connor, *Armor-Cavalry*, part 1, *Regular Army and Army Reserve*, Army Lineage Series (Washington, D.C.: Office of the Chief of Military History, U.S. Army, 1969), 71.

30. Ibid., 73.

31. The General Board, *Mechanized Cavalry Units*, no. 49: 7–8.

1—The Lessons of World War 1

1. *Cavalry Combat* (Harrisburg, PA: Telegraph Press, 1937), 507.

2. Alan Clark, *Suicide of Empires: The Battles of the Eastern Front* (New York: American Heritage Press, 1971), 44.

3. David E. Johnson, *Fast Tanks, Heavy Bombers: Innovation in the U.S. Army, 1917–1945* (Ithaca, New York: Cornell University Press, 1998), 27. Nine cavalry regiments remained on the Mexican border. John K. Herr and Edward S. Wallace, *The Story of the U.S. Cavalry, 1775–1942* (New York: Bonanza Books, 1953), 241–43.

4. Allen Millet and Peter Maslowski, *For the Common Defense: A Military History of the United States of America* (New York: The Free Press, 1984), 399.

5. Herr and Wallace, *The Story of the U.S. Cavalry*, 243; and *Cavalry Combat*, 73. Prior to the St. Mihiel offensive, the men of the 2nd CavRegt were performing construction.

6. John B. Wilson, *Maneuver and Firepower: The Evolution of Divisions and Separate Brigades* (Washington, D.C.: Center of Military History, 1998), 79.

7. Ibid., 83, 85.

8. Cavalry Board to Adjutant General, General Headquarters, American Expeditionary Forces, 24 April 1919.

9. Ibid., 3–4, 11, 20.

10. Ibid., 31.

11. Ibid., 4, 31.

12. Ibid., 4–6.

13. Ibid., 5–7.

14. Ibid., 8, 27–28.

15. Dominick Graham and Shelford Bidwell, *Fire-Power, British Army Weapons, and Theories of War, 1904–1945* (London: George Allen & Unwin, 1982), 205.

16. J.F.C. Fuller, *Memoirs of an Unconventional Soldier* (London: Ivor Nicholson and Watson, 1936), 361, 363.

17. Graham and Bidwell, *Fire-Power*, 205; and Robert M. Citino, *Armored Forces, History and Sourcebook* (Westport, CT: Greenwood Press, 1994), 51.

18. Wesley W. Yale, Isaac D. White, and Hasso E. von Manteuffel, *Alternative to Armageddon: The Peace Potential of Lightning War* (New Brunswick, NJ: Rutgers University Press, 1970), 70. For a concise description of actions at Tannenberg and the Masurian Lakes, see Winston Churchill's *Unknown War* (New York: Charles Scribner's Sons, 1931).

19. Heinz Guderian, *Panzer Leader* (New York: E.P. Dutton, 1952), 21, 24.

20. Griffith, *Men Wanted for the U.S. Army*, 1.

21. Memorandum, Major General W. A. Holbrook, Chief of Cavalry, to the Director, War Plans Division, "Duties and Responsibilities of the Chiefs of the Combatant Arms," 8 September 1920, file 323.362/316, Office of the Chief of Cavalry Correspondence, 1921–1942, box 7A, RG 177, cited in Dave E. Johnson, "Fast Tanks and Heavy Bombers: The United States Army and the Development of Armor and Aviation Doctrines and Technologies, 1917 to 1945" (Ph.D. diss., Duke University, 1990), 117.

22. Major Bertram C. Wright, *The 1st Cavalry Division in World War II* (Tokyo, Japan: Toppan Printing Company, 1947), 3.

23. Stubbs and Connor, *Armor-Cavalry*, 53; and Herr and Wallace, *The Story of the U.S. Cavalry*, 244.

24. "Plead to Maintain Cavalry Strength, Chief Wants Present 20,000 Retained When Army Total is Reduced, Points to Past Efficiency, and Necessity for Doing What Other Nations Consider Wise—Needed for Border Patrol," 17 April 1921, *The New York Times*.

25. George S. Patton, Jr., "The Cavalryman," in *Rasp* (Fort Riley, KS: The Cavalry School, Army of the United States, 1922), 166–67.

26. Hamilton S. Hawkins, "Why Cavalry is Indispensable?" in *Rasp* (Fort Riley, KS: The Cavalry School, Army of the United States, 1922), 164.

27. Ibid., 164–65.

28. Ibid. Open order refers to the amount of space between advancing soldiers as they attacked on line.

29. *Rasp*, 1922, p. 52.

30. Johnson, *Fast Tanks and Heavy Bombers*, 218; and Harold R. Winton, *To Change an Army: General Sir John Burnett-Stuart and British Armored Doctrine, 1927–1938* (Lawrence: University Press of Kansas, 1988), 82. Historian Timothy K. Nenninger, Military Records Archivist, National Archives II, writes in the notes section of "Organizational Milestones," in *Camp Colt to Desert Storm: The History of the U.S. Armored Forces*, ed. George Hofmann and Donn A. Starry (Lexington: University Press of Kentucky, 1999), 62.

31. J.F.C. Fuller, *The Army in My Time* (London: Rich and Cowan, 1935), 188.

32. Robert A. Miller, "The United States Army during the 1930s" (Ph.D. diss., Princeton University, 1973), 119–20.

33. Gillie, *Forging the Thunderbolt*, 24. The personal perseverance of Adna Chaffee, Jr., was instrumental in the creation of the Armored Force, and Chaffee's involvement is closely detailed by Gillie. Nenninger, "Organizational Milestones," in *Camp Colt to Desert Storm*, ed. Hofmann and Starry, 41. William J. Woolley, "Patton and the Concept of Mechanization," *Parameters* 15 (Autumn 1985): 75. Fort Leonard Wood in this instance is the contemporary Fort Meade, Maryland.

34. Vincent J. Tedesco, III. "'Greasy Automatons' and the 'Horsey Set': The U.S. Cavalry and Mechanization, 1928–1940" (Master's thesis, Pennsylvania State University, May 1995), 27.

35. Timothy K. Nenninger, "The Development of American Armor, 1917–1940" (Master's thesis, University of Wisconsin, 1968), 85–88; Gillie, *Forging the Thunderbolt*, 21–24; and Johnson, *Fast Tanks and Heavy Bombers*, 218–19.

36. "History of the 91st Reconnaissance Squadron, 15 February 1928–14 May 1941," folder CAVS-91-0.1, box 18231, entry 427, RG 407, NAII; *The Cavalry Journal* (hereafter cited as *CJ*) 32 (April 1928): 300–301; and Captain Harold G. Holt, "The 1st Armored Car Troop," *CJ* 32 (October 1928): 599.

37. *CJ* 32 (April 1928): 301; and Woolley, "Patton and the Concept of Mechanization," 75. *CJ* was published by the Office of the Chief of Cavalry, which was considered the "citadel of tradition."

38. Brigadier General Van Horn Mosley to Major General H.B. Crosby, 9 December 1927, folder 322.02, Office, Chief of Cavalry, box 7a, entry 39, RG 177, NAII, cited in Johnson, *Fast Tanks, Heavy Bombers,* 125.

39. Wilson, *Maneuver and Firepower,* 112.

40. Major George Dillman, "1st Cavalry Division Maneuvers," *CJ* 32 (January 1928): 47.

41. Anonymous student of the Cavalry School, "The Horse in the Palestine Campaign," *CJ* 32 (February 1928): 259–65. The article details desert operations and the impact they had on horses.

42. Captain Charles Cramer, "Portee Cavalry," *CJ* 32 (January 1928): 66–69. Porté referred to the use of trucks and or trucks and trailers to move men, horses, and equipment from one point to another.

43. Dillman, "1st Cavalry Division Maneuvers," 46–49, 52, 62–63.

44. Liberty Trucks were of the World War I vintage in terms of motorized transportation.

45. Cramer, "Portee Cavalry," 66–69.

46. Wilson, *Maneuver and Firepower,* 112–15.

47. George S. Patton Jr., "The 1929 Cavalry Division Maneuvers," *CJ* 39 (January 1930): 9.

48. Ibid., 10; Ryan, interview by Nenninger, 15 June 1967. Ryan holds a somewhat different view on mobility. He says he saw real potential in the armored car's ability not only to move greater distances cross-country but also to go faster than the horse.

49. Major E.C. McGuire, "Armored Cars in the Cavalry Maneuvers," *CJ* 39 (July 1930): 390–91; and Ryan, interview by Nenninger, 15 June 1967. At this time the troop was not equipped with "voice" radios. It relied on key-coded messages.

50. McGuire, "Armored Cars in the Cavalry Maneuvers," 392.

51. In Germany during the summer of 1932, Heinz Guderian deployed to Grafenwöhr with genuine armored reconnaissance vehicles much to the chagrin of school children and infantrymen alike. In the recent past, when only mock-up vehicles had been used, children satisfied their curiosity about the inter-workings of the vehicles by poking their pencils through the sides to see what was inside. Infantrymen had grown accustomed to using sticks, stones, and occasionally bayonets to repel the armored reconnaissance forces; both were now confounded with real armor. Guderian, *Panzer Leader,* 28–29.

52. Major George S. Patton, Jr., and Major C.C. Benson, "Mechanization and Cavalry," *CJ* 39 (April 1930): 235.

53. Ibid., 235–37.

54. Ibid., 239.

55. Major George S. Patton, Jr., "Motorization and Mechanization in the Cavalry," *CJ* 39 (July 1930): 331.

56. Ibid., 332. This was reminiscent of a lecture Patton gave in Boston in 1924 entitled "Cavalry Patrols." See *The Patton Papers,* vol. 1, *1885–1940,* ed. Martin Blumenson (New York: Houghton Mifflin, 1972), 781–83.

57. Patton, "Motorization and Mechanization in the Cavalry," 335.

58. Ibid., 336–37.

59. Ibid., 347.

60. Ibid., 330, 348.

61. Field Marshal Viscount E.H.H. Allenby to *CJ* 38 (January 1929): 2.

62. Photos accompany transcript of speech written by Hampson Gary, "Welcome to Field Marshal Lord Allenby," *CJ* 38 (January 1929): 4.

63. Patton read extensively on the desert campaigns during the interwar years. He gushed that *The Desert Mounted Corps* by R.M.P. Preston was "the greatest book" he had ever read. *Patton Papers,* ed. Blumenson, 1:762. See Roger H. Nye, *The Patton Mind: The Professional Development of an Extraordinary Leader* (Garden City Park, NY: Avery Publishing Group, 1993), 58–59, for a more extensive list of the books Patton read related to desert warfare.

64. "Progress and Discussion," *CJ* 39 (July 1930): 449, and (October 1930): 606–8.

2—The 1930s

1. *Patton Papers,* ed. Blumenson, 1:905.

2. Packed guns were those carried on the back of a horse. The inherent advantage was that a horse carrying a machine gun could go any place a mounted man could travel, unlike a towed weapon.

3. Mark Skinner Watson, *Chief of Staff: Prewar Plans and Preparations,* United States Army in World War II (Washington, D.C.: Army Historical Division, Dept. of the Army, 1950), 33.

4. The chief of staff of the army, GEN Malin Craig, formed a Modernization Board on 16 January 1936 to begin considering issues such as firepower, motorization, mechanization, and other matters that would have an impact should the army have to begin mobilizing. Like the Superior Board, the Modernization Board focused most of its attention on the organization of the infantry division. Wilson, *Maneuver and Firepower,* 127.

5. Stubbs and Connor, *Armor-Cavalry,* 55.

6. Charles P. Summerall, "Cavalry in Modern Combat," *CJ* 32 (October 1930): 491–93.

7. Ibid.

8. Ibid.

9. Johnson, *Fast Tanks, Heavy Bombers,* 270–71.

10. "General Principles to Govern in Extending Mechanization and Motorization throughout the Army." This note and citation are an interpretation of the aforementioned memo by MacArthur Members of the Office of the Chief of Cavalry held that General George Van Horn Mosley was the principal author of the revised mechanization policy. Memorandum, Robert W. Grow for COL Kent, 21 December 1936, Office of the Chief of Cavalry Correspondence, 1921–1942, file 322.02 folder, box 6, entry 39, RG 117, NAII.

11. Ibid.

12. Ibid.

13. Nenninger, "The Development of American Armor," 111.

14. Manuscript, autobiography folder, box 6, Papers of Guy V. Henry, Jr. [hereafter Henry Papers], MHI, 64.

15. Ibid., 65. The .50-caliber machine gun General Henry helped to field is still in use throughout the U.S. Army, especially in armor and cavalry units.

16. Ibid., 67.

17. Ibid., 65.

18. Ibid.

19. Memorandum by MG Guy V. Henry, Chief of Cavalry, for MAJ Dwight D. Eisenhower, 11 August 1933, with attachment entitled, "Trend of Tactics and Need of Cavalry Equipment," folder 322.02, box 7a, entry 39, RG 177, NAII.

20. Ibid.

21. Ibid.

22. Autobiography folder, box 6, Henry Papers, MHI, 68–70. In the past, when branch chiefs gave up their positions in Washington, they reverted from the rank of major general to their previous rank of colonel. When Henry departed as chief of cavalry, MacArthur saw to it that he was retained as a brigadier general, and Henry was sent to Fort Knox in May 1934 to command the 7th Cavalry Brigade.

23. Collier, "The Army and the Great Depression," 103. Gillie explains, in *Forging the Thunderbolt,* 46–47, that GEN Daniel Van Voorhis, who commanded the unit, resisted Chaffee's desire to see the new organization transferred to the control of Cavalry Branch. Van Voorhis appealed to GEN Henry to convince MacArthur to retain the force as it was. MacArthur would later claim he did everything he could to save the Mechanized Force, but related to the cost savings, he chose to maintain manpower rather than use the limited funds for new equipment. Nenninger, "The Development of American Armor," 116. See also *Patton Papers,* ed. Blumenson, 885. Miller, "The United States Army during the 1930s," 134–35.

24. Gillie, *Forging the Thunderbolt,* 37; and Arthur Wilson, "The Mechanized Force, Its Organization and Present Equipment," *CJ* 40 (March–April 1931): 7–10.

25. Arthur Wilson, "With the Mechanized Force on Maneuvers," *CJ* 40 (July–August 1931): 5–9.

26. Ibid.

27. Stubbs and Connor, *Armor-Cavalry,* 55.

28. War Dept., Adjutant General's Office, 5 April 1935, Subject: War Department Policies for Mechanization, citing AG 537.3 (5-13-31) Misc., Subject: Disposition of Mechanized Force, 322.02 2nd Cavalry folder, box 6, entry 39, RG 177, NAII.

29. Johnson, *Fast Tanks, Heavy Bombers,* 129.

30. Henry to 1st Cavalry Commander, Subject: "Mechanization of the 1st Cavalry," 17 November 1931, file 322.02 First Cavalry (Mechanized), box 6, RG 177, NAII, cited in Tedesco, "Greasy Automatons," 44.

31. Gillie, *Forging the Thunderbolt,* 58.

32. Ibid., 55.

33. Tedesco, "Greasy Automatons," 50.

34. Stubbs and Connor, *Armor-Cavalry,* 56.

35. "General Principles to Govern in Extending Mechanization and Motorization throughout the Army," 1 May 1931. MacArthur directed that, when employed with the infantry, armored vehicles would be referred to as "tanks" but when used with the cavalry, they would be called "combat cars."

36. COL Daniel Van Voorhis to Chief of Cavalry, Subject: Methods for the Tactical Employment and the Training of a Mechanized Cavalry Regiment, 30 OCT 1931, file 320.2 1st Cavalry (Mechz.), box 3, RG 177, NAII, cited in Tedesco, "Greasy Automatons," 51.

37. Ibid., 50.

38. H.W. Ketchum, Jr., "Communications of a Modern Motor Truck Convoy," *CJ* 42 (March–April 1933): 21–22.

39. Notes from the Chief of Cavalry, "Motor Trucks for Troop A, 1st Armored Car Squadron," *CJ* 42 (March–April 1933): 45.

40. By this time, the 1st CavDiv HQ was equipped with cars that had radios, but the horse-mounted troops and squadrons still lacked even a pack-radio set and were still dependent on mounted messengers.

41. James S. Corum, *The Roots of Blitzkrieg: Hans von Seeckt and German Military Reform* (Lawrence: University Press of Kansas, 1992), 184; and Robert Citino, *Armored Forces,* 54–55.

42. Gillie, *Forging the Thunderbolt,* 85.

43. Grow, interview by Nenninger, 10 June 1967.

44. Gillie, *Forging the Thunderbolt,* 89.

45. Ibid., 62.

46. Tedesco, "Greasy Automatons," 60. General George Van Horn Mosley, who had been instrumental in the development of the revised mechanized policy, also played a crucial role in shaping the first test of the newly mechanized regiment. He denied a request that the new regiment participate in maneuvers at Fort Benning in 1933. Not only would the new regiment have been largely untrained, given its recent arrival at Fort Knox and its work with the Civilian Conservation Corps (C.C.C.), Mosley recognized that the nature of the exercise proposed, combined with the terrain and distances at Fort Benning, favored horse cavalry. Rather than see the "first regiment (mechanized) killed off in a close-in tactical problem," he denied the request and thus delayed the first test of the mechanized regiment until 1934. Mosley to Van Voorhis, 16 February 1933, Washington, D.C., Office of the Chief of Cavalry, Correspondence 1921–1942, 322.012 folder, box 6, entry 39, RG 177, NA II.

47. War Department, by order of Secretary of War, 18 April 1934 reprinted in *CJ* 43 (May–June 1934): 44.

48. "Organization Activities," *CJ* 43 (May–June 1934): 61.

49. "The Cavalry Maneuvers at Fort Riley, Kansas, 1934," *CJ* 43 (July–August): 5–14. "Notional" enemy and friendly units were used to facilitate training. Robert W. Grow, "The Ten Lean Years, from the Mechanized Force (1930) to the Armored Force (1940)," Peter R. Mansoor and Kathy C. Garth, eds., *Armor* 96 (January–February 1987): 24.

50. "The Cavalry Maneuvers at Fort Riley," 5–14.

51. Ibid.

52. *Patton Papers,* ed. Blumenson, 1:906.

53. "The Cavalry Maneuvers at Fort Riley," 5–14.

54. Ibid.

55. Ibid.

56. MG Leon B. Kromer, "Lecture on Cavalry," delivered to Army War College, 17 September 1934, Fort Humphreys, Washington, D.C., folder Various Addresses and Speeches, box 3, Kromer Papers, MHI.

57. Ibid.

58. R.I. Sasse, "Course at the Army War College, 1933–1934, Memorandum for the Commandant, Subject: Mechanized Cavalry," 5 May 1934, Washington, D.C., AWC 407-73, MHI.

59. Ibid. See also George S. Patton, Jr., "Mechanized Forces," *CJ* 42 (September–October 1933): 5–8.

60. Patton, "Mechanized Forces," 5–8.

61. Kromer, "Address of MG Leon B. Kromer, Chief of Cavalry, at Fort Riley, Kansas, during the April–May Maneuvers," *CJ* 43 (May–June 1934): 44–46.

62. Ibid.

63. Collier, "The Army and the Great Depression," 104–5.

64. James V. Morrison, *F Troop, the Real One, through the Eyes of Trooper James V. Morrison* (Hiawatha, KS: R.W. Sutherland Printing, 1988), 40.

65. Collier, "The Army and the Great Depression," 107.

66. "Address of COL Bruce Palmer, First Cavalry (Mecz), October 12, 1936, before the Army War College," Speech File: Mechanized Tactics, 1938–1941 folder, box number unassigned, Isaac D. White Papers.

67. Draft memorandum for Kent from Grow, 21 December 1936.

68. *Cavalry Weapons and Material,* part 10, *Motor Vehicles* (Fort Riley, KS: The Cavalry School, 1935).

69. Ibid., 198.

70. Enclosure no. 1 to 3rd IND., HQ 13th CAV, 7 June 1935, RE: Report on Tentative Regulations—The Cavalry Scout Car, written by COL Charles L. Scott, folder 300.7, box 3, entry 39, RG 177, NAII.

71. Ibid.

72. CPT Wesley W. Yale, "The 1936 Maneuvers of the 1st Cavalry Division." *CJ* 45 (May–June 1936): 174–94.

73. COL Bruce Palmer, "Mechanized Cavalry in the Second Army Maneuvers," *CJ* 45 (November–December 1936): 461–78; and Draft memorandum for Kent from Grow. The Second Army consisted of V and VI Corps. 26,000 National Guard and regular troops composed these corps and participated in these exercises. *The Papers of George Catlett Marshall,* vol. 1, ed. Larry I. Bland and Sharon R. Ritenour (Baltimore, MD: The Johns Hopkins University Press, 1991), 500 [hereafter Marshall Papers]. Marshall commanded the 12th Brigade (Reinforced).

74. Palmer, "Mechanized Cavalry in the Second Army Maneuvers," 461–78.

75. *Marshall Papers,* ed. Bland and Ritenour, 499–501. Marshall to Frank R. McCoy, 16 August 1936, Bivouac North of Allegan, Michigan; and "Report on the Second Army Maneuvers," vol. 1, section 1, "Report of the Commanding General," folder 2nd AD and Armor, Miscellaneous Material, 1937–1942, box 33, White Papers.

76. *Marshall Papers,* ed. Bland and Ritenour, 499–501; and Marshall to McCoy, 16 August 1936.

77. Palmer, "Mechanized Cavalry in the Second Army Maneuvers," 470.

78. *Marshall Papers,* ed. Bland and Ritenour, 501–4. Report to the Commanding General Second Army Maneuvers, 21 August 1936, Camp Custer, Michigan.

79. Palmer, "Mechanized Cavalry in the Second Army Maneuvers," 474.

80. An after action report recognized a "need for a rifle organization to protect the flanks and rear from the sorties of hostile horse cavalry; to afford close-in protection to

artillery in position; to provide riflemen for dismounted patrolling and outpost duty; in short, to perform the many duties of combat best performed by riflemen." "Report on the Second Army Maneuvers," vol. 1, section 1, "Report of the Commanding General."

81. Ibid., 475–76.

82. Fort Humphreys is now Fort McNair. The Army War College has since moved to Carlisle Barracks, Pennsylvania.

83. COL Bruce Palmer, Commander 1st CavRegt (M), "The Cavalry," lecture delivered at the Army War College, Fort Humphreys, Washington, D.C., 12 October 1936, 6-3#10, 1937, MHI, 1.

84. Ibid., 2–3.

85. Ibid., 3.

86. Ibid., 9.

87. Ibid., 4–5.

88. Ibid., 10–11.

89. Ibid., 13.

90. Ibid., 4, of the Questions and Answers transcript pages that accompanied the text of the lecture.

91. I.D. White later attributed Palmer's fondness for the open-air scout in contrast to his lack of enthusiasm for the turreted armored car to claustrophobia. If this is true, one man's phobia had a lasting impact on mechanized reconnaissance equipment. I. D. White to Wes [Wesley W. Yale], 5 July 1967, [Honolulu, Hawaii], Correspondence with COL W.W. Yale and others, 1963–1969, RE: Alternative to Armageddon folder, box unassigned, White Papers.

92. Yale, "The 1936 Maneuvers," 175.

93. Ibid., 178.

94. Ibid., 177–78.

95. Ibid., 182.

96. Ibid., 181–82.

97. Ibid., 183.

98. Ibid.

99. Ibid., 188.

100. Ibid.

101. Ibid. p. 190.

102. Many hoped Bruce Palmer, commander of the 1st CavRegt(M) would be named the next chief of cavalry, which would have been a real boost for the mechanized portion of Cavalry Branch, but Herr was named instead. I.D. White to Wesley W. Yale [Honolulu], Hawaii , 5 July 1967; and Isaac D. White, General U.S.A.(ret.), interviewed by Colonel Charles S. Stodter, U.S. Army Military History Institute, Senior Officers Oral History Program, project 78-D, Carlisle Barracks, Carlisle, PA., 1978, 82 (hereafter referred to as White interview by Stodter).

103. COL Charles L. Scott, "Progress in Cavalry Mechanization: Scout Car Development," *CJ* 45 (July–August 1936): 281.

104. Ibid., 281.

105. Ibid., 282.

106. Ibid., 283.

107. Ibid., 284.

108. Ibid., 284.

109. Robert W. Grow became involved in the mechanization process while serving as the operations officer of the 12th Cavalry. COL Daniel Van Voorhis commanded the 12th and elected to take along his operations officer when he took command of the Mechanized Force being assembled at Fort Eustis, Virginia, in 1930. Grow saw extensive service at Fort Knox and at the Office of the Chief of Cavalry. During World War II he commanded the 6th AD. Grow, "The Ten Lean Years, from the Mechanized Force (1930) to the Armored Force (1940)," Peter R. Mansoor and Kathy C. Garth, eds., *Armor* 96 (January–February 1987), 23.

110. MAJ Robert W. Grow, "Military Characteristics of Combat Vehicles," *CJ* 45 (November–December 1936): 509.

111. Ibid.

112. Ibid.

113. Ibid.

114. Ibid., 511.

115. Grow served in the Supply and Budget Section of the Office of the Chief of Cavalry between 1936 and 1940. He used the opportunity to encourage General John K. Herr to save the branch by embracing the emerging mechanized combat force then under development. Grow, "The Ten Lean Years," 34–35.

116. Draft memorandum for Kent from Grow, 21 December 1946, pp. 2–3.

117. Ibid. This information is taken from an attached draft memorandum that was to be routed through the G3 to the Air Corps. The entire memorandum discussed the Second Army maneuvers., 6.

118. Ibid.

119. Draft memorandum for Kent from Grow, 21 December 1946, p. 4.

120. Ibid.

121. GEN Malin Craig, endorsement of article by LTC Alexander D. Surles, "The Cavalry and Mechanization, 1936," *CJ* 45 (January–February 1936): 6.

122. Ibid.

123. Surles, "The Cavalry and Mechanization, 1936," 6–7.

124. Kromer to Van Voorhis, Washington, 7 December 1936, folder 1936, box Correspondence 1935–1938, Crittenberger Papers, MHI, 1.

125. Ibid., 2.

126. Patton to Crittenberger, 27 April 1938, Fort Riley, Kansas, folder January–May 1938, Crittenberger Papers, MHI.

127. Carlo D'Este, *Patton, A Genius for War* (New York: Harper Collins, 1995), 368, cited in the Patton file at the Center of Military History, Fort McNair, Washington, D.C..

128. Killigrew, "The Impact of the Great Depression on the Army" (master's thesis, Indiana University, 1979), xv, 21–22, xv–31; and Russell F. Weigley, *The History of the United States Army* (New York: Macmillian, 1970), 415. Craig was a Pershing protégé. He had served as the chief of cavalry, as a corps commander, and as commandant of the War College. He also brought to his new post a general concern for the theoretical nature of the General Staff and lack of realism in planning.

129. Memo, Assistant Chief of Staff G3 to Chief of Staff, 10 April 1937, AG 381 (4-10-37), section 1, RG 94, NA, cited in Killigrew, "The Impact of the Great Depression on the Army," xv–30.

130. Johnson, *Fast Tanks, Heavy Bombers,* 115.

131. Watson, *Chief of Staff,* 29–30 and 42–43; and Weigley, *History of the United States Army,* 416.

132. Johnson, *Fast Tanks, Heavy Bombers,* 114.

133. Memo, Office of the Chief of Cavalry for Chief of Staff, subject: "Cavalry Requirements for the United States Army," 27 October 1938, folder 322.02 Cavalry, box 7b, entry 39, RG 177, NAII.

134. COL C.L. Scott to COL Guy W. Chipman, 17 March 1938, Fort Knox, folder January–May 1938, Crittenberger Papers, MHI.

135. MAJ Robert W. Grow, "Twenty Years of Evolution in American Cavalry Equipment," *CJ* 46 (March–April 1937): 127.

136. Memorandum, Grow to Miller, 10 January 1937, Office of the Chief of Cavalry, subject: "Thought for the Day," folder 322.02, box 7b, entry 39, RG 177, NAII.

137. Gillie, *Forging the Thunderbolt,* 101.

138. BG H.S. Hawkins, "Composition of Army Covering Forces and the Employment of Mechanized Force in This Role," *CJ* 46 (November–December 1937): 517.

139. Ibid., 518.

140. Ibid.

141. Ibid. 520.

142. BG Daniel Van Voorhis to MG Leon B. Kromer, Chief of Cavalry, 12 February 1937, Fort Knox, folder 1937, Crittenberger Papers, MHI, 2.

143. Ibid.

144. Ibid., 3.

145. Crittenberger to Senator Henry Cabot Lodge, 15 April 1937, Fort Knox, folder 1937, Crittenberger Papers, MHI, 2. Senator Henry Cabot Lodge, Jr., born 5 July 1902, was the grandson of Henry Cabot Lodge, Sr., who passed away in 1924. The younger Lodge entered the service as a reserve cavalry officer in 1925. Elected to the Senate in 1936 and again in 1942, he resigned his Senate seat when the war broke out. Lodge first saw service with the British Eighth Army in North Africa and went on to fight in Italy, southern France, and the Rhineland. The end of World War II found him serving in southern Germany, holding the rank of LTC.

146. Ibid., 3.

147. LTC Willis D. Crittenberger, "Fort Knox Maneuvers," *CJ* 46 (September–October 1937): 421–22.

148. Ibid., 425.

149. Ibid., 426.

150. Crittenberger to COL Adna R. Chaffee, 1 August 1937, Fort Knox, folder 1937, Crittenberger Papers, MHI, 1.

151. Ibid.

152. Ibid.

153. Ibid. Pack radios were carried on the back of a horse.

154. White, interview by Stodter, 75–76.

155. "Conference Delivered by CPT I.D. White, The Cavalry School, Fort Riley, Kansas, before Officers' Tactical School, Subject, Mechanized Reconnaissance Elements in Combat," Speech File: Mechanized Tactics, 1938–1941 folder, box unassigned, White Papers.

156. Ibid.

157. Ibid.

158. Ibid.

159. Ibid.

160. Ibid.

161. Ibid.

162. Ibid.

163. Ibid.

164. Ryan, interview by Nenninger, 15 June 1967.

165. MAJ F. During, trans., "The German Army Maneuvers, 1936," *The Command and General Staff School Quarterly* 17 (March 1937), 52; and Albert Seaton, *The German Army, 1933–1945* (London: Weidenfled and Nicolson, 1982), 69.

166. During, trans., "German Army Maneuvers, 1936," 54.

167. G. Bokis, "Certain Problems To Be Considered in the Training of Mechanized Forces," trans. Staff Sergeant Charles Berman, from *Auto Bronetankovyi Zhurnal*, Moscow, MHI Library UE147.B6513 1938, 2–3.

168. Ibid., 10.

169. MG Leon B. Kromer, "Cavalry," lecture delivered at the Army War College, 1 October 1937, Fort Humphreys, Washington, D.C., folder Various Addresses and Speeches, box 3, Kromer Papers, MHI, 2.

170. Ibid., 5.

171. Johnson, *Fast Tanks, Heavy Bombers,* 136; and LTC E.W. Taulbee to General Kromer, 9 February 1938, Fort Stotsenburg, Philippine Islands, folder Correspondence on Retirement from Chief of Cavalry, 1938, box 2, Kromer Papers, MHI. This personal letter also demonstrates the perception that Kromer too remained loyal to the horse cavalry community.

172. COL Charles L. Scott, "With the Mechanized Cavalry," *CJ* 47 (March–April 1938): 176.

173. "The Mechanized Cavalry Takes the Field," *CJ* 47 (July–August 1938): 291–92.

174. Ibid., 296.

175. Ibid., 297.

176. Ibid., 298–99.

177. Van Voorhis to LTC Crittenberger, 11 July 1938, Fort Knox, folder June–August 1938, Crittenberger Papers, MHI, 4.

178. Ibid.

179. Ibid.

180. BG Daniel Van Voorhis, "Organization of the Mechanized Cavalry Division," statement by commanding officer 7th CavBde(M) for the chief of cavalry. This was the edited result of Van Voorhis to Crittenberger, 11 July 1938.

181. Ibid.

182. Ibid.

183. LTC Willis D. Crittenberger, "Improvised Mechanized Reconnaissance," memorandum for the chief of cavalry, folder June–August 1938, Crittenberger Papers, MHI, 1.

184. Ibid., 3.

185. Van Voorhis to Crittenberger, 10 August 1938, Fort Knox, folder June–August 1938, Crittenberger Papers, MHI.

186. COL Charles L. Scott, "Armor for Cavalry Reconnaissance Vehicles Is Essential," *CJ* 47 (September–October 1938): 430.

187. Johnson, *Fast Tanks, Heavy Bombers*, 229.

188. "Notes from the Cavalry Board," *CJ* 47 (November–December 1938), 551. The original vehicle was introduced as the Bantam Chassis, 1/4 Ton. Early tests by the 2nd CavRegt recommended its use as the replacement for the sidecar motorcycle, but the regiment felt it lacked the requisite power required of tactical vehicles. Thomas Lynn, "Father of the Jeep," *Periodical Journal of America's Military Past* 24, no. 3 (Fall 1997): 25–30. Lynn suggests that his father invented a prototype "jeep" called the Dewblower at Jefferson Barracks, Missouri, in 1939, and that it was rejected by the officers of the 6th Infantry Regiment for being "too radical." Lynn's supposition seems flawed given the tests already underway at Fort Riley, but it does provide another example that the same kind of thinking was happening elsewhere at the same time, similar to how the evolution of mechanized warfare took place. True, there was a degree of cross-pollination, but there were also instances of nearly identical original thought going on simultaneously at different locations. The American Bantam Car Company delivered its first "jeeps" to the army in 1940. During the war, Ford and Willys-Overland assumed the contract, while the American Bantam Car Company manufactured aircraft and torpedo parts. There is some speculation as to why it was called the "jeep." One possibility is that the term derived from its abbreviated nomenclature, GPV (general-purpose vehicle). A more popular explanation is that it derived from the Popeye comic strip. Hereafter, the use of Jeep and Bantam are synonymous. Charles L. Scott later gave some attribution for the jeep's design to CPT Schauffle, who had served as a master mechanic in the 13th CavRegt(M) at Fort Knox during the interwar years. See letters exchanged between Scott and R.S. Demitz in 1944, folder October 1943–January 1944, box 2, Charles L. Scott Papers, LC.

189. Memorandum for Chief of Cavalry prepared by LTC Willis D. Crittenberger, Washington, subject: 1938 Cavalry Maneuvers at Fort Riley, folder June–August 1938, Crittenberger Papers, MHI, 4.

190. Ibid.

191. Ibid.

192. Crittenberger to Scott, 27 September 1938, Washington, D.C., folder September–December 1938, Crittenberger Papers, MHI.

193. Ibid.

194. Memorandum for Chief of Cavalry, 3 September 1938, Washington, D.C., subject: suggested changes to speech to be delivered at the Army War College, folder September–December 1938, Crittenberger Papers, MHI, 1.

195. Ibid., 2.

196. Ibid.

197. MG H.B. Fiske (Ret.) to MG John K. Herr, 14 September 1938, San Diego, folder 322.02, box 7, entry 39, RG 177, NAII. This letter was written in response to Herr's letter of 6 September 1938.

198. Ibid. Ralph Parker to Johnny [John K. Herr], 13 October 1940, C*R*O*Z*A*L, Canal Zone, box 6, Herr Papers, USMA, West Point.

199. Memorandum for Chief of Staff from Office of the Chief of Cavalry, 27 October 1938, subject: Cavalry Requirements for the United States Army, appendix G, folder 322.02, box 7b, entry 39, RG 177, NAII.

200. "Remarks RE Conference with Assistant Chief of Staff, G-3," folder 322.02 Chief of Cavalry, box 6, entry 39, RG 177, NAII.

201. Memorandum for Chief of Staff from Office of the Chief of Cavalry, 27 October 1938, pp. 1, 3.

202. Ibid., 2.

203. Ibid., 3.

204. Ibid., 2.

205. Ibid., 4.

206. Ibid., appendix F.

207. Ibid., 2.

208. Ibid., appendix C, p. 2.

209. Ibid., 1. See also Scott to COL Guy W. Chipman, 17 March 1938, Fort Knox, folder January–May 1938, Crittenberger Papers, MHI, 7.

210. Memorandum for Chief of Staff from Office of the Chief of Cavalry, 27 October 1938, p. 2.

211. Ibid., appendix C, 2.

212. Ibid., appendix H.

213. Albert Seaton, *The German Army, 1933–1945,* 60–61.

214. Crittenberger to Scott, 27 September 1938, Washington, D.C., folder September–December 1938, Crittenberger Papers, MHI. The comments are attributed to MAJ Little of the British War Office.

215. Fred W. Merten, "Cavalry in the War in Spain," *CJ* 47 (November–December 1938): 520.

216. Scott, "With the Mechanized Cavalry," 175–76; and Gillie, *Forging the Thunderbolt,* 102.

217. LTC Klemens Rudnicki, "The Role of Cavalry in a Future War," condensed from *Bellona* (August 1938, Warsaw), trans. MAJ E.M. Benitez, *The Command and General Staff School Quarterly* 18 (December 1938), 38.

218. LTC P.R. Davison and MAJ E.M. Benitez, *The Command and General Staff School Quarterly* 18 (September 1938), 5.

219. Ibid., 14.

220. Morrison, *F Troop,* 40.

221. Collier, "The Army and the Great Depression," 107.

3—The Big Maneuvers and War

1. Hanson Baldwin, *Battles Lost and Won: Great Campaigns of World War II* (New York: Konecky & Konecky, 1966), 20.

2. George Fielding Elliot, "The Future of American Cavalry," *CJ* 48 (May–June 1939): 195.

3. MG John K. Herr, "What of the Future?" *CJ* 48 (January–February 1939): 3–4. In this case Herr was not delusional. The advent of long-range strategic bombers during the 1930s placed the coast of Brazil within range of the west coast of Africa. The reality of the bombers, combined with the belief that foreign/European interlopers might inject themselves into South and Central America, fed some of the concern and hence attention paid to the development of the ways and means of defending the Western Hemisphere in support of the Monroe Doctrine. To this end, the Army War College responded to the needs of the War Department, War Plans Division. Henry G. Gole, *The Road to Rainbow: Army Planning for Global War, 1934–1940* (Annapolis, MD: Naval Institute Press, 2003), 83, 92–93.

4. "Cavalry Affairs Before Congress" [MG Herr testifies to Congress], *CJ* 48 (March–April 1939): 130.

5. Ibid., 131, 135. Herr provided to the committee taking his testimony an interesting

example whose conclusion was that a body of mounted men with the proper amount of dispersion could charge across open ground and capture a machine gun before it would have time to engage all of them. He used a tabletop to illustrate his reasoning.

6. Ibid., 130, 134.

7. Bland, *Marshall Papers,* ed. Bland and Ritenour, 707–8; BG George C. Marshall to BG Lesley McNair, 4 March 1939, Washington, D.C.

8. LTC Robert W. Grow, "New Developments in the Organization and Equipment of Cavalry," *CJ* 48 (May–June 1939): 205–7.

9. LTC E.E. Schwien, "Cavalry Division Maneuvers," *CJ* 48 (January–February 1939): 464–65.

10. Ibid., 468.

11. Ibid.

12. Jean R. Moenk, *A History of Large-Scale Maneuvers in the United States, 1935–1964* (Fort Monroe, VA: Headquarters, U.S. Continental Army Command, 1969), 23–25. "Square" infantry divisions contained two brigades, each of which had two regiments, and of which the regiments each had four battalions. "Triangular" divisions were initially formed with three regiments, each containing three battalions. It was believed this smaller, more streamlined force would have additional agility compared to the larger "square" divisions.

13. *Patton Papers,* ed. Blumenson, 1:939–41.

14. Gillie, *Forging the Thunderbolt,* 127.

15. Ibid., 134–35.

16. Johnson, *Fast Tanks, Heavy Bombers,* 176. Although MG Herr would later accuse Marshall and McNair of being opposed to horses, it is clear he had an affinity for equines and certain members of the Cavalry Branch. Soon after George C. Marshall was named chief of staff of the army, Terry Allen, then stationed at Fort Riley, Kansas, selected two horses to be shipped to Washington for the new chief's use. Marshall found his daily rides therapeutic, and they enabled him to keep the demands and challenges of his new job in perspective. Gerald Astor, *Terrible Terry Allen, Combat General of World War II: The Life of an American Soldier* (New York, Ballantine Books, 2003), 86. Patton invited Marshall to live with him until the chief's quarters were prepared. Writing to his wife, Patton quipped, "I have just consumated (sic) a pretty snappy move. Gen George C Marshall is going to live at our house!!! He and I are batching it. I think that once I can get my natural charm working I wont (sic) need any letters from John J.P. [Pershing] or any one else." Patton also made sure that when Marshall was sworn in, he had eight sterling silver stars to affix to his shoulder loops. *Patton Papers,* ed. Blumenson, 1:939, 944.

17. Blitzkrieg represented far more than tactics and weapons; it included operational and political action, all of which suited the needs of its "parent" Adolph Hitler. The fast-paced action of blitzkrieg presented its opponents and the world a fait accompli. Graham and Bidwell, *Fire-Power,* 205.

18. "Training of Modern Cavalry for War, Polish Cavalry Doctrine," *CJ* 48 (July–August 1939): 300–5.

19. Ibid., 300.

20. Ibid., 301–5.

21. Baldwin, *Battles Lost and Won,* 15; and Gillie, *Forging the Thunderbolt,* 137.

22. Johnson, *Fast Tanks, Heavy Bombers,* 138.

23. M. Kamil Dziewanowski, "Last Great Charge of the Polish Cavalry," *Army* 20, no. 4 (April 1970), 41–43. The Poles also had an assortment of light armored vehicles in their horse cavalry units, intended to deal with enemy mechanization. These in combination with improvised gasoline grenades allowed Dziewanowski's unit to destroy 31 armored vehicles in a week of fighting. By the end of the week, the unrelenting action transformed his unit's horses into walking skeletons. Operating deep in the rear of the growing German penetration, the Poles had limited success. Not unlike American peacetime maneuvers, the Poles used their armored cars to the flanks on high-speed avenues of approach, unlimbered their heavy machine guns to provide a base of fire, and on one occasion, operating on terrain conducive to horse cavalry, conducted a mounted charge against an unsupported German infantry battalion marching in the open. With sabers and lances, two squadrons (U.S.

equivalent of two troops) descended upon the Germans. When the charge was over the Poles had only lost three men killed, but 30 to 40 of their mounts lay dead on the battlefield.

24. Herr and Wallace, *The Story of the U.S. Cavalry, 1775-1942,* 252.

25. BG Adna R. Chaffee, "Mechanized Cavalry," lecture delivered to the Army War College, 29 September 1939, Fort Humphreys, G3#12, 1940, copy 1, MHI.

26. Ibid., 11.

27. Ibid., 16.

28. Ibid., 17.

29. Terry Allen, *Combat Communication for Regiments and Smaller Units of Horse Cavalry* (Harrisburg, PA: Military Service Publishing Co., 1939), 6-7.

30. Ibid., 9, 36. Pack radios could only be monitored on the march.

31. Ibid., 17-18.

32. Ibid., 26.

33. Yale, White, and Manteuffel, *Alternative to Armageddon,* 78.

34. The assistant chief of staff, G3, MG Frank M. Andrews, was in no rush to make any radical changes in the organization of the horse cavalry division in 1940. While operations in Europe attracted the attention of many, and reinforced the position of those bent on further mechanization, Andrews held some reservations about eliminating the horse entirely from the army's array of capabilities. Not knowing definitely where the army would fight but with full knowledge that it might be asked to fight in places where vehicles could not travel, he was unwilling to make sweeping changes. Andrews did not completely lack initiative in regard to the future of the cavalry, and he did want to explore the possibility that horse cavalry might effectively cooperate with mechanized forces. *Report of the Secretary of War, 1941,* 60, Letter from the Adjutant General to the Commanding General 1st CavDiv, 24 April 1940, subject: Report on the Corps and Army Maneuvers, AGO 320.2 (4-15-40) file, RG 407, NARA, cited in Wilson, *Maneuver and Firepower,* 145.

35. MAJ Thomas J. Heavey, "The Horse-Mechanized Regiment," *CJ* 44 (September–October 1940): 426.

36. Ibid.

37. LTC John A. Considine to LTC Willis D. Crittenberger, 20 January 1940, Fort Leavenworth, Kansas, folder January–March 1940, Crittenberger Papers, MHI. Considine writes that the Horse-Mechanized is "one of the best ideas that has come out of the cavalry in years.

38. Memorandum for Chief of Cavalry, 23 January 1949, folder 320.02 Mech Cav, box 8, entry 39, RG 177, NAII.

39. Memorandum for Chief of Cavalry, 26 February 1940, folder January–March 1940, Crittenberger Papers, MHI.

40. LTC Willis D. Crittenberger to LTC John B. Coulter, 9 February 1940, Washington, D.C., folder January–March 1940, Crittenberger Papers, MHI. At the time of the letter Coulter commanded the 4th CavRegt (H-M).

41. CPT M.S. Biddle to LTC Willis D. Crittenberger, 9 March 1940, Fort Bliss, Texas, folder January–March 1940, Crittenberger Papers, MHI. Biddle, then stationed with the 1st CavDiv, raised the issue of escort and firepower in his letter.

42. LTC Willis D. Crittenberger to CPT M. S. Biddle, 12 March 1940, Washington, D.C., folder January–March 1940, Crittenberger Papers, MHI.

43. LT Tom E. Matlack, "Reconnaissance Troop Officer's Diary," *CJ* 44 (November–December 1940): 532–34.

44. Ibid., 532.

45. Ibid., 534.

46. *Patton Papers,* ed. Blumenson, 1:943.

47. LTC E.E. Schwien, "Intensive Training of the First Cavalry Division," *CJ* 44 (March–April 1940): 115–16.

48. *Patton Papers,* ed. Blumenson, 1:946–48.

49. "Cavalry Affairs in Congress," *CJ* 44 (May–June 1940): 206.

50. Ibid.

51. Ibid.

52. Robert S. Cameron, "Americanization of the Tank: U.S. Army Administration and Mechanized Development within the Army, 1917–1943" (Ph.D. diss., Temple University, Philadelphia, 1994), 486. This statement refers only to the poor reconnaissance performance.

53. *Final Report, Third Army Maneuvers*, vol. 3, annex 21, *Final Critques*, cited in Moenk, *A History of Large-Scale Maneuvers*, 32–33.

54. Stubbs and Connor, *Armor-Cavalry*, 57; and Tedesco, "Greasy Automatons," 102, cited in LTG Alvan C. Gillem, Jr. (Ret.), interview by Timothy Nenninger, 27 September 1967, Atlanta, Georgia.

55. Christopher R. Gabel, *U.S. Army GHQ Maneuvers, 1941* (Washington, D.C.: Center of Military History, 1991), 23; Stubbs and Connor, *Armor-Cavalry*, 63; and Peter R. Mansoor, *The G.I. Offensive in Europe: The Triumph of American Infantry Divisions, 1941–1945* (Lawrence: University Press of Kansas, 1999), 17.

56. Gillie, *Forging the Thunderbolt*, 163–65; Gabel, *U.S. Army GHQ Maneuvers*, 23; and Nenninger, "The Development of American Armor," 185.

57. D'Este, *Patton, Genius for War*, 378. Patton, ever trying to keep his career hopes alive, had supplied the eight silver stars to Marshall when he was promoted to general in 1940. He had also supplied the stars for the promotion of Joyce to 1st CavDiv Commander the same year.

58. Memorandum for Assistant Chief of Staff, G3, 3 June 1940, folder 320.02 Mech Cav, box 8, entry 39, RG 177, NAII.

59. Stubbs and Connor, *Armor-Cavalry*, 58. The new Armored Corps was built on the existing foundations of the 7th Cavalry Brigade (Mechanized), which immediately became the 1st Brigade, 1st AD. The Provisional Tank Brigade at Fort Benning became the 2nd AD. In an especially astute Army/Branch political move, Chaffee cross-pollinated his new force by bringing the infantryman, Magruder, to Fort Knox and sending Scott to Fort Benning. The word "armored" was specifically chosen to help the new force divorce itself from its roots. "Mechanized" was associated with cavalry branch, and "tank" had been the term used by infantry branch. Nenninger, "The Development of American Armor," 3.

60. Cameron, "Americanization of the Tank," 498, 883. Cameron places greater importance on the rapid destruction of the French army in 1940 as the catalyst for the major changes effected by the creation of the Armored Corps.

61. D'Este, *Patton, Genius for War*, 382.

62. COL George S. Patton, Jr., to MG Herr, 10 September 1940, Headquarters 2nd Armored Brigade, Fort Benning, Georgia, box 7, Herr Papers, USMA.

63. Ibid.

64. Ibid.

65. Ibid.

66. Morrison, *F Troop*, 76–77.

67. MG John K. Herr to MG Innis P. Swift, 9 June 1941, Office of the Chief of Cavalry, Washington, D.C., box 6, Herr Papers, USMA.

68. *Patton Papers*, ed. Blumenson, 1:949.

69. Gabel, *U.S. Army GHQ Maneuvers*, 5.

70. Moenk, *History of Large-Scale Maneuvers*, 69; and Cameron, "Americanization of the Tank," 708 and 725–26. Cameron speaks specifically to problems associated with the horse-mechanized regiments.

71. Mansoor, *The G. I. Offensive in Europe*, 25–26.

72. Wright, *The First Cavalry Division in World War II*, 3.

73. "Editorial Comment," *CJ* 50 (November–December 1941): 26–28.

74. Weigley, *History of the United States Army*, 432.

75. Memorandum for Commanding General, Field Forces, 31 January 1942, Subject: Cavalry Corps Reconnaissance Regiment, folder 322.02, box 44, RG 337, NAII.

76. U.S. Army. *Training Circular*, no. 32, 8 May 1941, *Employment of the Cavalry Regiment Horse and Mechanized (Corps Reconnaissance Regiment)*. The National Guard regimental commanders shared the same prejudices. COL Maxwell A. O'Brien, commander of the 113th CavRegt (H-M), commented that his horses performed better at night because his mechanized troops "were compelled to halt to await daylight and for the purpose of

servicing equipment. "113th Cavalry (H-Mecz) in Louisiana Maneuvers," *CJ* 50 (November–December 1941): 37.

77. COL John Millikin to MG Herr, 2 May 1940, HQ 6th Cavalry, folder 320.3, box 3, entry 39, RG 177, NAII.

78. "Fourth Cavalry (H-Mecz) in Fourth Army Maneuver," *CJ* 44 (September–October 1940): 444.

79. COL John Millikin to MG Herr, 2 May 1940, HQ 6th Cavalry, folder 320.3, box 3, entry 39, RG 177, NAII, 3–4.

80. COL Maxwell A. O'Brien to Commanding General, VIII Army Corps, Brownwood, Texas, 30 October 1941, Camp Bowie, folder 320.3 Revision of Table of Organization, box 5, entry 39, RG 177, NAII.

81. "113th Cavalry (H-M) in Louisiana Maneuvers," *CJ* 50 (November–December 1941): 38.

82. O'Brien to Commanding General, VIII Army Corps, Brownwood, Texas, 30 October 1941; "Report of 106th Cavalry, June to October Maneuvers, Louisiana, 1941," 12 October 1941, Camp Livingston, Louisiana, folder 320.3 Revision of Table of Organization, box 5, entry 39, RG 177, NAII; and long quotation attributed to COL John Millikin (Millikin to Herr, 2 May 1940, HQ 6th Cavalry, folder 320.3, box 3, entry 39, RG 177, NAII).

83. "Report of 106th Cavalry, June to October Maneuvers, Louisiana, 1941," 12 October 1941.

84. MAJ Thomas J. Heavey, "The Horse-Mechanized Regiment," *CJ* 44 (November–December 1940): 427–28; and Matlack, "Reconnaissance Troop Officer's Diary," 535.

85. "113th Cavalry (H-M) in Louisiana Maneuvers," *CJ* 50 (November–December 1941): 39.

86. "Report of 106th Cavalry, June to October Maneuvers, Louisiana, 1941," 12 October 1941.

87. LTC Charles R. Johnson, Jr., "106th Cavalry, Last Phase, Louisiana Maneuvers," *CJ* 50 (November–December 1941): 45.

88. Millikin to Herr, 2 May 1940; Matlack, "Reconnaissance Troop Officer's Diary," 535; and "113th Cavalry (H-M) in Louisiana Maneuvers," *CJ* 50 (November–December 1941), 37, a call for the replacement of the SCR-203 pack radio with a SCR-245, modified for use as a pack radio or in a jeep.

89. "The Sixth Cavalry in the IV Corps Maneuvers," *CJ* 49 (May–June 1940): 194. Until the spring of 1941, all units trained their own men in regimental schools at their home stations.

90. Millikin to Herr, 2 May 1940.

91. CPT Bruce Palmer, Jr., "Turn 'em Over!" *CJ* 44 (November–December 1940): 542.

92. CPT John F. Franklin, Jr., "The 6th Cavalry, Maneuvers in Louisiana," *CJ* 50 (November–December 1941): 65–66.

93. In a brief, but telling, thank-you note, McNair provided real insight into his understanding of modern cavalry. Swift had sent him a series of photographs of the entire division assembled on a large review field. These photos moved McNair to comment, "[S]ince I have no very clear idea of what the modern Cavalry division looks like," the pictures, "the big one especially," was appreciated. MG L.J. McNair to BG I.P. Swift, 3 March 1941, General Headquarters, U.S. Army, Army War College, Washington, D.C., box 6, Herr Papers.

94. Adjutant General to Commanding General, Second Army, 8 October 1941, Washington, folder 322.02 Cavalry, box 44, RG 337.

95. War Department, Operations and Training Division, G3, brief, 30 December 1941, folder 322.02 Cavalry, box 44, RG 337.

96. Millikin to Herr, 4 October 1941, Headquarters 2nd CavDiv, Camp Funston, Kansas, box 6, Herr Papers, USMA.

97. "Further Notes on Narrative Record to Date of the G-3 Effort to Substitute Mechanized for Horse and Mechanized Corps Cavalry Regiments," 20 October 1941, Washington, D.C., folder 320.2, box 7c, entry 39, RG 177, NAII.

98. "Narrative Record to Date of the G-3 Effort to Substitute Mechanized for Horse and Mechanized Corps Cavalry Regiments," 11 October 1941.

99. Ibid. COL Considine's remarks may have had some bearing on his service during

the remainder of World War II as a military attaché to Paraguay. It should be noted that he served as a BG in the Guatemalan army from 1930 to 1935.

100. MG John K. Herr to MG Innis P. Swift, 11 October 1941, Office of the Chief of Cavalry, Washington, D.C., box 6, Herr Papers, USMA.

101. Ibid. See also Memorandum to Herr, 7 October 1941, Washington, D.C., record of the conversation conducted between COL J.T. Duke, Office of the Chief of Cavalry and COL Campbell, representative of the Army Chief of Staff, G3, folder 320.2, box 7c, entry 39, RG 177, NAII.

102. Herr to Swift, 11 October 1941.

103. Herr to Swift, 11 October 1941; and Orders thru Chief of Cavalry to MAJ Wesley W. Yale (0-14969) Cavalry, 21 October 1941, issued by Adjutant General, folder 320.2, box 7c, entry 39, RG 177, NAII. Yale had written a number of articles for *CJ*, which were generally positive in the sense of supporting the horse. Yale went on to serve in the 11th AD with Herr's son-in-law, Willard Holbrook, himself the son of a former chief of cavalry and future father-in-law of George S. Patton IV.

104. "Observer's Report, Test of Cavalry Reconnaissance Units, First Army Maneuvers, 1941," for Chief of Cavalry, undated, appendix A, p. 1, folder 322.02, box 7c, entry 39, RG177, NAII.

105. Ibid.

106. Ibid., 10; and appendix A:4.

107. Ibid., 11.

108. Ibid., 5.

109. Ibid., 4.

110. Ibid., and appendix A:6.

111. Ibid., 5.

112. Ibid., 2.

113. "Report of Cavalry Reconnaissance Units," Headquarters Sixth Cavalry, 4 December 1941, Fort Oglethorpe, Georgia, file 3203, box 3, entry 39, RG 177, NA II.

114. Ibid.

115. Ibid.

116. Kent Roberts Greenfield, Robert R. Palmer, and Bill I. Wiley, *The Army Ground Forces, the Organization of Ground Combat Troops* (Washington, D.C.: Dept. of the Army, Historical Division, 1947), 308–9.

117. Staff writer, "The Sixth Cavalry at the Fourth Corps Maneuvers," *CJ* 48 (May–June 1939): 200.

118. Memorandum for Chief of Cavalry from War Department, G3 Training, 27 June 1940, folder 322.02 Infantry-Division Reconnaissance Troop, box 7a, entry 39, RG 177, NAII.

119. Memorandum for Assistant Chief of Staff, G3, 2 July 1940, folder 322.02 Infantry-Division Reconnaissance Troop.

120. Memorandum for Chief of Staff, 16 August 1940, folder 322.02 Infantry-Division Reconnaissance Troop.

121. Chief of Cavalry to the Adjutant General, 22 August 1940, Washington, D.C., folder 322.02 Infantry-Division Reconnaissance Troop.

122. Ibid.

123. Ibid.

124. War Department, Adjutant General's Office, to the Chief of Cavalry, 3 September 1940, folder 322.02 Infantry-Division Reconnaissance Troop.

125. COL K.S. Bradford to Commandant, Cavalry School, 2 December 1940, Washington, D.C., folder 322.02 Infantry-Division Reconnaissance Troop.

126. Ibid.

127. U.S. Army, *Training Circular*, no. 18, 15 March 1941 (Washington, D.C.: Government Printing Office).

128. *The Infantry School Mailing List* 22 (May 1941), 102.

129. Ibid.

130. "Organization of New American Armored Corps," *CJ* 50 (July–August 1940): 317.

131. Gillie, *Forging the Thunderbolt*, 179.

132. MAJ I.D. White, "The Armored Force Reconnaissance Battalion, Armored Division," *CJ* 50 (May–June 1941): 48–49.

133. Ibid., 51.

134. Ibid.

135. Ibid., 52. Program for "Reconnaissance in Force Demonstration for U.S.M.A. Cadets, 82nd Armored Reconnaissance Battalion in Collaboration with the Infantry School," 18 June 1942, Fort Benning, Georgia, Miscellaneous Material, 1937–1942 folder, box 33, White Papers.

136. Cameron, "Americanization of the Tank," 789.

137. Patton's critique of the Tennessee Maneuvers, delivered to the entire 2nd AD, 8 July 1941, box 54, Patton Papers, LC.

138. I.D. White later recounted this episode in a letter to Wesley W. Yale, stating that he only released Drum after checking with Patton. Drum criticized White in the formal after action review of the maneuver, when he "derided the efforts of 'a little lieutenant scampering around trying to find his way back home with his odd collection of 'farm implements!'" White was surprised that any of them [early mechanized advocates] had survived and was grateful that "many of them [those behind the times] were too old to do much in WWII." White to Yale, 16 January 1968, Honolulu, Hawaii, Correspondence with COL W.W. Yale and others, 1963–1969, RE Alternative to Armageddon folder, box unassigned, White Papers.

139. LTC H.H.D. Heiberg, "Armored Reconnaissance, *CJ* 51 (May–June 1942): 66, 69.

140. Taking elements from one unit and assigning them to another unit within the parent organization is often referred to as "cross-attachment." Cross-attachment is a means of building small combined-arms organizations.

141. D.W. Dean and C.H. Hulse, *The 91st Cavalry Reconnaissance Squadron and WW II* (Self-published for unit reunion, August 1993), 5–6, 10.

142. MG John K. Herr to MG I.P. Swift, 31 May 1941, Office of the Chief of Cavalry, Washington, D.C., box 6, Herr Papers, USMA.

143. Ibid.

144. Herr to Swift, 9 June 1941.

145. Ibid.

146. Ibid.

147. Ibid.

148. Swift to Herr, 10 June 1941, Headquarters, First Cavalry Division, Fort Bliss, Texas, box 6, Herr Papers, USMA.

149. Ibid.

150. Ibid.

151. Herr to Swift, 16 June 1941, Office of the Chief of Cavalry, Washington, D.C., box 6, Herr Papers, USMA.

152. Ibid.

153. Swift to Herr, 30 June 1941, Headquarters, First Cavalry Division.

154. Swift to Herr, 7 July 1941.

155. Ibid.

156. Ibid.

157. MG McNair arrived at the same conclusion and developed the tank destroyer force with its own reconnaissance agencies. See Christopher Gabel, *Seek, Strike, and Destroy: U.S. Army Tank Destroyer Doctrine in World War II* (Fort Leavenworth, KS: Combat Studies Institute, 1985), for a detailed history of tank destroyer development and the army's failure to employ this force in accordance with its doctrine.

158. Swift to Herr, 10 January 1941, Headquarters, Second Cavalry Brigade, Fort Bliss, Texas, box 6, Herr Papers, USMA; and Swift to Herr, 17 March 1941, Headquarters, First Cavalry Division, Office of the Commanding General.

159. Swift to Herr, 2 July 1941, Headquarters, First Cavalry Division.

160. Swift to Herr, 2 October 1941, in camp vicinity, Many, Louisiana, box 6, Herr Papers, USMA.

161. MG John K. Herr, "A Survey of Our Cavalry," *CJ* 44 (November–December 1940): 483.

162. COL K.S. Bradford to Miss Louise Hoffer, 15 February 1941, Washington, D.C., folder 350.051 Military Information Dissemination, box 16, entry 39, RG 177, NAII.

163. Grow, "The Ten Lean Years," 35.

164. "Notes from the Chief of Cavalry," *CJ* 44 (November–December 1940): 511.

165. LT George M. White, "Cavalry's Iron Pony," *CJ* 50 (March–April 1941): 85.

166. Ibid., 86–87.

167. Ibid., 87.

168. CPT W.F. Damon, Jr., "9th Reconnaissance Troop on Maneuvers," *CJ* 50 (September–October 1941): 87.

169. CPT Bruce Palmer, Jr., "The Bantam in the Scout Car Platoon," *CJ* 50 (March–April 1940): 89–92.

170. Greenfield, Palmer, Wiley, *Organization of Ground Combat Troops,* 33, 46.

171. Gabel, *U.S. Army GHQ Maneuvers,* 5. Reference is only to the army's expansion.

172. Herr and Wallace, *The Story of the U.S. Cavalry,* 250.

173. Ibid.

174. Edwin Price Ramsey and Stephen J. Rivele, *Lieutenant Ramsey's War: From Horse Soldier to Guerrilla Commander* (Washington, D.C.: Potomac Books, 1996), 66.

175. "History of the 91st Reconnaissance Squadron, 1928–20 July 1945." Folder CAVS-91-0, box 18231, entry 427, RG 407, NA II, 1928–20 July 1945.

176. Morrison, *F Troop,* 76–77.

177. LTC John A. Hettinger to LTC George I. Smith, 3 December 1941, Fort Jackson, South Carolina, box 5, entry 39, records group 177, NAII.

178. Ibid.

179. Ibid.

180. Ibid.

181. Ibid.

182. Ibid.

183. G2 is the intelligence officer for the division. During World War II, the U.S. Army did not have a military intelligence branch; rather, officers were detailed into the job from their basic branch of service.

184. MG O.W. Griswold to the Adjutant General, Subject "Test of Cavalry Reconnaissance Units," 9 December 1941, unknown, box 5, entry 39, records group 177, NAII.

185. Ibid.

186. Ibid.

187. Ibid.

188. Ibid.

189. MG John K. Herr to the Assistant Chief of Staff, G-3, Subject: Recommendations for the Setup of a Corps Cavalry Reconnaissance Regiment, Mechanized, Based on Recommendations of MG O.W. Griswold, 29 December 1941, Washington, D.C., box 5, entry 39, RG 177, NAII.

190. Ibid.

191. Herr to the Adjutant General, Subject: Cavalry Reconnaissance Units, 23 January 1942, Washington, D.C., box 7a, entry 39, RG 177, NAII.

192. Ibid.

193. Ibid.

194. Herr to the Adjutant General, Subject: Training Cavalry Divisions and Separate Brigades, 28 January 1942, Washington, D.C., box 7a, entry 39, RG 177, NAII.

195. Herr to Swift, 3 February 1942, Office of the Chief of Cavalry, Washington, D.C., box 6, Herr Papers, USMA.

196. Memorandum, Chief of Cavalry, 5 February 1942, file 322.02, Office of the Chief of Cavalry, Correspondence, 1921–1942, box 8, RG 177, NAII.

197. Ray S. Cline, *The War Department: Washington Command Post, The Operations Division,* United States Army in World War II (Washington, D.C.: Office of the Chief of Military History, 1951), 90; and Herr and Wallace, *The Story of the U.S. Cavalry,* 252.

198. Cline, *The War Department,* 92–93.

199. Ibid., 93.

200. Herr and Wallace, *The Story of the U.S. Cavalry,* 252.

201. Herr had a strange ally in the Soviets. A group of Soviet military attachés touring American factories in May 1942 remarked that the U.S. cavalry had become "over mechanized and too road bound"; it lacked sufficient "mounted combat training," and the "use of sabres was disliked." The Soviets regarded American cavalry as "potentially excellent," yet "unnecessarily limited to minor reconnaissance functions." Extract of Memorandum for ACOS G3, 12 May 1942, appendix H, "Review of Tactical Doctrine," prepared by Organization and Doctrine Branch, G3, 28 August 1942, folder Record of Review of Tactical Doctrine, box 957, entry 461, RG 337, NAII.

202. Ramsey and Rivele, *Lieutenant Ramsey's War,* 60–71.

203. MAJ Spelman Downer, "Cavalry Commandos," 51 (September–October 1942): 36–39.

204. Shelby L. Stanton, *Order of Battle, U.S. Army, World War II* (Presidio: Novato, CA, 1984), 23; and Stubbs and Connor, *Armor-Cavalry,* 70–72.

205. Stubbs and Connor, *Armor-Cavalry,* 72. These included: 101st (Ohio), 102nd (New Jersey), 104th (Pennsylvania), 106th (Illinois), 107th (Ohio), 113th (Iowa), and the 115th (Wyoming).

206. Table of Organization, 2-71, 1 April 1942.

207. Stubbs and Connor, *Armor-Cavalry,* 60–61.

208. Senator Henry Cabot Lodge, Jr., to LTC Willis D. Crittenberger, 15 August 1940, Washington, D.C., folder April–August 1940, Crittenberger Papers, MHI.

209. "Report on the Organization of a Mechanized Reconnaissance Regiment as the Cavalry Component of an Army Corps," 14 February 1942, Fort Riley, Kansas, folder 322.02, Mechanized Reconnaissance Regiment, box 8, entry 39, RG 177, NAII.

210. On 20 March 1942, BG Terry Allen, then the assistant division commander of the 36th Infantry Division and a notorious cavalry officer, signed an endorsement requesting the replacement of the Scout Car because of "its great lack of cross country mobility" and "limited maneuverability." He said the car should be replaced with a half-track. Army Ground Forces responded that the T22, which would become the M8 light armored car, "may not be expected prior to September 1942," and that the M3A1 Scout Car would continue to serve as a substitute until the T22 arrived in units. "Deficiencies in Equipment of Division Reconnaissance Troop under Proposed Table of Organization," 19 March 1942, Camp Blanding, Florida; and Memorandum, 15 April 1942, AGF to Commanding General, II Army Corps, folder 475.2 Cavalry Equipment, General HQ USA, General Correspondence 1940–1942, 475–78, box 97, RG 337, NAII.

211. Ibid.

212. William S. Biddle Papers. Biddle believed this proved the organization was better organized and equipped to fight for information and maintain contact until larger forces could arrive.

213. U.S. War Department, *FM 2-15, Cavalry Field Manual, Employment of Cavalry* [hereafter *FM 2—15, Employment of Cavalry*] (Washington, D.C.: Government Printing Office, 1941), 148.

214. Extract of Memorandum for ACOS G3, 12 May 1942, appendix I, "Review of Tactical Doctrine," prepared by Organization and Doctrine Branch, G3, 28 August 1942; and "Report on the Organization of a Mechanized Reconnaissance Regiment as the Cavalry Component of an Army Corps," 14 February 1942.

215. Christopher R. Gabel, introduction to September 1991 reprint of *FM 100-5, Operations.* The U.S. Army expanded eightfold between 1939 and 1941, and it had also "embarked on a modernization program that affected virtually every facet of military activity." As Michael Doubler asserts in his book, *Closing with the Enemy* (Lawrence: University Press of Kansas, 1994), the doctrine contained in *FM 100-5, Operations* was extremely important because it was generally sound. The U.S. did not have to find an entirely new doctrine after it was committed to combat. Rather, it only had to make relatively minor adjustments to the existing doctrine by applying lessons learned.

216. Ibid., 6–7.

217. *FM 100-5, Operations,* 7.

218. Ibid., 40–41.

219. Ibid., 258–59.

220. Ibid.

221. Ibid., 40–41.

222. Ibid., 43.

223. Ibid.

224. Ibid., 44.

225. Ibid.

226. Ibid., 45.

227. Ibid.

228. Ibid.

229. Ibid.

230. Ibid.

231. Ibid., 263.

232. Ibid.

233. Ibid.

234. *FM 2-15, Employment of Cavalry,* 1. *FM 2-15* reflected almost no difference from *FM 2-5, Cavalry Field Manual, Horse Cavalry* (Washington, D.C.: Government Printing Office, 1940) in regard to the proper use and employment of mechanized forces. I have selected the later field manual for description and analysis, since it represents the last organizational document speaking to the issue prior to actual combat.

235. *FM 2-15, Employment of Cavalry,* 5.

236. Ibid., 1.

237. Ibid.

238. Ibid., 2.

239. Ibid.

240. Ibid.

241. Ibid.

242. Ibid., 87–88.

243. Ibid., 94. "Transportation facilities" probably means the horse trailers used to move the horse over great distances at considerable speed to save "horse flesh." Doctrinally, this was only possible if the horse squadron's axis of reconnaissance was supported by a detachment of mechanized elements clearing the road in advance of the trailers.

244. Ibid., 95.

245. Ibid., 100.

246. Ibid., 102.

247. Ibid.

248. Ibid., 112.

249. U.S. War Department, *FM 17-22, Armored Force Field Manual, Reconnaissance Battalion* [hereafter *FM 17-22, Reconnaissance Battalion*] (Washington, D.C.: Government Printing Office, 1942).

250. *FM 100-5, Operations,* 263.

251. Ibid., 264; and Officers' School Armored Force Replacement Training Center, *Instructional Bulletin* [hereafter Officer's School, *Instructional Bulletin*], no. 8, 19 February 1942, Fort Knox, Kentucky, Papers of John Wesley Castle, Jr., binder Officer's School Armored Force Replacement Training Center, box 2, USMA.

252. *FM 100-5, Operations,* 263.

253. *FM 17-22, Reconnaissance Battalion,* 4–10.

254. *FM 100-5, Operations,* 263; and *FM 17-22, Reconnaissance Battalion,* 39.

255. Officer's School, *Instructional Bulletin,* no. 8; and *FM 17-22, Reconnaissance Battalion,* 3.

256. Ibid.

257. Ibid.

258. *FM 100-5, Operations,* 266.

259. Officer's School, *Instructional Bulletin,* no. 8.

260. "Report on the Organization of a Mechanized Reconnaissance Regiment as the

Cavalry Component of an Army Corps," 14 February 1942.

261. Ibid.

262. "Review of Tactical Doctrine" prepared by Organization and Doctrine Branch, G3, 28 August 1942.

263. T/O 2-51 and T/O 2-71.

264. "Review of Tactical Doctrine," prepared by Organization and Doctrine Branch, G3, 28 August 1942.

265. George S. Patton, Jr., "Notes on Tactics and Techniques of Desert Warfare (Provisional)," 30 July 1943, Headquarters Desert Training Center, Camp Young, California, box 3, Patton Papers, USMA, 1.

266. Ibid., 4.

267. Ibid., 12.

268. MG Charles L. Scott, "Armored Reconnaissance," *CJ* 51 (November–December 1942): 21–22.

269. Ibid., 22.

270. Cameron, "Americanization of the Tank," 498, 883. Cameron places greater importance on the rapid destruction of the French army in 1940 as the catalyst for the major changes effected by the creation of the Armored Corps.

271. D'Este, *Patton, Genius for War*, 382.

272. Gabel, *U.S. Army GHQ Maneuvers*, 5. Reference only to the army's expansion.

273. Herr and Wallace, *The Story of the U.S. Cavalry*, 250.

274. Ibid.

4—War in the Mediterranean

1. "Troop History Ninth Reconnaissance Troop for the Year 1942," 309-CAV-0.1 folder, box 7452, entry 427, RG 407, NAII. Corporal Arnold F. Buechler and Private First Class Peter P. Silver both earned the Silver Star for Gallantry for their actions in Safi Harbor.

2. Headquarters Cavalry Replacement Training Center, Fort Riley, Kansas, "Increase in Horse Type Trainees," 8 June 1943, folder AFG 475 Cavalry, box 1057, entry 55, RG 337, NAII.

3. Notes sent to Dr. George F. Howe, Robert M. Marsh Papers, MHI; George F. Howe, *Northwest Africa: Seizing the Initiative in the West*, U.S. Army in World War II, Mediterranean Theater of Operations (Washington, D.C.: Center of Military History, 1993), 61; and "Troop History Ninth Reconnaissance Troop for the Year 1942." The Secon2nd AD contributed critical elements to Task Force Brushwood, which landed at Safi, south of Casablanca.

4. George F. Howe, *The Battle History of the 1st Armored Division, "Old Ironsides"* (Washington, D.C.: Combat Forces Press, 1954), 21–22.

5. At this time, armored regiments still included a reconnaissance company, which demonstrates the continued recognition for the need to echelon the reconnaissance effort.

6. Howe, *The Battle History of the 1st Armored Division*, 32–33.

7. FM 17-22, *Reconnaissance Battalion*, 39.

8. "Troop History Ninth Reconnaissance Troop for the Year 1942."

9. First Reconnaissance Troop to Commanding General 1st ID, "Report of Battle," 21 November 1942, folder 301-CAV-.03, box 5876, entry 427, RG 94, NA II. Things did not always go smoothly for the members of 1st Reconnaissance Troop on their initial day in combat. The first thing Paul Skogsberg recalled was the ramp going down on his landing craft and his motorcycle disappearing beneath the water as he tried to make his way to the beach. Folder Paul Lester Skogsberg, Box 1st ID HQ and Other Unidentified Units, MHI.

10. Cavalry School, *Cavalry Reconnaissance*, no. 7, *Operations of a Reconnaissance Company in Tunisia* ([Fort Riley, KS]: Cavalry School, [1943]), 1–5.

11. Howe, *Battle History of the 1st Armored Division*, 118; and Rick Atkinson, *Army at Dawn* (New York: Henry Holt, 2002), 306. The British had made the same mistake of dispersing their forces when they arrived in the African desert. Dispersion had led to defeat in detail, similar to the fate of the 1st AD. Graham and Bidwell, *Fire-Power*, 222.

12. Howe, *Battle History of the 1st Armored Division*, 119. Ward proposed on 24 January

1943 a two-pronged attack on Maknassy with one column traveling from the vicinity of Sidi-bou-Zid and another from Gafsa. Since Sened Station lay halfway between Gafsa and Maknassy, Ward feared the raid would provide the Germans an indication of his next move. Fredendall wanted a "hit-and-run operation" to build confidence.

13. Marsh Papers, MHI.

14. Howe, *Battle History of the 1st Armored Division*, 123. The main elements carrying out the initial attack were drawn from the 21st Panzer Division.

15. Ibid., 126.

16. Martin Blumenson, "Kasserine Pass," in *America's First Battles, 1776–1965*, ed. Charles E. Heller and William A. Stofft (Lawrence: University Press of Kansas, 1986), 246–47. Information on the initial movements of the 81st ARB is taken from Frank Sears, "Supply Operations in Combat," 1 May 1948, Curricular Files, Armor School Library, Fort Knox, Kentucky; and *Cavalry Reconnaissance*, no. 7: 21.

17. *Cavalry Reconnaissance*, no. 7: 24; and Headquarters, 81st ARB, "Operations Sidi Bou Zid, February 4–17," Inclusive, Historical Documents of World War II, 1st AD, 81st ARB, 31 JAN 1943–SEPT 1944, microfilm reel box 2189, USMA, and Sears, "Supply Operations in Combat."

18. Headquarters, 81st ARB, "Operations Sidi Bou Zid, February 4–17"; and Sears, "Supply Operations in Combat." Report based on patrolling conducted on 12 and 13 February, the same day that A Company relieved B Company.

19. Headquarters, 81st ARB, "Operations Sidi Bou Zid, February 4–17."

20. Ibid.

21. Ibid.

22. Marsh Papers, MHI.

23. Headquarters, 81st ARB, "Operations Sidi Bou Zid, February 4–17"; and Sears, "Supply Operations in Combat."

24. "The Attack on Sidi Bou Zid, by The Second Battalion, First A.R. [armored regiment] 15 February 1943," box 1, Alger Papers, USMA. The officers of the 2nd Battalion, 1st Armored Regiment, 1st AD, prepared a report of their action at Sidi Bou Zid in which they were "annihilated on 15 February 1943." Most of their remarks were collected from the perspective of the battalion headquarters, while the authors were held in German captivity at Oflag 64 in Szubin, Poland.

25. Ibid.

26. Headquarters, 81st ARB, "Operations Sidi Bou Zid, February 4–17"; and Sears, "Supply Operations in Combat." Report based on patrolling conducted on 12 and 13 February, the same day that A Company relieved B Company.

27. *Cavalry Reconnaissance*, no. 7, 26–28.

28. Ibid.

29. Marsh Papers, MHI.

30. Marsh Papers, MHI; and *Cavalry Reconnaissance*, no. 7: 29–30.

31. Hamilton H. Howze Papers, folder correspondence and related documents, 1943–1946, 308-02-08, box 7, MHI.

32. Sears, "Supply Operations in Combat"; and "Operations Sidi Bou Zid, February 4–17." D Company was temporarily disbanded to recreate 3 companies.

33. "Operations from February 28–March 13, 1943; and *Cavalry Reconnaissance*, no. 7: 41–52.

34. Sears, "Supply Operations in Combat."

35. "Maknassy Operations 14–28 March," Historical Documents of World War II, 1st AD, 81st ARB, 31 JAN 1943–SEPT 1944, microfilm reel box 2189, USMA.

36. Headquarters 81st ARB, "El Guettar Operation, 29 March to 8 April 1943," Historical Documents of World War II, 1st AD, 81st ARB, 31 JAN 1943–SEPT 1944, microfilm reel box 2189, USMA; and Howe, *The Battle History of the First Armored Division*, 214–18.

37. "History of the 91st Reconnaissance Squadron, 15 February 1928–11 October 1943."

38. Ibid. See also CPT Jack H. Ficklen, "A Reconnaissance Squadron in Tunisia," *CJ* 52 (July–August, 1943): 16–18.

39. Fred H. Salter, *Recon Scout* (Kalispell, Montana: Scott Publishing, 1994), 49–50.

40. Salter, *Recon Scout*, 46; and Ficklen, "A Reconnaissance Squadron," 16–18.

41. Salter, *Recon Scout,* 56; and Keith Royer, interview by Matthew Morton, 13 January 2001, Reichelt Oral History Program, Florida State University, 12. Hereafter cited as Royer interview.

42. "History of the 91st Reconnaissance Squadron, 15 February 1928–11 October 1943." During its detached service, the light tank company, E Troop, lost 4 of 5 tanks while conducting a reconnaissance patrol on 27 April 1943.

43. Royer interview, 12. A "ditty bag" was a small backpack.

44. Salter, *Recon Scout,* 60–66, Royer interview, 13–14. Royer and Salter served in the same platoon throughout the war. Their accounts of the events are remarkably similar, with Royer being credited in Salter's memoir for naming the unknown hill "About-Face Hill."

45. Royer interview, 14.

46. Headquarters 81st ARB, "Report on Combat Experience and Battle Lessons for Training Purposes," 9 June 1943, folder 601-CAV-0.3, box 14820, entry 427, RG 407, NAII.

47. Ibid. Maps used by members of the 1st Reconnaissance Troop in North Africa were printed in black and white, and all the words were written in French. Troopers remarked it was very hard to locate their position with any precision. COL Sidney S. Haszard and Paul Skogsberg, interview by John Votaw, 25 August 1990, Louisville, Kentucky, Robert R. McCormick Research Center, 25 [hereafter Votaw interview]. German forces operating in North Africa prized the maps they were able to capture from the British, because the quality of the Italian maps was lacking. The Germans credited the British Long Range Desert Group for collecting the detailed information contained on the maps through its use of patrols that penetrated deeply into the German and Italian rear areas. Alfred Toppe, *Desert Warfare, German Experiences in World War II,* trans. H.Heitman, ed. H. Heitman, Reviewer Captain N. E. Devereux (Historical Division European Command, June 1952) 3.

48. LTC Bruce Palmer, Jr., "Battle Lessons on Reconnaissance," *CJ* 52 (Sept.–Oct.) cited in *Modern Reconnaissance: A Collection of Articles from the Cavalry Journal* (Harrisburg, PA: Military Service Publishing Company, 1944), 113. Palmer drew the material for his article from Hoy.

49. Mrs. Wendell Sharp to Mr. John Sharp, no date, New York, NY, Wendell C. Sharp Papers, MHI.

50. Headquarters 81st ARB, "Report on Combat Experience and Battle Lessons for Training Purposes," 9 June 1943.

51. Ibid.

52. Ibid.

53. *FM 17-12, Reconnaissance Battalion,* 10.

54. Headquarters 81st ARB, "Report on Combat Experience and Battle Lessons for Training Purposes," 9 June 1943.

55. Ibid.

56. Ibid.

57. MG Ernest N. Harmon, "Notes on Combat Experience during the Tunisian and African Campaigns," curricular files, Armor School Library, Fort Knox, Kentucky, 5.

58. Ibid., 4, 13.

59. A review of the unit summaries for the 1st and 9th Reconnaissance Troops while they served in North Africa reveals that these units conducted a host of missions. The following entries make up only a partial list of the missions for which they were responsible: route reconnaissance, liaison with U.S. units (ranging from regiment to corps), liaison with Allied units, training of Allied personnel on U.S. equipment, patrolling the Spanish-Moroccan border, and extensive mounted and dismounted patrolling. Annex no. 18, Operations Report, 1st U.S. Infantry Division, Unit—1st Reconnaissance Troop, 19 Jan–8 April 1943, folder 301-CAV-.03, box 5876, entry 427, RG 407, NAII; and "Troop History of the Ninth Reconnaissance Troop for the Year 1943," box 7452, entry 427, RG 407, NAII. Note: Dispersion. 1st ID dispatched LT Paul Skogsberg's platoon to the Spanish-Moroccan border, where Skogsberg secured an intelligence team that was attempting to gauge the feelings of the local inhabitants. While he performed that mission, another platoon patrolled the Mediterranean coastline. Veterans characterized their service in North Africa as being platoon-oriented rather than that of operating as a troop. Haszard and Skogsberg, Votaw interview, 14–15.

60. "Summary of Discussions, Conference of G-2s of U.S. Army in North Africa, Held

at Allied Force Headquarters, 23 May 1943 to 26 May 1943," reproduced at HQ AGF, 21 July 1943, enclosure 1 in "Observers' Report," Team 3, Army Ground Forces, 22 August 1943, p. 7, file 319.1/35, curricular files, Armor School Library, Fort Knox, Kentucky.

61. Ibid.

62. Ibid., 11.

63. Palmer, "Battle Lessons on Reconnaissance," 117.

64. "Summary of Discussions, Conference of G-2s of U.S. Army in North Africa, Held at Allied Headquarters, 23 May 1943 to 26 May 1943," 11.

65. Dennis J. Vetock, *Lessons Learned: A History of U.S. Army Lessons Learning* (Carlisle Barracks, PA: U.S. History Institute, 1988), 56–61.

66. "Visit to NATO, 15 April–5 May 1943," folder 314.81, HQSCG General Decimal File, HQ AGF, RG 337, NAII. McNair's visit took place in late April 1943 as Allied forces were making their final push into Tunis. On the day McNair was wounded he was visiting the 26th Infantry Regiment of the 1st ID. He had little positive to say about the regiment, remarking that all he found was "100% lethargy."

67. "Field Operations of a Maintenance Battalion," circa 1943, W.I. Rossie Papers, MHI.

68. Report of LTC Albert B. Crowther and MAJ Barton E. Miles, 13 March 1943, and report of LTC T.A. Seely, 19 February 1943, in Observer Board (Mediterranean Theater of Operations), U.S. Army, Army Ground Forces, *Reports of the AGF Observer Board (MTO)*, 2 vols. (Washington, D.C.: AGF, 1942–1945), 1:3.

69. LTC E.A. Russel, Executive Officer, CC B, 1st AD, cited in report of Seely, 19 February 1943, commenting on the utility of light tanks for reconnaissance, *Reports of the AGF Observer Board (MTO)*, 1:2; report of MAJ Allerton Cushman, 29 March 1943; and report of COL Richard J. Werner, 29 March 1943, *Reports of the AGF Observer Board (MTO)*, 2:12. Werner commented on the utility of the tracked Bren gun carrier used by the Derbyshire Yeomanry to keep II Corps informed.

70. Cushman, report of 5 May 1943, *Reports of the AGF Observer Board (MTO)*, 2:5. The Germans also abandoned their motorcycles and even their half-tracked version of the motorcycle for Volkswagens. Toppe, *Desert Warfare*, 33.

71. Marsh Papers, MHI. Robert Marsh recalled that on 2 May a tank rolled over six Teller mines stacked on top of one another. The tank's hull "split" as a result of the blast and only one crew member survived.

72. Cushman, report of 5 May 1943, and Werner, report of 29 March 1943, *Reports of the AGF Observer Board, (MTO)*, 2:5–6. Folder Paul L. Skogsberg, MHI; and "Annex No. 18, Operations Report, 1st U.S. Infantry Division, Unit—1st Reconnaissance Troop, Period 19 January–8 April 1943."

73. Cushman, 5 May 1943, *Reports of the AGF Observer Board (MTO)*, 2:5.

74. Robert M. Citino, *The Evolution of Blitzkrieg Tactics: Germany Defends Herself against Poland, 1918–1933* (Westport, CT: Greenwood Press, 1987), 80; and Graham and Bidwell, *Fire-Power*, 235.

75. These are the same *Cavalry Reconnaissance* documents that have been used previously in this chapter.

76. Charles J. Hoy, "The Last Days in Tunisia" *CJ* 53 (January–February 1944): 8.

77. Ibid., 12.

78. Ibid., 13.

79. Ibid.

80. "Observers' Report," Team 3, Army Ground Forces, 22 August 1943, p. 5, enclosure to "Summary of Discussions, Conference of G-2s of U.S. Army in North Africa Held at Allied Force Headquarters, 23 May 1943 to 26 May 1943."

81. Ibid., 4.

82. Harry W. Candler, "91st Reconnaissance Squadron in Tunisia," *CJ* 53 (March–April 1944): 22.

83. Ibid. Even though LTC Candler commented favorably on the effectiveness of the 37mm cannon, it must be remembered that all the direct-fire weapons found in the mechanized reconnaissance units were grossly overmatched by the Germans. Only American Grant and Sherman tanks, not found in the reconnaissance units, faced a chance against the

German panzers. George B. Jarrett, an Ordnance Observer to the British Eighth Army, had observed, "The tank gunner who needed a better gun during the battle is all too often very, very dead, killed in his own tank before he got in a good shot at his enemy." Jarrett blamed the American's inability to field a "hard hitting cannon" on years of "stinting." George B. Jarrett, "Achtung Panzer," manuscript, George B. Jarrett Papers, MHI, 35, 178.

84. Ibid.

85. Toppe, *Desert Warfare*, 40.

86. Report of C.P. Hall, 7 May 1943, *Reports of the AGF Observer Board (MTO)*, 2:6.

87. Martin Blumenson, *Salerno to Cassino*. The United States Army in World War II, the Mediterranean Theater of Operations (Washington, D.C.: Center of Military History, 1969), 234.

88. LTC Albert N. Garland and Howard McGaw Smyth, assisted by Martin Blumenson, *Sicily and the Surrender of Italy*. The United States Army in World War II, the Mediterranean Theater of Operations (Washington, D.C.: Center of Military History, 2002), 52.

89. Hamilton H. Howze, *A Cavalryman's Story, Memoirs of a Twentieth Century Army General* (Washington: Smithsonian Institution Press, 1996), 83–84 .

90. Garland and Smyth, *Sicily and the Surrender of Italy*, 53. This reference is specific to Sicily, but the same can be said of the terrain in Italy and the impact it had on military operations.

91. Garland and Smyth, *Sicily and the Surrender of Italy*, 95–96.

92. *Patton Papers*, vol. 2, *1940–1945*, ed. Martin Blumenson (Boston: Houghton Mifflin Company, 1974), 275.

93. *Cavalry Reconnaissance, no. 5, Operations of the Third Reconnaissance Troop in Sicily* ([Fort Riley, Kansas]: Cavalry School [1943 or 1944]), 1–6. CPT Alvin C. Netterblad, Jr., commanded the Third Cavalry Troop during the Sicilian campaign and prepared this selection—with its numerous footnoted references to doctrine—for educational purposes at Fort Riley, Kansas. See also LTC Paul A. Disney, "Reconnaissance in Sicily: Activities of a Reconnaissance Battalion from the Landings at Licata and Gela to the Entry into Palermo," *CJ* 53 (May–June 1944): 11–16; and "Report of Operations, 82nd ARB, 1–31 July 1943," Hell on Wheels Historical Records, 1963–1965, the 1941–1945 folder, box 7, White Papers. LTC Disney commanded the 82nd ARB throughout the campaign.

94. *Patton Papers*, ed. Blumenson, 2:285.

95. Ibid., 286.

96. Ibid., 287; and Garland and Smyth, *Sicily and the Surrender of Italy*, 224.

97. Lucien K. Truscott, Jr., *Command Missions, A Personal Story* (New York: E.P. Dutton and Company, 1954), 221.

98. *Cavalry Reconnaissance*, no. 5: 10–11.

99. Garland and Smyth, *Sicily and the Surrender of Italy*, 194; and LTC E.A. Trahan, *A History of the Second United States Armored Division* (Atlanta, GA: Albert Love Enterprises, 1946), chap. 2: 9.

100. Disney, "Reconnaissance in Sicily," 10, 17.

101. Garland and Smyth, *Sicily and the Surrender of Italy*, 250; and Disney, "Reconnaissance in Sicily," 17–22.

102. Disney, "Reconnaissance in Sicily," 21–22; Trahan, *A History of the Second United States Armored Division*, 11–12; and "Operations of 1st Platoon—Company 'C,' 82nd ARB, 10–25 July, 1943," Hell on Wheels Historical Records, 1963–1965, the 1941–1945 folder, box 7, White Papers.

103. Donald E. Houston, *Hell on Wheels: The 2nd Armored Division* (Novato, CA: Presidio, 1977), 174–75.

104. Garland and Smyth, *Sicily and the Surrender of Italy*, 254.

105. Houston, *Hell on Wheels*, 176.

106. *Patton Papers*, ed. Blumenson, 2:296.

107. Ibid., 296–97.

108. Operation Report [of the 91st Cavalry Reconnaissance Squadron], 10 July 1943–March 1944, folder CAVS-91-0.3, box 18231, entry 427, RG 94 NAII .

109. Garland and Smyth, *Sicily and the Surrender of Italy,* 311, 314.

110. Operation Report of the 91st Cavalry Reconnaissance Squadron, 10 July 1943–March 1944; and Salter, *Recon Scout,* 129–30.

111. Salter, *Recon Scout,* 117, 124, 129; and Operation Report of the 91st Cavalry Reconnaissance Squadron, 10 July 1943–March 1944.

112. LTG George S. Patton, Jr., "Notes on the Sicilian Campaign," [no date], Headquarters Seventh Army, APO 758, box 3, Patton Papers, USMA.

113. Ibid. In the same document, Patton offered that armies should have two corps each, and each corps in turn should have three divisions. In addition, each army should have two armored divisions, to exploit success when armies or corps achieved "break through[s]" or to envelop a flank uncovered "by the Infantry." Patton wrote the same message about army composition and the continued utility of horse cavalry to the chief architect of the American army during World War II, Lesley McNair. Patton to McNair, 4 September 1943, HQ Seventh Army, folder McN-McQ, box 33, Patton Papers, LC.

114. Edward L. Bimberg, *The Moroccan Goums, Tribal Warriors in a Modern War* (Westport, CT: Greenwood Press, 1999), 39–43.

115. Ibid., 121–23. The Goums were noted for their patrolling abilities in Italy also. Observer Report, 110, MAJ Harold E. Miller, 9 March to 6 April 1944, 110 folder, box 54, entry 15A, RG 337, NA II.

116. Garland and Smyth, *Sicily and the Surrender of Italy,* 107.

117. Edgar R. Raines, Jr., *Eyes of Artillery: The Origins of Modern U.S. Army Aviation in World War II* (Washington, D.C.: Center of Military History, 2000), 138.

118. Ibid., 161–63.

119. Truscott, *Command Missions,* 230–31.

120. Ibid., 255. Observer Report 81, MAJ A.J. Crist, December 1943–March 1944, 81 folder, box 53, entry 15A, RG 337, NAII.

121. CPT A.T. Netterblad, "Tactical Employment of a Provisional Mounted Squadron," *CJ* 54 (March–April 1945): 68–69. In a different *CJ* article, "Are Horses Essential," the editorial staff made sure its readers knew that Generals Patton, Eddy, and Truscott all cited the need for horse cavalry. Truscott wrote, "I am firmly convinced that if one squadron of horse cavalry and one pack troop of 200 mules had been available to me at San Stefano on August 1, they would have enabled me to cut off and capture the entire German force opposing me along the north coast road, and would have permitted my entry into Messina at least 48 hours earlier." Cited in MAJ P.D. Eldred's "Are Horses Essential," *CJ* 53 (May–June 1944): 4.

122. A review of the unit histories for the 91st CavReconSqdn and the 81st ARB reveals frequent stints in the line serving as infantry. Example: "14 October 1944 . . . Squadron was attached to CC 'A' and relieved the 91st Rcn Sqdn in their sector . . . 15 October: A, B, and C troops were in position along the front, F troop in reserve and in position to support line troops with artillery fire . . . 16 October: . . . moved in the sector to the immediate East and relieved the 6th Infantry in that sector." Headquarters, 81st Reconnaissance Squadron, History October 1944, Historical Documents, World War II, microfilm reel box 2189, USMA.

123. COL Willard A. Holbrook, Jr., to Czar [John K. Herr], 9 December 1943, Los Angeles, CA, box 6, Herr Papers, USMA.

124. MG C.H. Gerhardt to MG John K. Herr (Ret.), January 1944, V Mail return address New York, box 6, Herr Papers, USMA.

125. Hamilton H. Howze to Hamilton Hawkins, 10 February [1945], somewhere in Italy, Family Papers, CC A, 1st AD, MG Hamilton S. Hawkins folder, box 7, Papers of Hamilton H. Howze, MHI.

126. Ibid.

127. Summary of Interviews, section 2, pp. 32–33, box 1, Howze Papers MHI.

128. Howze to Hawkins, 10 February [1945].

129. Ibid.

130. Ibid.

131. Observer Report 110, Miller, 9 March to 6 April 1944.

132. Ibid.

133. Harmon, "Notes on Combat Experience," 20.

134. LTC Charles A. Ellis, "Demolition Obstacles to Reconnaissance," *CJ* 54 (May–June 1945): 29.

135. Ibid.

136. Harmon, "Notes on Combat Experience," 20.

137. Observer Report 81, Crist, "Report on the Italian Campaign," December 1943 to March 1944. Crist specifically observed the 34th and the 36th IDs. See also, Observer Report 110, Miller, 9 March to 6 April 1944; Observer Report 85, MAJ Elias C. Townsend, nd, folder 85, box 53, entry 15A, RG 337, NAII; "Collection and Dissemination of Information and the Employment of Reconnaissance Units," 23 May 1944, Headquarters Army Ground Forces, folder 142, box 56, entry 15A, RG 337, NAII; and Observer Report 90, LTC M.S. Cralle, Observer to Fifth Army, 24 December 1943 to 8 March 1944, folder 90, box 53, entry 15A, RG 337, NAII.

138. Observer Report 93, MAJ H.F. Suffield on the Italian Campaign, 23 December 1943 to 5 March 1944, 91 folder, box 83, entry 15A, RG 337, NAII.

139. Ibid.

140. Observer Report 110, Miller, 9 March to 6 April 1944.

141. Harmon, "Notes on Combat Experience," 20.

142. Ibid., 18; and "Report of Operations, 81st ARB, June 1944," Historical Documents, World War II, microfilm reel box 2189, USMA.

143. Harmon, "Notes on Combat Experience," 21.

144. Ibid., 5.

145. Ibid.

146. Committee 17, Officers Advanced Class, 1949–1950, "Operation of CAV RCN SQ Integral to the Armored Division" (Fort Knox, KY: Armored School, 1950), 39.

147. Robert R. Palmer, Bill I. Wiley, and William R. Keast, *Procurement and Training of Ground Combat Troops*, United States Army in World War II (Washington, D.C.: Historical Division, U.S. Army, 1948), 489–93.

148. MG John K. Herr to MG Innis P. Swift, 16 June 1941, Office of the Chief of Cavalry, Washington, D.C., box 6, Herr Papers, USMA.

149. Greenfield, Palmer, and Wiley, *Army Ground Forces*, 365.

150. Ibid., 365; and Stanton, *Order of Battle*, 6.

151. Greenfield, Palmer, and Wiley, *Army Ground Forces*, 353.

152. Ibid., 286–91.

153. Stubbs and Connor, *Armor-Cavalry*, 60–61.

154. Greenfield, Palmer, and Wiley, *Army Ground Forces*, 256–57.

155. This renaming of each of the squadrons can lead to some confusion. For example, 113th CavRegt with its 1st and 2nd squadrons became the 113th Cavalry Group (CavGrp) with 113th CavReconSqdn and the 125th CavReconSqdn. The army assembled other cavalry groups from the separate mechanized reconnaissance squadrons that had served the motorized infantry divisions, none of which saw service as motorized divisions as a result of McNair's reorganization of the army in 1943.

156. The General Board, *Mechanized Cavalry Units*, no. 49, appendix 15; and William Eagen, *The Man on the Red Horse* (Portland, OR: Metropolitan Printing Co., 1975), 63.

157. Greenfield, Palmer, and Wiley, *Army Ground Forces*, 356–57.

158. Stubbs and Connor, *Armor-Cavalry*, 70–72.

159. *FM 100-5, Operations* (1944), 8.

160. *FM 100-5, Operations* (1944), 8.

161. Ibid., 9.

162. Ibid.

163. *FM 100-5, Operations* (1944), 9; *FM 100-5, Operations* (1941), 8.

164. *FM 100-5, Operations* (1944), 9–10.

165. Ibid., 10.

166. Ibid., 11.

167. Ibid., 10.

168. Cavalry Board to the Adjutant General, General Headquarters American Expeditionary Forces, 24 April 1919.

5—D-Day to VE-Day

1. Ernest E. Epps, *Fourth Cavalry Group: The History of the Fourth Cavalry Reconnaissance Squadron, European Theatre of Operations* (Frankfurt: Gerhard Blümlein, 1945), 1.

2. The General Board, *Mechanized Cavalry Units,* no. 49, appendix 6. The board drew 3 examples from the June–November 1944 period, 4 from December 1944–March 1945, and 7 from March–April 1945.

3. "Troop History of the 4th Reconnaissance Troop, 1 June 1944 to 30 June 1944" and "Troop History . . . 1 July 1944 to 31 July 1944," Historical Documents, World War II, box 150, 4th ID (CAV) 4th Recon Troop (Mecz) Combat History, 4 August 1940–30 August 1945, USMA microfilm collection.

4. Haszard and Skogsberg, Votaw interview, 34.

5. Epps, *Fourth Cavalry,* 1. During the same period, Troop C supported the 101st Airborne Division in a similar fashion.

6. Ibid., 2.

7. Ibid., 3.

8. U.S. Army, General Staff, *St-Lô (7 July–19 July 1944),* American Forces in Action Series (Washington, D.C.: War Dept., 1947), 1–6.

9. Ibid., 7; and Operations Order, 113th Cavalry Group, 4 July 1944, France, unmarked folder, box 5, Biddle Papers, MHI.

10. U.S. Army, *St-Lô,* 9.

11. Ibid., 19; and Observer Report C-156, "Story of the Operations of the 113th Squadron Cavalry Mechanized in Normandy for the Period 7 July–10 July," 7 August 1944, France, in Observer Board (ETO), U.S. Army, Army Ground Forces, *Reports of the AGF Observers Board (ETO),* 6 vols. (Carlisle Barracks, PA: U.S. Army Military History Institute, 1944–1945), vol. 2, MHI, 1–3.

12. Observer Report C-156.

13. U.S. Army, *St-Lô,* 19–22.

14. Ibid., 23; and Observer Report C-156.

15. U.S. Army, *St-Lô,* 36–37; and Observer Report C-156.

16. U.S. Army, *St-Lô,* 37–38.

17. Omar N. Bradley, *A Soldier's Story* (New York: Rand McNally and Company, 1951), 318.

18. Ibid., 337.

19. Ibid., 348–49.

20. Statement prepared by General Charles L. Scott, 27 July 1944, [Fort Knox], folder July 1944–September 1944, box 3, Scott Papers, LC. Scott and McNair had been cadets in the same company at West Point.

21. White interview by Stodter, 245-246.

22. Martin Blumenson, *Breakout and Pursuit,* The U.S. Army in World War II, ETO (Washington, D.C.: Center of Military History, 1989), 634.

23. "Report of Combat Operations, 1 August–5 November 1944, 2nd Cavalry Group (Mecz)," file CAVG-2-0.3, box 17942, RG 407, NA II.

24. Blumenson, *Breakout and Pursuit,* 349.

25. "15th Cavalry Group: An Ambush in Brittany," part 1, *CJ* 54 (September–October 1945): 3.

26. Ibid., 4; and MG Adrian St. John (Ret.); Reichelt Oral History Program, interview conducted by Matthew Morton, 14 June 2000.

27. G.J. Dobbins and Thomas Fiori, "Cavalry and Infantry at St. Malo," *CJ* 54 (November–December 1945): 15–16; Blumenson, *Breakout and Pursuit,* 641; and "15th Cavalry Group: An Ambush in Brittany," 2.

28. Blumenson, *Breakout and Pursuit,* 428–29.

29. Ibid., 436–39.

30. Ibid., 497.

31. Thomas J. Howard et al., eds., *The 106th Cavalry Group in Europe, 1944–1945* (Augsburg, Germany: J.P. Himmer, 1945), 42, 49, 51, 59; and James W. Cocke, "Battle

Reconnaissance," *CJ* 54 (May–June 1945): 18–23.

32. "3rd Cavalry Group, Metz Operations, 10 August–1 November 1944," 1–44, Combat interviews folder 321, box 17953, entry 427, RG 407, NAII. The 3rd Cavalry Group only arrived in France on 9 August and was committed to combat on 10 August. Hugh M. Cole, *The Lorraine Campaign*, U.S. Army in World War II, ETO (Washington, D.C.: Center of Military History), 117–18.

33. "Old Nicomus in Western Europe," box 5, Benjamin A. Dickson Papers, USMA.

34. Joseph Bleich, "Thirty Men at Thionville," Combat interviews, folder 321, box 17953, entry 427, RG 407, NAII.

35. Cole, *The Lorraine Campaign*, 119–24.

36. Christopher R. Gabel, *The 4th Armored Division in the Encirclement of Nancy* (Fort Leavenworth, KS: Combat Studies Institute, 1986), 10.

37. Ibid., 16.

38. Ibid., 18–19.

39. Headquarters, 2nd Cavalry Group (Mecz), APO 403, 5 November 1945, Report of Combat Operations 1 August–5 November 1944, Section II, Narrative, box 17942, entry 427, RG 407, NAII 5–9; and "2nd Cavalry Group Report of Combat Operations, Period 1 September – 23 December 1944, box 17942, entry 427, RG 407, NAII Combat interview conducted by 2LT Charles Howards, folder 320, p. 2.

40. Ibid.

41. Ibid. Report of Operations, 9–11, and Combat interviews, 5–7.

42. Ibid. Report of Operations, 11, and Combat interviews, 8; and Cole, *The Lorraine Campaign*, 221.

43. Lewis Hawkins, AP War Correspondent, "Mechanized Cavalry Retains Tradition of Slashing Advance," cited in *CJ* 53 (November–December 1944): 11.

44. Blumenson, *Breakout and Pursuit*, 349–50.

45. Manteuffel to White, 25 March 1967, Diessen am Ammersee, Germany, Correspondence between I.D. White and General Hasso von Manteuffel, 1967–1976 folder, box unassigned, White Papers. Manteuffel went on to write about Patton, "His preparations and transmissions respectively of orders—we say his technics in issue of orders—is of the same kind we cavalrymen used!" Manteuffel arrived at his use of "cowboy" on his own. According to White, "he called them his cowboys with subordinate units." White, interview by Stodter, 254.

46. Lyman C. Anderson, "Third Army Reconnaissance," *CJ* 54 (January–February 1945): 20–23.

47. Ibid. See also War Department, *After Action Report, Third U.S. Army, 1 August 1944–9 May 1945*, vol. 2, *Staff Section Reports* (Department of History Library, USMA: Army Information Service, G3, August [1945]), 12.

48. War Department, *After Action Report, Third U.S. Army, 1 August 1944–9 May 1945*, 12.

49. Ibid., 12.

50. Robert D. Sweeney, "How Patton Kept Tabs on His Third Army," *Armored Cavalry Journal* 58 (March–April 1949): 53.

51. "The 6th Cavalry Group, Attack to Seize L'Hôpital and Clear Karlsbrunn Forest, December 2–5, 1944," *CJ* 54 (May–June 1945): 12–13.

52. War Department, *After Action Report, Third U.S. Army, 1 August 1944–9 May 1945*, 25.

53. Blumenson, *Breakout and Pursuit*, 350, gives 28 battle casualties, 30 traffic accident casualties.

54. Bradley, *A Soldier's Story*, 384.

55. Leclerc to Third Army HQ, 15 August 1944 [France], folder La-Lec, box 34, Patton Papers, LC. Within Patton's Third Army, Leclerc initially served in Wade H. Haislip's XV Corps. Haislip was fluent in French and had attended the École de Guerre. Leclerc wrote to Patton around the time he was detached from Haislip and assigned to General Gerow's V Corps. Henry Maule, *Out of the Sand: The Epic Story of General Leclerc* (London: Odhams Books, 1966), 173, 183.

56. Maule, *Out of the Sand*, 185.

57. Ibid., 205–6.

58. Harold J. Samsel, "Operational History of the 102nd Cavalry Regiment (Group) 'Essex Troop' 38th CavReconSqdn, 102nd CavReconSqdn, World War II," Samsel folder, box 3, 102nd Cavalry Regiment, WWII Veterans Survey, MHI.

59. "The Liberation of Paris," statement prepared by Sergeant Robert Schreil, B Troop [102nd CavReconSqdn], 102nd Cavalry Group, cited in Samsel, "Operational History of the 102nd Cavalry Regiment (Group)."

60. Colonel Paul Willing to MAJ Matthew Morton, 6 October 2003, Paris, France, possession of author.

61. Ibid.

62. Samsel, "Operational History of the 102nd Cavalry Regiment (Group)."

63. Ibid.

64. Blumenson, *Breakout and Pursuit,* 682, 693–94.

65. Epps, *Fourth Cavalry,* 13–20.

66. The General Board, *Mechanized Cavalry Units,* no. 49, appendix 6: 8.

67. Eagen, *The Man on the Red Horse,* 93–95.

68. [Summary of Operations, 113th Cavalry Group, July 1944–May 1945], unmarked folder, box 5, Biddle Papers, MHI; and Blumenson, *Breakout and Pursuit,* 679–80.

69. "113th Cavalry Group in Mission of Reconnaissance," p. 1, box 5, Biddle Papers, MHI; and Eagen, *The Man on the Red Horse,* 108.

70. "113th Cavalry Group in Mission of Reconnaissance," p. 3, box 5.

71. Ibid., 4–5.

72. Charles H. Corlett, *Cowboy Pete: The Autobiography of Major General Charles H. Corlett,* ed. William Farrington (Santa Fe, NM: Sleeping Fox, 1974), 101.

73. [Summary of Operations, 113th Cavalry Group, July 1944–May 1945]; and Eagen, *The Man on the Red Horse,* 116–17.

74. [Summary of Operations, 113th Cavalry Group, July 1944–May 1945]; and Corlett, *Cowboy Pete,* 101–2.

75. Jeffrey J. Clarke and Robert Ross Smith, *Riviera to the Rhine,* United States Army in World War II, ETO (Washington, D.C.: Center of Military History, 1993).

76. Clarke, *Riviera to the Rhine,* 132; and Frederic B. Butler, "Southern Exploits of Task Force Butler," part 1, *The Armored Cavalry Journal* 57 (January–February 1948): 13.

77. Butler, "Southern Exploits of Task Force Butler," part 1: 13.

78. Ibid.

79. Ibid., 15–18.

80. Butler, "Southern Exploits of Task Force Butler," part 2, *The Armored Cavalry Journal* 57 (March–April 1948): 33; and Clarke, *Riviera to the Rhine,* 146–47.

81. Butler, "Southern Exploits of Task Force Butler," part 2: 34.

82. "History of the 117th Cavalry Reconnaissance Squadron (Mecz), from 1 August 1944 to 31 August 1944," 8–9, Unit History Summary folder, Charles J. Hodge Papers, MHI.

83. Clarke, *Riviera to the Rhine,* 170, 223.

84. "History of the 117th Cavalry Reconnaissance Squadron (Mecz), Federal Service During World War II, January 6, 1941 – November 25, 1945," 22, Robert C. Lutz folder, Non-Divisional Cavalry Units box, MHI.

85. "Report No. 5, Conversations with General J. Lawton Collins," transcribed by MAJ Gary Wade (Fort Leavenworth, KS: Combat Studies Institute, [1983]), 9.

86. Epps, *Fourth Cavalry,* 22.

87. "Role of Task Force Polk (3rd Cavalry Group Reinforced) in the Metz Operation," 3–4 January 1945, revised copy, box 17953, entry 427, RG 407, NAII. Task Force Polk was named after its commander, Colonel James Polk, or "Jimmy," as he was known within the old cavalry community. While assigned to the 8th Cavalry Regiment under the command of Colonel "Jingles" Wilson, himself a Medal of Honor winner from the Philippine Insurrection, the spirited Polk and other single officers of the command had gone to Mexico. Upon their return they raised so much hell that the 2nd Squadron commander wanted to prefer court martial charges. On the advice of Ernest N. Harmon, a horse cavalryman who had gained more experience in the command of armored divisions than any other commander during World War II, nothing came of the incident. "Personal Memoirs of Major General

E. N. Harmon, U.S.A. Retired," 39–40, Special Collections, Norwich University, Northfield, Vermont.

88. "Role of Task Force Polk (3rd Cavalry Group Reinforced) in the Metz Operation," 3–4 January 1945, revised copy, 1–5.

89. "Role of Task Force Polk (3rd Cavalry Group Reinforced) in the Metz Operation," 3–4 January 1945, revised copy, pp. 8–15; and Cole, *The Lorraine Campaign,* 485.

90. Willard A. Holbrook to Helen Herr Holbrook, November–December 1944, France, folder World War II letters, box 41, Willard Holbrook, Jr., Papers, USMA.

91. Hamilton S. Hawkins, "Cavalrymen—Mounted, Dismounted and Mechanized," *CJ* 53 (September–October 1944): 29–30.

92. Ibid.

93. 12th Army Group, *Report of Operations (Final After Action Report),* vol. 11, *Antiaircraft Artillery, Armored Artillery, Chemical Warfare, and Signal* (Germany, 1945) 41.

94. Joe Holly to Charles L. Scott, 5 September 1944, HQ Communications Zone (FWD), folder General Holly, box 12, Scott Papers, LC.

95. Ibid.; and Hanson W. Baldwin, *Tiger Jack* (Fort Collins, CO: Old Army Press, 1979).

96. Observer Report 169, "Extracts from Various Observers," 18 November 1944, box 56, entry 15a, RG 337, NAII.

97. White, interview by Stodter, 245.

98. Ibid., 240–51.

99. Observer Report 157, "Armored Reconnaissance in the European Theater of Operations," [23 August 1944], box 56, entry 15a, RG 337, NA II; and Kenneth T. Barnaby, "Face-Lifting a Cavalry Squadron," *CJ* 55 (November–December 1946): 8, 17.

100. Barnaby, "Face-Lifting a Cavalry Squadron," 8–9.

101. Observer Report 385, "Cavalry and Armored Report," 27 November 1944 *Report of Observers,* 3:9, MHI; and Observer Report C-479, "Interview with Colonel S.N. Dolph, Commanding Officer, 102nd Cavalry Group," 31 December 1944 [Belgium], in Observer Board (ETO), U.S. Army Ground Forces, *Reports of the AGF Observers Board, ETO,* 3:1–4, MHI; and Barnaby, "Face-Lifting a Cavalry Squadron," 10.

102. Observer Report 157, "Armored Reconnaissance in the European Theater of Operations," [23 August 1944].

103. 12th Army Group, *Report of Operations (Final After Action Report),* vol. 11, *Antiaircraft Artillery, Armored Artillery, Chemical Warfare, and Signal,* 38, 58.

104. Willing to Morton, 6 October 2003.

105. Holly to Scott, 5 September 1944.

106. *12th Army Group, Report of Operations (Final After Action Report),* vol. 11, *Antiaircraft Artillery, Armored Artillery, Chemical Warfare, and Signal,* 43. James P. Hart to Charles L. Scott, 30 September 1944, HQ Ninth U.S. Army [France], folder LTC J.P. Hart, box 12, Scott Papers, LC.

107. Ibid., 58.

108. Memorandum prepared by Charles H. Reed for Commanding General Third U.S. Army, Subject: Authority For Assignment of Two (2) M5 or M5A1 Light Tanks to HQ, 2nd Cav Gp (Mecz), 12 August 1944, [France], file CAVG-2-0.1, box 17942, entry 427, RG 407, NAII.

109. Samsel, "Operational History of the 102nd Cavalry Regiment (Group)," 17.

110. White, interview by Stodter, 96–97.

111. Robert D. Sweeney, "How Patton Kept Tabs on His Third Army," *Armored Cavalry Journal* 58 (March–April 1949): 53.

112. "Notes from Combat," *CJ* 53 (November–December 1944): 17.

113. Willing to Morton, 6 October 2003.

114. Observer Report C-479, "Interview with Colonel S.N. Dolph, Commanding Officer, 102nd Cavalry Group," 31 December 1944.

115. Ibid.

116. Ibid.

117. Ibid.

118. Haszard and Skogsberg, Votaw interview, 30, 37–38.

119. Ibid., 36.

120. Observer Report 157, "Cavalry Notes," 23 December 1944.

121. Observer Report 186, "Armored Reconnaissance in the European Theater of Operations" [23 August 1944], box 56, entry 15a, RG 337, NAII.

122. "Notes observed on active front by an armored division and sent to Armored Center for its information," enclosure to memorandum prepared by Charles L. Scott, 23 November 1944, Fort Knox, Kentucky, folder Charles L. Scott, classified, box 11, Scott Papers, LC. While awaiting combat, MG Kilburn, commanding officer of the 11th AD, dispatched his officers to the active front to gather information. Among the observers was Colonel Wesley W. Yale.

123. Ibid.

124. Ibid.

125. Memorandum prepared by Scott for Commanding General, "Army Ground Forces, Subject: Reconnaissance Training in Armored Divisions," 21 August 1944, Fort Knox, Kentucky, folder July 1944–September 1944, box 3, Scott Papers, LC.

126. Ibid.

127. Bill Yenne, *"Black '41": The West Point Class of 1941 and the American Triumph in World War II* (New York: John Wiley and Sons, 1991), 166.

128. Observer Report C-479, "Interview with Colonel S.N. Dolph, Commanding Officer, 102nd Cavalry Group," 31 December 1944.

129. Ibid..

130. Observer Report C-483, "Notes on the 4th Cavalry Group," 29 December 1944 [Belgium], in Observer Board (ETO), U.S. Army, Army Ground Forces, *Reports of the AGF Observers Board (ETO)*, 3:1-5, MHI.

131. Bradley, *A Soldier's Story*, 444.

132. "History of the 117th Cavalry Reconnaissance Squadron (Mecz), from 1 August 1944 to 31 August 1944."

133. Patton to McNair, 4 September 1943, HQ Seventh Army, folder McN-McQ, box 33, Patton Papers, LC.

134. Summary of Operations of the 14th Cavalry Group, 16 December–24 December 1944, prepared by LT Shea, 1 February 1945, Villers l'Eveque, Belgium, 14th Cavalry Group, p. 34, folder 14th Cav Group, box 5, Charles B. MacDonald Papers, MHI.

135. "G-2 (AIR) Daily Summary Covering Period 1 Dec.–15 Dec. (INCL)," 6 March 1945, First Army Headquarters, APO 230, Ardennes folder, box 2, Dickson Papers, USMA.

136. Hugh M. Cole, *The Ardennes: Battle of the Bulge*, U.S. Army in World War II, ETO (Washington, D.C.: Center of Military History, 1994), 51–53.

137. LTG George S. Patton, Jr., "Notes on Bastogne Operation," Headquarters, Third United States Army, APO 403, 16 January 1945, folder 11, box 4, Patton Papers, USMA.

138. Ibid.

139. Patton diary, 16 December 1944, vol. 7, 1 August 1944 to 23 March 1945, box diary vols. 7 and 8; and Gay diary (vol. 1), Patton Papers, USMA.

140. J.D. Morelock, *Generals of the Ardennes: American Leadership in the Battle of the Bulge* (Washington, D.C.: National Defense University Press, 1993), 2–5.

141. Charles B. MacDonald, *A Time for Trumpets: The Untold Story of the Battle of the Bulge* (New York: Quill, 1985), 26; and Cole, *The Ardennes*, 78.

142. Colonel M.A. Devine, Jr., to MG John K. Herr, 8 December 1944, somewhere in Luxembourg, box 6, Herr Papers, USMA.

143. Ibid.

144. Ibid.

145. Cole, *The Ardennes*, 137–39.

146. Summary of Operations of the 14th Cavalry Group, 16–24 December 1944.

147. Cole, *The Ardennes*, 139–40.

148. Ibid., 142–45.

149. Summary of Operations of the 14th Cavalry Group, 16–24 December 1944, 2-5; and Cole, *The Ardennes*, 146–47.

150. Summary of Operations of the 14th Cavalry Group, 16–24 December 1944, 6; and Cole, *Ardennes*, 146–47.

151. Ralph G. Hill to Roger Cirillo, 10 November 1992, Wyomissing, Pennsylvania,

p. 5, Historical Correspondence concerning Events on the Northern Shoulder of the Bulge folder, box 1, Ralph G. Hill Papers, MHI.

152. "Transcript of the 32nd CAV RCN SQDN UNIT History, 15 December–30 DEC 1944"; and Summary of Operations of the 14th Cavalry Group, 16–24 December 1944, 21–23.

153. Hill to Cirillo, 10 November 1992, p. 5. Hill insists 14th Cavalry Group Executive Officer MAJ Lawrence J. Smith reported to him in a letter that he spent the entire day of 16 December trying to get additional artillery support.

154. Summary of Operations of the 14th Cavalry Group, 16–24 December 1944, p. 22; and Lawrence J. Smith to Charles B. MacDonald, 22 October 1983, Lake Odessa, Michigan, 14th Cavalry Group folder, box 5, Charles B. MacDonald Papers, MHI.

155. Ibid.; and Cole, *The Ardennes*, 261. Members of Troop A continued to drift back into Allied lines until mid-January.

156. Summary of Operations of the 14th Cavalry Group, 16–24 December 1944, pp. 28–29.

157. Ibid.

158. Ibid., 39–41; and "TRANSCavRegtIPT OF THE 32nd CAV RCN SQDN UNIT HISTORY, 15 DEC–30 DEC 1944."

159. Summary of Operations of the 14th Cavalry Group, 16–24 December 1944, p. 59.

160. First United States Army, *Report of Operations, 1 August 1944–22 February 1945*, annex no. 5: 40; and "14th Cavalry at Remagen Bridgehead," *CJ* 54 (May–June 1945): 15.

161. *Conquer: The Story of Ninth Army, 1944–1945* (Washington, D.C.: Infantry Journal Press, 1947), 117.

162. *Patton Papers*, ed. Blumenson, 2:595.

163. *The Battle at St. Vith, Belgium, 17–23 December 1944* (Fort Knox, KY: U.S. Army Armor School, nd), 4–5. The Malmedy Massacre occurred along the eastern route of march assigned to the 7th AD. Battery B, 285th Field Artillery Observation Battalion infiltrated the 7th AD's eastern route of march and was destroyed by Kampfgruppe Peiper.

164. Ibid., 9; and Cole, *The Ardennes*, 275–76.

165. *The Battle at St. Vith*, 12.

166. Ibid., 28.

167. Ibid., "Editor's Note," comments regarding German objectives attributed to the Fifth Panzer Army commander, Hasso von Manteuffel.

168. Patton, "Notes on Bastogne Operation."

169. Patton diary, 19 December 1944.

170. Patton, "Notes on Bastogne Operation."

171. Patton diary, 23 December 1944.

172. Ibid.

173. Robert Willoughby Williams, "With the Unicorn West of Bastogne," *CJ* 54 (March–April 1945): 7–8; Patton, "Notes on Bastogne Operation."

174. Williams, "With the Unicorn West of Bastogne," 6.

175. Yenne, *Black '41*, 192–93.

176. Michael J. Greene, "Contact at Houffalize," *The Armored Cavalry Journal* 58 (May–June 1949): 37–38.

177. Ibid., 38.

178. Ibid. LTC H.M. Foy was not pleased to learn, later, that his young executive officer had taken the majority of his squadron on the mission without Foy's knowledge. Greene was only commissioned as a second lieutenant in June 1941.

179. Greene, "Contact at Houffalize," 41.

180. Ibid., 42–43.

181. Patton, "Notes on Bastogne Operation." See also, "11th Armored Division After Action Report, Narrative, 23 December 1944–31 January 1945," and "11th Armored Division After Action Report, 1–31 January 1945," AAR 11th AD, December–March folder, box 1, Holmes E. Dager Papers, MHI.

182. Robert S. Allen, *Lucky Forward: The History of Patton's Third U.S. Army* (New York: Vanguard Press, 1947), 237.

183. Patton diary, 7 February 1945.

184. "Report of Operations, 2nd Cavalry Group (Mecz), 24 December 1944–28 February 1945," file CAVG-2-0.3, box 17943, entry 427, RG 407, NAII; and Cole, *The Ardennes,* 494.

185. Patton diary, 7 February 1945.

186. The General Board, *Mechanized Cavalry Units,* no. 49, appendix 6: 10–11.

187. "After Action Review for Period 1–28 February 1945," 5 March 1945, Headquarters 3rd Cavalry Group (M), box 17953, entry 427, RG 407, NAII.

188. "The 316th Provisional Mechanized Cavalry Brigade in Combat," *CJ* 55 (May–June 1946): 8.

189. Ibid., 9.

190. Willard A. Holbrook to John K. Herr, 4 February 1945, folder "Letters, Hunk and John K. Herr," box 41, Holbrook Papers, USMA.

191. Colonel W.W. Yale to MG J.K. Herr, 13 February 1945, V Mail return address New York, box 6, Herr Papers, USMA.

192. Holbrook to Czar [Herr], 12 March 1945, Germany, box 6, Herr Papers, USMA. The American occupation force in Germany, of which Herr and Holbrook were both a part, had been stationed at Koblenz, just east of Mayen.

193. Holbrook to Czar [Herr], 12 March 1945.

194. Yale to Herr, 24 March 1945.

195. "General Hawkins' Notes, Horse-Armor for U.S. Army?" *CJ* 54 (March–April 1945): 42.

196. Ibid.

197. Ibid.

198. Charles B. MacDonald, *The Last Offensive,* U.S. Army in World War II, ETO (Washington, D.C.: Government Printing Office, 1984), 231–32; and Edward N. Bedessem, *Central Europe* (Washington, D.C.: Center of Military History, nd), 33.

199. The General Board, *Mechanized Cavalry Units,* no. 49, appendix 6: 12–13.

200. Holbrook to Czar [Herr], 27 April 1945.

201. "11th Armored Division After Action Review, 1–31 March 1945."

202. Holbrook to Czar [Herr], 27 April 1945.

203. Ibid.

204. Ibid.

205. Ibid.

206. See Robert A. Doughty's *The Breaking Point: Sedan and the Fall of France, 1940* (Hamden, CT: Archon Books, 1990) for a detailed military history of General Guderian's drive to the Meuse and beyond in 1940.

207. "11th Armored Division After Action Review, 1–31 March 1945."

208. Holbrook to Czar [Herr], 27 April 1945.

209. George L. Haynes and James C. Williams, *The Eleventh Cavalry from the Roer to the Elbe, 1944–1945* (Nurnberg: Entwurf, Druck Union-Werk, 1945), 74; and The General Board, *Mechanized Cavalry Units,* no. 49, appendix 6: 9–10.

210. Haynes and Williams, *The Eleventh Cavalry,* 74.

211. Ibid., 75.

212. Ibid., 74.

213. George S. Patton, Jr., to Kenyon A. Joyce, 24 April 1945, Headquarters Third Army, duplicate Patton Correspondence folder, box 1, Papers of Kenyon A. Joyce, MHI.

214. The General Board, *Mechanized Cavalry Units,* no. 49, appendix 6: 3–4.

215. Howard et al., *The 106th Cavalry Group,* 127–28; and MacDonald, *The Last Offensive,* 437.

216. Howard et al., *The 106th Cavalry Group,* 126–31.

217. Charles H. Reed, "The Rescue of the Lippizanner Horses," 4 November 1970, appended to a communication from Reed to William S. Biddle, 14 February 1973, Richmond, Virginia, 113th CavGrp Correspondence and papers concerning WWII service, history folder, box 22, Biddle Papers, MHI.

218. Ibid. CPT Thomas M. Stewart of Tennessee accompanied CPT Lessing.

219. Ibid.

220. Ibid.

221. Reed to Biddle, 14 February 1973.

222. Barbra Burke Berntsen to CPT Matt Morton, 7 March 2002, California, citing material from her father's personal papers as executive officer of the 102nd Cavalry Group.

223. "Observations of CPT Harold Wassell, Pilot and Air Observer—117th Cavalry Reconnaissance Squadron," 4 June 1982, Springfield, Illinois, complete statement included in Samsel's "Operational History of the 117th Cavalry Reconnaissance Squadron (Mecz.) World War II."

224. Bradley to Patton, 9 February 1945, Twelfth Army Group, HQ, folder Be-Bu, box 34, Patton Papers, LC.

225. William S. Biddle to William Eagen, 16 April 1974, Alexandria, Virginia, 113th CavGrp Correspondence and papers concerning WWII service, history folder, box 22, Biddle Papers, MHI.

226. Observer Report C-775, "Mechanized Cavalry Notes," 28 March 1945, Observer Board (ETO), U.S. Army, Army Ground Forces, *Reports of the AGF Observers Board, ETO*, 3:1–2, MHI.

227. Ibid.

228. Ibid. Attached is the after action report, "Action against the Enemy, Report After," 1 January 1945, Headquarters, 28th Cavalry Reconnaissance Squadron.

229. Holbrook to Czar [Herr], 27 April 1945.

230. "After Action Review for Period 1–28 February 1945," 5 March 1945, Headquarters Third Cavalry Group (M), box 17953, entry 427, RG 407, NA II.

231. "William Shepard Biddle, MG, U.S. Army (Ret.) and Commandant Pennsylvania Military College, Chester, Pennsylvania," 113th Cavalry Group: Correspondence and papers concerning WWII service, history of 1960s folder, box 22, Biddle Papers, MHI.

232. First U.S. Army, *Report of Operations, 1 August 1944–22 February 1945*, annex no. 5: 48–49; and "A Report on United States vs. German Armor," prepared for General of the Army Dwight D. Eisenhower by MG I.D. White, 20 March 1945, box 3, White Papers, Special Collections, Norwich University.

233. Holbrook to Czar [Herr], 27 April 1945.

234. R.P. Hunnicutt, *Armored Car: A History of American Wheeled Combat Vehicles* (Novato, CA: Presidio, 2002), 138–39, 144–45.

235. Observer Report C-775, "Mechanized Cavalry Notes," 28 March 1945. These comments were directly attributed to Colonel Edward Fickett, commander of the 6th Cavalry Group and Colonel James Polk, commander of the 3rd Cavalry Group.

236. Ibid.

237. Memorandum from LTG George S. Patton, Jr., to Commanding General, Twelfth Army Group, Subject: Post-War Army, 14 February 1945, folder Bradley, box 34, Patton Papers, LC.

6—The Last Cavalry War

1. LTG L.K. Truscott, Jr. (Ret.), to MG John K. Herr, 29 May 1951, Germany, box 7, Herr Papers, USMA. Herr inserted this comment into the text he had prepared based on the letter from Truscott.

2. The General Board, *Mechanized Cavalry Units*, no. 49, appendix 2: 1.

3. Herr (Ret.) to Patton, 24 May 1945, [Washington], box 7, Herr Papers, USMA.

4. Ibid.

5. Herr (Ret.) to Truscott, 24 May 1945, [Washington], box 7, Herr Papers, USMA.

6. [BG Willard Holbrook] Willard to Herr, 13 June 1945, Headquarters 11th AD, APO 261, box 7, Herr Papers, USMA. "I agree entirely with your analysis of armored employment of cavalry. Mechanized cavalry especially is specialized for country with a considerable number of metalized roads. However, our mechanized cavalry is not properly equipped. It can not perform all missions required of cavalry adequately. It is road bound, especially in mountainous or marshy country. . . . Streams which horse cavalry would not have given a thought to fording become a major obstacle requiring several hours to negotiate. . . . Yale

has started a letter, supported by me, and approved by General Dager which recommends the inclusion of porté cavalry as an organic part of each armored division. . . . Still intended to `bridge the gap in mobility' between the man on foot and the man mounted on a vehicle . . . I'm not sure that our present methods of operation could be applied in China even if we were able to move across the northern portion of that country."

7. Herr to Truscott, 24 May 1945.

8. Patton to Herr, 7 July 1945, Headquarters, Third United States Army, APO 403, box 7, Herr Papers, USMA.

9. Ibid.

10. Ibid.

11. Ibid.

12. Ibid.

13. Ibid.

14. Ibid.

15. Ibid.

16. Patton to Tom T. Handy [Deputy Chief of Staff, War Department], 2 June 1945 [Germany], folder Han 1945, box 35, Patton Papers, LC.

17. Ibid.

18. Ibid.

19. Patton to Herr, 7 July 1945.

20. Herr to Patton, Jr., 8 August 1945, Washington, box 7, Herr Papers, USMA.

21. Ibid.

22. Ibid.

23. Ibid.

24. Ibid.

25. Ibid.

26. Ibid.

27. Patton to Herr (Ret.), 19 August 1945, Headquarters Third Army, APO 403, box 7, Herr Papers, USMA.

28. Ibid.

29. Ibid.

30. Truscott, Jr., to COL T.Q. Donaldson, Army Ground Forces Board, 14 July 1945, Headquarters, Fifth Army, APO 464, box 7, Herr Papers, USMA.

31. Ibid.

32. Ibid.

33. Ibid.

34. Ibid.

35. Herr (Ret.) to Truscott, Jr., 26 July 1945 [Washington], box 7, Herr Papers, USMA.

36. MG E.N. Harmon to Herr (Ret.), 25 July 1946, Headquarters, U.S. Constabulary, APO 46, box 7, Herr Papers, USMA.

37. Ibid.

38. Ibid.

39. Ibid.

40. Ibid.

41. Ibid.

42. Herr (Ret.) to Harmon, 2 August 1946, Washington, box 7, Herr Papers, USMA.

43. Ibid.

44. Ibid.

45. MG John K. Herr (Ret.), testimony before the U.S. Senate Armed Services Committee, "Question of Ownership of Captured Horses," 80th Congress, 1st Session, 1947: 302–7.

46. Patton diary, 8 August 1945.

47. *Patton Papers,* ed. Blumenson, 2:795–96.

48. "The General Board—European Theater," 1–3, folder 1946–1947, 11th Constabulary Regiment and ETO General Board, box 16, Biddle Papers, MHI.

49. Ibid., 7.

50. Ibid., 8–9.

51. Ibid., 10–11. Biddle and his staff drew heavily on the First Army's *Report of Operations*. Divisional troop commanders were particularly hard to track down.

52. First U.S. Army, *Report of Operations, 1 August 1944–22 February 1945,* annex no. 5: 55.

53. The General Board, *Mechanized Cavalry Units,* no. 49: 7. Biddle and his staff determined through their study of after action reports that the ratio of dismounted to mounted combat was 1.8:1.

54. First U.S. Army, *Report of Operations, 1 August 1944–22 February 1945,* annex no. 5: 55.

55. Herr to the Adjutant General, Subject: Cavalry Reconnaissance Units, 23 January 1942, Washington, D.C., box 7a, entry 39, records group 177, NAII.

56. The General Board, *Mechanized Cavalry Units,* no. 49: 9.

57. Ibid., 13.

58. Ibid., 20.

59. The General Board, *Mechanized Cavalry Units,* no. 49, appendix 7.

60. Ibid., 20.

61. First U.S. Army, *Report of Operations, 1 August 1944–22 February 1945,* annex no. 5: 56.

62. Patton to Handy, 2 June 1945.

63. Ibid.

64. Ibid.

65. The General Board, *Mechanized Cavalry Units,* no. 49: 21.

66. "The General Board—European Theater," 13, folder 1946–1947.

67. Herr to the Adjutant General, Subject: Cavalry Reconnaissance Units, 23 January 1942.

68. Patton to Handy, 14 July 1945.

69. Patton to Herr, 7 July 1945, Headquarters, Third United States Army, APO 403, box 7, Herr Papers, USMA.

70. "The General Board—European Theater," 13, folder 1946–1947.

71. Ibid., 14.

72. Ibid.

73. Ibid., 14–15.

74. Ibid., 14.

75. Ibid., 15; and The General Board, *Mechanized Cavalry Units,* no. 49: 21.

76. LTC John A. Hettinger to LTC George I. Smith, 3 December 1941, Fort Jackson, South Carolina, box 5, entry 39, RG 177, NA II.

77. Herr to the Adjutant General, Subject: Cavalry Reconnaissance Units, 23 January 1942.

78. Ibid.

79. "The General Board—European Theater," 17–18, folder 1946–1947.

80. Ibid., 18.

81. Ibid.

82. Ibid.

83. Ibid.

84. The General Board, *Mechanized Cavalry Units,* no. 49: 21.

85. Ibid. Appendix 21 contains the original questions and the tabular data derived from them. Appendix 15 and appendix 16 contain excerpts from these conferences.

86. Ibid., 18.

87. Ibid., appendix 16.

88. Ibid., appendix 15.

89. Ibid., appendix 16.

90. Ibid., 18.

91. Although the board made this recommendation, the army never adopted the measure.

92. The General Board, *Mechanized Cavalry Units,* no. 49: 18–19.

93. Ibid., 19; and Memorandum for Commanding General, Army Ground Forces,

prepared by Willard A. Holbrook, Jr., 3 June 1945, HQ 11th AD, box 7, Herr Papers, USMA.

94. "The General Board—European Theater," 20, folder 1946–1947.

95. *Patton Papers,* ed. Blumenson, 2:830–31.

96. "The General Board—European Theater," 20, folder 1946–1947.

97. Roy W. Cole, Jr., "Let's Face Facts," *CJ* 54 (September–October 1945): 35.

98. Ibid.

99. Hamilton Hawkins, "Hawkins' Notes," *CJ* 54 (September–October 1945): 32.

100. Ibid.

101. Obituary of Hamilton Smith Hawkins, *Assembly* 10 (April 1951), 47–48.

102. John K. Herr to the Editor of *The Cavalry Journal, CJ* 55 (May–June 1946): 35–40.

103. Fred W. Koester, "The Horse's Place in Our Future," *CJ* 54 (May–June 1945): 33.

104. Ibid.

105. Ibid.

106. Ibid.

107. C.L. Scott to I.D. White, 15 July 1945, Fort Knox, Kentucky, correspondence and personal business, 1942–1945 folder, box unassigned, White Papers, Special Collections, Norwich University.

108. Tom Herren to I.D. White, 25 July 1945, HQ Seventieth Division, correspondence and personal business, 1942–1945 folder, box unassigned, White Papers, Special Collections, Norwich University.

109. White, interview by Stodter, 341; and Jacob L. Devers to I.D. White, 31 October 1946, AGF HQ, Fort Monroe, Virginia, correspondence and personal business, 1946–1949 folder, box unassigned, White Papers, Special Collections, Norwich University.

110. "Remarks made by MG I.D. White, Commandant of the Cavalry School to the National Horse Association Dinner, New York City, 3 November 1946," folder of same name, box unassigned, White Papers, Special Collections, Norwich University.

111. Ibid.

112. Ibid.

113. "Minutes of the Special Meeting of the United States Cavalry Association," 8 July 1946, Washington, *The Armored Cavalry Journal* 55 (July–August 1946): 33.

114. Ibid.

115. Patton to Handy, 2 June 1945.

116. "Minutes of the Special Meeting of the United States Cavalry Association," 8 July 1946, 33.

117. John K. Herr to Verne Mudge, 7 May 1946, Washington, Cavalry testimony folder, box 7, Herr Papers, USMA. Mudge finished World War II as the commander of the 1st CavDiv, which had fought the entire war as a dismounted (infantry) division.

118. Mudge to Herr, 9 May 1946. Herr's note is written and initialed at the bottom of this letter.

119. "Mobilitate Vigemus," *The Armored Cavalry Journal* 55 (July–August 1946): 33.

120. Ibid.

121. Table of Contents, *The Armored Cavalry Journal* 55 (July–August 1946).

122. James A. Sawicki, *Cavalry Regiments of the U.S. Army* (Dumfries, VA: Wyvern Publications, 1985), 124.

123. Quoted from the Army Organization Act of 1950, cited in "From Horse to Horsepower . . . the Mobile Arm Becomes Armor," *Armor* 59 (July–August 1950): 14.

124. Ibid.

125. Sawicki, *Cavalry Regiments of the U.S. Army,* 124.

126. Of all people, COL Biddle and COL Reed had insisted on the inclusion of horses in the headquarters of the constabulary regiments used to patrol occupied Germany. "The Horse Platoon of the 11th Constabulary Regiment," folder 1946–1947, 11th Constabulary Regiment and ETO General Board, box 16, Biddle Papers, MHI; and Reed to Biddle, 14 February 1973, Richmond, Virginia, folder 113thCavGrp correspondence and papers concerning World War II service, box 22, Biddle Papers, MHI.

127. Obituary for John K. Herr, *Assembly* 14 (October 1955), 57–58.

128. BG R.L. Esmay, Adjutant General, to MG John Knowles Herr, 17 September 1951,

Cheyenne, Wyoming, box 7, Herr Papers, USMA.

129. John K. Herr to Honorable Carl Vinson, 28 June 1951, Washington, D.C., cavalry testimony folder, box 7, Herr Papers, USMA.

130. Vinson to Herr, 29 June 1951, Washington, D.C., cavalry testimony folder, box 7, Herr Papers, USMA.

131. Vinson to Herr, 2 July 1951, Washington, D.C., cavalry testimony folder, box 7, Herr Papers, USMA; and H.R. 5156, 14 August 1951, 82nd Congress, 1st Session.

132. H.R. 3338, 20 March 1951, 82nd Congress, 1st Session.

133. "The Cause for the Cavalry," *The Camden* [South Carolina] *Chronicle,* 23 October 1951.

134. Jonathan M. Wainwright, foreword to *The Story of the U.S. Cavalry, 1775–1942,* by Herr and Wallace (New York: Bonanza Books, 1984), vii–viii.

135. Statement prepared by John K. Herr upon receiving note from J. Nelson Tribly, clerk of the Armed Services Committee, 25 August 1953, Washington, D.C., cavalry testimony folder, box 7, Herr Papers, USMA.

136. Obituary for William Shepard Biddle, III, *Assembly* 42 (March 1984), 126–27.

Conclusion

1. Harold Kennedy, "Special Ops Equipment: Newest—and Oldest," *National Defense Magazine* 86 (February 2002), accessed 3 February 2003 at http://www.nationaldefensemagazine.org/article.cfm?Id=725.

2. Major General John K. Herr to Major General Innis P. Swift, 9 June 1941, Office of the Chief of Cavalry, Washington, D.C., box 6, Herr Papers, USMA.

3. Ibid.

4. Ibid.

5. Phil Bolte and Randy Myers, "The 'Cavalry' Rides Again!" *CJ* 26 (December 2002): 6–7.

6. Harold A. Buhl, Jr., "The Future of Scout and Cavalry Systems," *Armor* 112 (March–April 2003): 20.

7. Robert H. Scales, Jr., *United States Army in the Gulf War: Certain Victory* (Washington, D.C.: Office of the Chief of Staff, U.S. Army, 1993), 1. Colonel H.R. McMaster prepared a monograph, "Crack in the Foundation: Defense Transformation and the Underlying Assumptions of Dominant Knowledge in Future War," while serving as an Army War College Fellow at the Hoover Institution, Stanford University.

8. Scales, *United States in the Gulf War,* 237–38, 261–62.

9. Peter A. Wilson, John Gordon IV, and David E. Johnson, "An Alternative Future Force: Building a Better Army," *Parameters* 33, no. 4 (Winter 2003–2004): 21, 26–27.

10. Matthew Cox, "Front-line Training for Rear Area Troops," *Army Times* (15 December 2003), 12.

11. Email exchange with John Gordon (RAND Corp.) regarding ongoing study of Operation Enduring Freedom, 11 February 2004, copy in the possession of Matthew Morton. The same report cites improved situational awareness at division and corps level. Better communication links between these higher headquarters and units conducting reconnaissance may, under certain conditions, help reconnaissance units avoid "bumping into the enemy."

Selected Bibliography

PRIMARY SOURCES
Museums and Archives
Library of Congress (LC)

> The George S. Patton, Jr., Papers
> The John J. Pershing Papers
> The Charles L. Scott Papers

National Archives II at College Park, Maryland (NAII)

> Record Group 177. Records of the Chiefs of Arms
> Record Group 337. Army Ground Forces
> Record Group 407. Office of the Adjutant General

United States Army, Military History Institute, Carlisle Barracks, Carlisle, Pennsylvania (MHI)

> Papers of William S. Biddle
> Papers of Willis D. Crittenberger
> Papers of Holmes E. Dager
> Papers of Guy V. Henry, Jr.
> Papers of Ralph Hill
> Papers of Charles J. Hodge
> Papers of Hamilton H. Howze
> Papers of George B. Jarrett
> Papers of Kenyon A. Joyce
> Papers of Leon B. Kromer
> Papers of Charles B. MacDonald
> Papers of Robert M. Marsh
> Papers of W. L. Rossie
> Papers of Charles E. Rousek
> Papers of Wendell C. Sharp
> Curricular Files of the Army War College
> Observer Board (European Theater of Operations), U.S. Army, Army Ground Forces. *Reports of the AGF Observers Board, ETO*. 6 vols. Carlisle Barracks, PA: Military History Institute, 1944–1945.
> Observer Board (Mediterranean Theater of Operations), U.S. Army, Army Ground Forces. *Reports of the AGF Observers Board*. 2 vols. Carlisle Barracks, PA: Military History Institute, 1942–1945.

United States Military Academy, West Point, New York (USMA)

Papers of J. D. Alger
Papers of John Wesley Castles
Papers of Benjamin A. Dickson
Papers of John K. Herr
Papers of Willard A. Holbrook, Jr.
Papers of George S. Patton, Jr.
Historic Documents of World War II, Microfilm Collection

Norwich University, Northfield, Vermont

Papers of Ernest J. Harmon
Papers of Isaac D. White
Isaac D. White, General, U.S. Army, retired, interviewed by Colonel Charles S. Stodter, *U.S. Army Military History Institute, Senior Officers Oral History Program,* project 78-D. Carlisle Barracks, Carlisle, PA, 1978. Special Collections, Norwich University.

United States Armor Center Library, Fort Knox, Kentucky

Curricular Files
"Operations of CAV RCN SQ Integral to the Armored Division." Committee 17, Officers Advanced Class, 1949–1950. Fort Knox, KY: Armored School, 1950.
Alfred Toppe, *Desert Warfare, German Experiences in World War II,* trans. H.Heitman, ed. H. Heitman, Reviewer Captain N. E. Devereux, Historical Division European Command, c. 1952.

Combined Arms Research Library, Fort Leavenworth, Kansas (CARL)

S. D. Slaughter. "The Cavalry Group as an Economy Force (4th Cavalry Group in World War II, 19 December–31 December 1944): A Research Report." Fort Knox, KY: Armor School, 1950.

Interviews
Interviews and Questionnaires conducted by Timothy K. Nenninger, 1967.

LTG John L. Ryan (Ret.)
LTG Robert W. Grow (Ret.)
LTG Alvan C. Gillem, Jr. (Ret.)

Colonel Robert R. McCormick Research Center, First Division Museum, Cantigny, Wheaton, Illinois.

Interview with Colonel Sidney "Hap" Haszard and Paul Skogsberg by John Votaw, 25 August 1990.

Reichelt Oral History Program, Florida State University, Tallahassee, FL, interviews conducted by Matthew Morton.

Major General Adrian St. John (Ret.)
Keith Royer

Government Reports

ETO. U.S. Army. *Order of Battle of the U.S. Army, World War II: European Theater of Operations, Divisions*. Paris: 1945.
First U.S. Army. Report of Operations, 1 August 1944–22 February 1945
Seventh Army. Report of Operations (Final After Action Report)
Third Army. Report of Operations
12th Army Group. Report of Operations (Final After Action Report)
U.S. Army. Forces in the European Theater. The General Board. *Mechanized Cavalry Units*. Study No. 49. Washington, D.C.: [May 1946].

United States Army and War Department Publications

U.S. Army. *Field Manual 3-20.96. Cavalry Squadron (RSTA)*. Washington, D.C.: Headquarters Department of the Army, 2002.
———. *Field Manual 17-98-1. Scout Leader's Handbook*. Washington, D.C.: Department of the Army, 1990.
———. *FM 100-5. Field Service Regulations, Operations*. Washington, D.C.: Government Printing Office, 1941.
———. *FM 100-5. Field Service Regulations, Operations*. Washington, D.C.: Government Printing Office, 1944.
———. *St-Lo (7 July–19 July 1944)*. American Forces in Action Series. Washington, D.C.: War Department, 1947.
———. *Table of Organization No. 2-51. Cavalry Regiment, Horse and Mechanized*. Washington, D.C.: Government Printing Office, 1940.
———. *Table of Organization No. 2-55. Cavalry Squadron, Horse Regiment, Horse and Mechanized*. Washington, D.C.: Government Printing Office, 1940.
———. *Table of Organization No. 2-71. Cavalry Regiment, Mechanized*. Washington, D.C.: Government Printing Office, 1942.
———. *Table of Organization No. 2-25. Cavalry Squadron, Mechanized*. Washington, D.C.: Government Printing Office, 1 April 1942.
———. *Table of Organization and Equipment No. 2-25. Cavalry Reconnaissance Squadron, Mechanized*. Washington, D.C.: Government Printing Office, 1943.
———. *Training Circular. No. 42. Employment of Cavalry Mechanized Reconnaissance Elements*. Washington, D.C.: Government Printing Office, 1942.
———. *Training Circular. No. 107. Employment of Mechanized Cavalry Units*. Washington, D.C.: Government Printing Office, 1943.
U.S. War Department. *FM 2-15. Cavalry Field Manual, Employment of Cavalry*. Washington, D.C.: Government Printing Office, 1941.
———. *FM 2-20. Cavalry Field Manual, Reconnaissance Troop Mechanized*. Washington, D.C.: Government Printing Office, 1944.

Service School Publications

Cavalry School, U.S. Army. *Rasp*. Fort Riley, KS: Cavalry School, 1922.
———. *Mechanized Cavalry*. Fort Riley, KS: Cavalry School, 1936.
———. *Cavalry Combat*. Harrisburg, PA: Telegraph Press, 1937.
———. *Cavalry Reconnaissance*. No. 4. *Operations of the 91st Cavalry Reconnaissance Squadron, Mechanized, from Mateur to Bizerte (Northern Tunisia)*. [Fort Riley, KS]: Cavalry School, [1943].
———. *Cavalry Reconnaissance*. No. 5. *Operations of Third Reconnaissance Troop in Sicily*. [Fort Riley, KS]: Cavalry School, [1943 or 1944].
———. *Cavalry Reconnaissance*. No. 7. *Operations of a Reconnaissance Company in Tunisia*. [Fort Riley, KS]: Cavalry School, [1943].

———. *Cavalry Weapons and Material*. Pt. 10. *Motor Vehicles*. Fort Riley, KS: Cavalry School, 1935.

General Service School. U.S. Army. *Tactics and Technique of Cavalry*. Fort Leavenworth, KS: General Service School Press, 1921.

Books

Allen, Terry. *Combat Communication for Regiments and Smaller Units of Horse Cavalry*. Harrisburg, PA: The Military Service Publishing Company, 1939.

Bland, Larry I., and Sharon R. Ritenour, eds. *The Papers of George Catlett Marshall*. Vol. 1. Baltimore: The Johns Hopkins University Press, 1991.

Blumenson, Martin, ed. *The Patton Papers, 1885–1945*. 2 vols. Boston: Houghton Mifflin Company, 1972–1974.

Bradley, Omar N. *A Soldier's Story*. New York: Rand McNally, 1951.

Brown, Vernon H. *Mount Up! We're Moving Out! A World War II Memoir of D Troop, 94th Cavalry Reconnaissance Squadron [Mechanized] of the 14th Armored Division*. Bennington, VT: Merriam Press, 2002.

Butcher, Harry C. *My Three Years with Eisenhower*. New York: Simon & Schuster, 1946.

Corlett, Charles H. *Cowboy Pete: The Autobiography of Major General Charles H. Corlett*. Edited by William Farrington. Santa Fe, NM: Sleeping Fox, 1974.

Dean, D.W., and C.H. Hulse. *The 91st Cavalry Reconnaissance Squadron and WW II*. Self published for unit reunion, August 1993.

Epps, Ernest E. *Fourth Cavalry Group: The History of the Fourth Cavalry Reconnaissance Squadron, European Theater of Operations*. Frankfurt: Gerhard Blümlein, 1945.

———. *The Army in My Time*. London: Rich and Cowan, 1935.

Fuller, J.F.C. *Memoirs of an Unconventional Soldier*. London: Ivor Nicholson and Watson, 1936.

Guderian, Heinz. *Panzer Leader*. New York: E.P. Dutton, 1952.

Harmon, Ernest Nason. *Combat Commander*. Englewood Cliffs, NJ: Prentice-Hall, 1970.

Haynes, George L., and James C. Williams. *The Eleventh Cavalry from the Roer to the Elbe, 1944–1945*. Nurnberg: Entwurf, Druck Union-Werk, 1945.

Howard, Thomas J., et al., eds. *The 106th Cavalry Group in Europe, 1944–1945*. Augsburg: J.P. Himmer, 1945.

Howze, Hamilton H. *A Cavalryman's Story, Memoirs of a Twentieth Century Army General*. Washington: Smithsonian Institution Press, 1996.

Morrison, James V. *F Troop, the Real One, through the Eyes of Trooper James V. Morrison*. Hiawatha, KS: R.W. Sutherland Printing, 1988.

Ramsey, Edwin Price, and Stephen J. Rivele. *Lieutenant Ramsey's War: From Horse Soldier to Guerrilla Commander*. Washington, D.C.: Potomac Books, 2005.

"Report No. 5, Conversations with General J. Lawton Collins." Transcribed by Gary Wade. Fort Leavenworth, KS: Combat Studies Institute [1983].

Salter, Fred. *Recon Scout*. Kalispell, MT: Scott Publishing, 1994.

Truscott, Lucian K., Jr. *Command Missions, A Personal Story*. New York: E.P. Dutton, 1954.

———. *The Twilight of the U.S. Cavalry: Life in the Old Army*. Lawrence: University Press of Kansas, 1989.

Wright, Bertram C. *The 1st Cavalry Division in World War II*. Tokyo, Japan: Toppan Printing Company, 1947.

SECONDARY SOURCES
Books

Allen, Robert S. *Lucky Forward: The History of Patton's Third U.S. Army*. New York: Vanguard Press, 1947.

Anzio Beachhead, 22 January–25 May 1944. Washington, D.C.: Center of Military History, 1990.

Astor, Gerald. *Terrible Terry Allen, Combat General of World War II: The Life of an American Soldier.* New York: Ballantine Books, 2003.

Atkinson, Rick. *Army at Dawn.* New York: Henry Holt, 2002.

Bailey, Charles M. *Faint Praise: American Tanks and Tank Destroyers during World War II.* Hamden, CT: Archon Book, 1983.

Baldwin, Hanson, *Battles Lost and Won: Great Campaigns of World War II.* New York: Konecky & Konecky, 1966.

———. *Tiger Jack.* Fort Collins, CO: Old Army Press, 1979.

The Battle of St. Vith, Belgium, 17–23 December 1944. Fort Knox, KY: Armor School [nd].

Bedessem, Edward N. *Central Europe.* Washington, D.C.: Center of Military History [nd].

Bimberg, Edward L. *The Moroccan Goums, Tribal Warriors in a Modern War.* Westport, CT: Greenwood Press, 1999.

Birtle, Andrew J. *U.S. Army Counterinsurgency and Contingency Operations Doctrine, 1860–1941.* Washington, D.C.: Center of Military History, 1998.

Blumenson, Martin. *Breakout and Pursuit.* U.S. Army in World War II, European Theater of Operations. Washington, D.C.: Government Printing Office, 1984.

———. *Salerno to Cassino.* U.S. Army in World War II, Mediterranean Theater of Operations. Washington, D.C.: Center of Military History, 1969.

Bonn, Keith E. *When the Odds Were Even.* Novato, CA: Presidio, 1994.

Churchill, Winston. *The Unknown War.* New York: Charles Scribner's Sons, 1931.

Citino, Robert M. *Armored Forces: History and Sourcebook.* Westport, CT: Greenwood Press, 1987.

———. *The Evolution of Blitzkrieg Tactics: Germany Defends Herself against Poland, 1918–1933.* Westport, CT: Greenwood Press, 1994.

Clark, Alan. *Suicide of Empires: The Battles of the Eastern Front.* New York: American Heritage Press, 1971.

Clarke, Jeffrey J., and Robert Ross Smith. *Riviera to the Rhine.* U.S. Army in World War II, ETO. Washington, D.C.: Center of Military History, 1993.

Cline, Ray S. *The War Department: Washington Command Post, The Operations Division.* U.S. Army in World War II. Washington, D.C.: Office of the Chief of Military History, 1951.

Cole, Hugh M. *The Ardennes: Battle of the Bulge.* U.S. Army in World War II, ETO. Washington, D.C.: U.S. Government Printing Office, 1965.

———. *The Lorraine Campaign.* U.S. Army in World War II, ETO. Washington, D.C.: Center of Military History, 1984.

Conquer: The Story of the Ninth Army, 1944–1945. Washington, D.C.: Infantry Journal Press, 1947.

Corum, James S. *The Roots of Blitzkrieg: Hans von Seeckt and German Military Reform.* Lawrence: University Press of Kansas, 1992.

D'Este, Carlo. *Fatal Decision: Anzio and the Battle for Rome.* New York: Harper Collins, 1991.

———. *Patton, A Genius for War.* New York: Harper Collins, 1995.

Doubler, Michael. *Closing with the Enemy.* Lawrence: University Press of Kansas, 1994.

Doughty, Robert A. *The Breaking Point: Sedan and the Fall of France, 1940.* Hamden, CT: Archon Books, 1990.

Eagen, William. *The Man on the Red Horse.* Portland, OR: Metropolitan Printing, 1975.

Fisher, Ernest F. *Cassino to the Alps.* U.S. Army in World War II, MTO. Washington, D.C.: Center of Military History, 2002.

Gabel, Christopher R. *The Fourth Armored Division in the Encirclement of Nancy.* Fort Leavenworth, KS: Combat Studies Institute, 1986.

———. *Seek, Strike, and Destroy: U.S. Army Tank Destroyer Doctrine in World War II.* Fort Leavenworth, KS: Combat Studies Institute, 1985.

———. *U.S. Army GHQ Maneuvers, 1941.* Washington, D.C.: Center of Military History, 1991.

Garland, Albert N., and Howard McGaw Smyth, assisted by Martin Blumenson. *Sicily and the Surrender of Italy.* U.S. Army in World War II, MTO. Washington, D.C.: Center of Military History, 2002.

Gillie, Mildred Hanson. *Forging the Thunderbolt: A History of the Development of the Armored Force*. Harrisburg, PA: The Military Service Publishing Company, 1947.

Gole, Henry G. *The Road to Rainbow: Army Planning for Global War, 1934–1940*. Annapolis, MD: Naval Institute Press, 2003.

Graham, Dominick, and Shelford Bidwell. *Fire-Power: British Army Weapons and Theories of War, 1904–1945*. London: George Allen & Unwin, 1982.

Greenfield, Kent Roberts. *American Strategy in World War II, A Reconsideration*. Malabar, FL: Robert E. Krieger Publishing, 1963.

Greenfield, Kent Roberts, Robert R. Palmer, and Bill I. Wiley. *The Army Ground Forces: The Organization of Ground Combat Troops*. Washington, D.C.: Department of the Army, Historical Division, 1947.

Griffith, Robert K., Jr. *Men Wanted for the U.S. Army: America's Experience with an All-Volunteer Army between the World Wars*. Westport, CT: Greenwood Press, 1982.

Harrison, Gordon A. *Cross-Channel Attack*. U.S. Army in World War II, ETO. Washington, D.C.: Government Printing Office, 1951.

Heller, Charles E., and William A. Stofft, eds. *America's First Battles, 1776–1965*. Lawrence: University Press of Kansas, 1986.

Herr, John K., and Edward S. Wallace. *The Story of the U.S. Cavalry, 1775–1942*. New York: Bonanza Books, 1953.

Hofmann, George, and Donn A. Starry, eds. *Camp Colt to Desert Storm: The History of the U.S. Armored Forces*. Lexington: University Press of Kentucky, 1999.

Houston, Donald E. *Hell on Wheels: The 2nd Armored Division*. Novato, CA: Presidio, 1977.

Howard, Michael. *The Franco-Prussian War: The German Invasion of France, 1870–1871*. New York: Dorset Press, 1990.

Howe, George F. *The Battle History of the 1st Armored Division, "Old Ironsides."* Washington, D.C.: Combat Forces Press, 1954.

———. *Northwest Africa: Seizing the Initiative in the West*. U.S. Army in World War II, MTO. Washington, D.C.: Center of Military History, 1993.

Hunnicutt, R.P. *Armored Car: A History of American Wheeled Combat Vehicles*. Novato, CA: Presidio, 2002.

Johnson, David E. *Fast Tanks, Heavy Bombers: Innovation in the U.S. Army, 1917–1945*. Ithaca, NY: Cornell University Press, 1998.

Jones, James Pickett. *Yankee Blitzkrieg: Wilson's Raid through Alabama and Georgia*. Lexington: University Press of Kentucky, 2000.

MacDonald, Charles B. *The Last Offensive*. U.S. Army in World War II, ETO. Washington, D.C.: Center of Military History, 1993.

———. *The Siegfried Line Campaign*. U.S. Army in World War II, ETO. Washington, D.C.: Government Printing Office, 1984.

———. *A Time for Trumpets: The Untold Story of the Battle of the Bulge*. New York: Quill, 1985.

Mansoor, Peter R. *The G. I. Offensive in Europe: The Triumph of American Infantry Divisions, 1941–1945*. Lawrence: University Press of Kansas, 1999.

Maule, Henry. *Out of the Sand: The Epic Story of General Leclerc*. London: Odhams Books, 1966.

Millet, Allen, and Peter Maslowski. *For the Common Defense: A Military History of the United States of America*. New York: The Free Press, 1984.

Moenk, Jean R. *A History of Large-Scale Maneuvers in the United States, 1935–1964*. Fort Monroe, VA: Headquarters, U.S. Continental Army Command, 1969.

Morelock, J. D. *Generals of the Ardennes: American Leadership in the Battle of the Bulge*. Washington, D.C.: National Defense University, 1993.

Nye, Roger H. *The Patton Mind: The Professional Development of an Extraordinary Leader*. Garden City Park, NY: Avery Publishing Group, 1993.

Palmer, Robert R., Bell I. Wiley, and William R. Keast. *Procurement and Training of Ground Combat Troops*. U.S. Army in World War II. Washington, D.C.: Historical Division, U.S. Army, 1948.

Raines, Edgar R., Jr. *Eyes of Artillery: The Origins of Modern U.S. Army Aviation in World War II*. Washington, D.C.: Center of Military History, 2000.

Register of Graduates and Former Cadets of the United States Military Academy, West Point, New York. West Point, NY: Association of Graduates, 1999.

Sawicki, James. *Cavalry Regiments of the U.S. Army*. Dumfries, VA: Wyvern Publications, 1985.

Scales, Robert H., Jr. *United States Army in the Gulf War, Certain Victory*. Washington, D.C.: Office of the Chief of Staff, U.S. Army, 1993.

Seaton, Albert. *The German Army, 1933–1945*. London: Weidenfeld and Nicolson, 1982.

Stanton, Shelby L. *Order of Battle: U.S. Army, World War II*. Novato, CA: Presidio, 1984.

Stubbs, Mary Lee, and Stanley Russell Connor. *Armor-Cavalry*. Part 1. *Regular Army and Army Reserve*. Army Lineage Series. Washington, D.C.: Office of the Chief of Military History, U.S. Army, 1969.

To Bizerte with the II Corps, 23 April–13 May 1943. Washington, D.C.: Center of Military History, 1990.

Trahan, E. A. *A History of the Second United States Armored Division*. Atlanta, GA: Albert Love Enterprises, 1946.

Vetock, Dennis J. *Lessons Learned: A History of U.S. Army Lessons Learning*. Carlisle Barracks, PA: U.S. Army History Institute, 1988.

Watson, Mark Skinner. *Chief of Staff: Prewar Plans and Operations:* United States Army in World War II. Washington, D.C.: Historical Division, Department of the Army, 1950.

Weigley, Russell F. *Eisenhower's Lieutenants: The Campaign of France and Germany, 1944–1945*. Bloomington: Indiana University Press, 1981.

——. *History of the United States Army*. New York: Macmillan, 1970.

Wilson, Dale E. *Treat 'Em Rough! The Birth of American Armor, 1917–1920*. Novato, CA: Presidio, 1989.

Wilson, John B. *Maneuver and Firepowers: The Evolution of Divisions and Separate Brigades*. Washington, D.C.: Center of Military History, 1998.

Winton, Harold R. *To Change an Army: General Sir John Burnett-Stuart and British Armored Doctrine, 1927–1938*. Lawrence: University Press of Kansas, 1988.

Yale, Wesley W., Isaac D. White, and Hasso E. von Manteuffel. *Alternative to Armageddon: The Peace Potential of Lightning War*. New Brunswick, NJ: Rutgers University Press, 1970.

Yenne, Bill. *"Black '41": The West Point Class of 1941 and the American Triumph in World War II*. New York: John Wiley and Sons, Inc., 1991.

Journals

Armor
Armored Cavalry Journal
Army
Assembly
The Camden [South Carolina] Chronicle
The Cavalry Journal
The Command and General Staff School Quarterly
The Infantry School Mailing List
The Journal of Military History
Military Review
Parameters: Journal of the U.S. Army War College
Periodical Journal of America's Military Past

Theses and Dissertations

Cameron, Robert S. "Americanization of the Tank: U.S. Army Administration and Mechanized Development within the Army, 1917–1943." Ph.D. diss., Temple University, 1994.

DiMarco, Louis A. "The U.S. Army's Mechanized Cavalry Doctrine in World War II." Master's thesis, U.S. Army Command and General Staff College, Fort Leavenworth, KS, 1995.

Johnson, Dave E. "Fast Tanks and Heavy Bombers: The United States Army and the Development of Armor and Aviation Doctrines and Technologies, 1917 to 1945." Ph.D. diss., Duke University, 1990.

Killigrew, John W. "The Impact of the Great Depression on the Army." Master's thesis, Indiana University, 1979.

Miller, Robert A. "The United States Army during the 1930s." Ph.D. diss., Princeton University, 1973.

Nenninger, Timothy K. "The Development of American Armor, 1917–1940." Master's thesis, University of Wisconsin, 1968.

Powell, James. "The Making of a Combat Unit: A National Guard Regiment Goes to War," Master's thesis, Texas A and M University, 2002.

Runde, Richard J., Jr. "The Intelligence and Reconnaissance Platoon, 1935–1965: Lost in Time." Master's thesis, U.S. Army Command and General Staff College, Fort Leavenworth, KS, 1994.

Tedesco, Vincent J., III. "'Greasy Automatons' and the 'Horsey Set': The U.S. Cavalry and Mechanization, 1928–1940." Master's thesis, The Pennsylvania State University, 1995.

Tully, John N. "Doctrine, Organization, and Employment of the Fourth Cavalry Group during World War II." Master's thesis, Command and General Staff College, Fort Leavenworth, KS, 1994.

Movies

Colonel Tom Gillis. "A Year on a Cavalry Post." Silent movie, 1938.

Index